Mathematical Models for Structural Reliability Analysis

CRC Mathematical Modelling Series

Titles included in the series:

Movchan/Movchan: ***Mathematical Modelling of Solids with Nonregular Boundaries***

Brandimarte/Villa: ***Advanced Models for Manufacturing Systems Management***

Kusiak/Bielli: ***Designing Innovations in Industrial Logistics Modelling***

Kliemann/Namachchivaya: ***Nonlinear Dynamics and Stochastic Mechanics***

Bellomo: ***Modelling Mathematical Methods and Scientific Computation***

Mathematical Models for Structural Reliability Analysis

Edited by
Fabio Casciati
Brian Roberts

CRC Press
Boca Raton New York London Tokyo

Library of Congress Cataloging-in-Publication Data

Catalog record available from the Library of Congress.

Direct all inquiries to CRC Press, Inc., 2000 Corporate Blvd., N.W., Boca Raton, Florida 33431.

International Standard Book Number 0-8493-9631-x
Printed in the United States of America 1 2 3 4 5 6 7 8 9 0
Printed on acid-free paper

Contents

Preface

We are most grateful to the Series Editor, Professor N. Bellomo of the Polytechnic of Turin, Italy, for inviting us to make this contribution. Following Professor Bellomo's suggestion our aim has been to provide a coverage of important developments in the application of probabilistic models to the field of Structural Reliability. Much of the work in this area was initiated in the Seventies, was developed in the Eighties, and has achieved maturity in the Nineties. It is therefore timely to bring together, within a single volume, treatments of some of the most important aspects of this activity. It was clear, at the outset, that it would be impossible to invite all the scientists and engineers who are active in this field to contribute to the book. However, we believe that the following chapters do reflect clear milestones in the development of reliability methods for engineering applications.

To maintain a uniform style of treatment and character in the book all the individual contributors have striven to explain both theoretical modeling models and practical applications in as clear, concise, and readable a fashion as possible. A reasonable level of mathematical rigour has been achieved throughout whilst avoiding the inherent dangers of becoming too formal. Thus, it is hoped that the book will be accessible and useful to a wide range of researchers and practising engineers.

All the authors acknowledge the contributors of co-workers and/or supporting funding agencies. For Chapter 1 the authors jointly thank the Italian Ministry of University and Scientific and Technological Research (MURST) for supporting the research co-ordinated at a national level by Professor Casciati. The author of Chapter 2 thanks H. Ugur Köylüoglu for valuable comments, S. Randrup-Thomsen for producing the computer drawings, and Annette Baerentzen for the TeX typing. The work has been financially supported by the Danish Technical Research Council. The author of Chapter 3 acknowledges the contribution of Dr. Sadeghi in producing the experimental results and wishes to thank Linda Nutley for the TeX typing. The co-operation of the authors of Chapter 5 was facilitated by funding from the Human Capital Mobility Network on Stochastic Mechanics, of which Professor Faravelli is the coordinator (European Union grant ERBCHXCT940565). In particular, a period of study by the first author at Pavia occurred within this exchange program. It is worth remarking here that this network has been of value in a more general sense, helping other contributors to the book to interact more closely than would otherwise have been possible. Part of the work described in Chapter 6 was funded by the UK Engineering and Physical Research Council (EPSRC) and the Marine Technology Directorate; the authors gratefully acknowledge this support.

The editors of this book also wish to thank all the authors for their effort and time in producing their contributions and the CRC staff, Ms. Nora Konopka and Dr. C.L. Tondo, for their pleasant and friendly cooperation.

F. Casciati J.B. Roberts

1

Stochastic Process Models

F. Casciati
University of Pavia, Italy

M. Di Paola
University of Palermo, Italy

1.1 Introduction

Mechanical systems are studied in an analytical way through the identification of a mathematical model and the solution of its governing equations. A satisfactory level of generality is achieved by writing a set of ordinary differential equations:

$$\frac{d}{dt}x_j(t) = f_j(\boldsymbol{x}(t), t) + \sum_k F_{jk}(\boldsymbol{x}(t), t)\xi_k(t) \tag{1.1}$$

or in matrix notation

$$\dot{\boldsymbol{x}} = \boldsymbol{f}(\boldsymbol{x}(t), t) + \boldsymbol{F}(\boldsymbol{x}(t), t)\boldsymbol{\xi}(t) \tag{1.2}$$

A higher order equation can be replaced by an equivalent set of first order equations and, hence, it is still represented by Eq. (1.1). In this equation, $\boldsymbol{x}(t)$ is a vector of system response variables, $\boldsymbol{\xi}(t)$ is a vector of excitations, and $\boldsymbol{f}$ and $\boldsymbol{F}$ are a vector and a matrix, respectively, of either linear or nonlinear functions. Indeed, greater generality would require a set of partial differential equations in the time t and in the spatial coordinates. Continuous systems of finite size, however, are regarded as discretized within a finite element scheme or by other standard procedure: this rearranges the space differentiation in an algebraic form, i.e., partial differentiation into ordinary derivatives with respect to t.

Uncertainty in the modeling of the system means that $\boldsymbol{f}$ and $\boldsymbol{F}$ are random functions of t and $\boldsymbol{x}$. The response variables would show in this case a random nature. This random nature still remains, when $\boldsymbol{f}$ and $\boldsymbol{F}$ are deterministic, if at least one of the excitations possesses a random character. Any excitation ξ_k is multiplicative or additive depending on whether or not $\boldsymbol{F}$ is a function of $\boldsymbol{x}$. A static problem is achieved when all the ξ_k are constant in time and the system is time invariant.

This chapter analyses the possible mathematical idealizations for ξ_k. The simplest cases do not require more than a short mention:

a constant-in-time excitation is simply modeled by a **random variable**;

a time variant excitation that can be modeled through its maximum in a reference period is simply modeled by a **random variable of the extreme type**;

a time variant excitation with constant-in-time moments (see Appendix) is a **stationary random process** and may be studied by Fourier analysis.

The last category has a preliminary value for the understanding of non-stationary random processes. It is discussed in the next two Sections.

1.2 The Orthogonal-Increment Model

Let $Z(\omega)$ be a complex-valued stochastic process defined on (a, b) and let the ensemble average $E[\cdot]$ of its squared modulus be bounded. Given two non-overlapping intervals (ω_1, ω_2) and (ω_3, ω_4), the process is said:

with *uncorrelated increments* when

$$E[(Z(\omega_2) - Z(\omega_1))\ (Z^*(\omega_4) - Z^*(\omega_3))] =$$

$$E[Z(\omega_2) - Z(\omega_1)]\ \ E[Z^*(\omega_4) - Z^*(\omega_3)] \tag{1.3}$$

(where the star denotes the complex conjugate);

with *orthogonal increments* when they are uncorrelated and the r.h.s. of Eq. (1.3) is zero.

It can be shown [18] that, given a process $Z(t)$ with orthogonal increments:

- the following relation holds:

$$E[(Z(\omega_3) - Z(\omega_1))(Z^*(\omega_4) - Z^*(\omega_2))] = E[|Z(\omega_3) - Z(\omega_2)|^2] \quad (1.4)$$

- for any ω_0 on the ω axis the function

$$\psi(\omega) = \text{sign}(\omega - \omega_0)E[|Z(\omega) - Z(\omega_0)|^2] \quad (1.5)$$

  is non-decreasing. For $\omega_2 \geq \omega_1$, in fact:

$$E[|Z(\omega_2) - Z(\omega_1)|^2] = \psi(\omega_2) - \psi(\omega_1) \quad (1.6)$$

- for $\omega_1 = \omega$ and $\omega_2 = \omega_1 + d\omega$, Eq. (1.6) becomes:

$$E[|dZ(\omega)|^2] = d\psi(\omega) \quad (1.7)$$

In summary, for a stochastic process $Z(\omega)$ with orthogonal increments and for $\omega \neq \omega'$ one writes:

$$E[dZ(\omega)dZ^*(\omega')] = 0$$

but, for $\omega = \omega'$,

$$E[dZ(\omega)dZ^*(\omega')] = d\psi(\omega) \quad (1.8)$$

Such orthogonal-increment relations can be re-written in the more compact form:

$$E[dZ(\omega)dZ^*(\omega')] = \Delta_K(\omega - \omega')d\psi(\omega)$$

$\Delta_K(\omega)$ being the Kronecker delta ($\Delta_K(\omega) = 1$ if $\omega = 0$; $\Delta_K(\omega) = 0$ otherwise).

Some models for the stochastic process $\xi(t)$ in Eq. (1.1) can easily be constructed on these processes with orthogonal increments and are discussed in Sections 1.3 and 1.4.

1.3 The Correlation-Stationary Model

A *correlation-stationary* stochastic process $\xi(t)$ shows a correlation function $E[\xi(t_1)\xi(t_2)]$ which only depends on the time difference $\tau = t_2 - t_1$:

$$E[\xi(t_1)\xi(t_2)] = R_\xi(\tau) \quad (1.9)$$

Under the assumptions that $\xi(t)$ is a real-valued process, $E[\xi(t)] = 0$ and t is a continuous parameter, it turns out that a necessary and sufficient condition for $\xi(t)$ to be a correlation stationary stochastic process is to possess the integral representation:

$$\xi(t) = \int_{-\infty}^{\infty} e^{i\omega t} dZ_\xi(\omega) \tag{1.10}$$

where $Z_\xi(\omega)$ is an orthogonal increment process. The proof requires to introduce $\xi(t-\tau)$ in its integral form with $dZ_\xi^*(\omega)$. Then it results:

$$R_\xi(\tau) = \int_{-\infty}^{\infty} e^{i\omega\tau} d\psi_\xi(\omega) \tag{1.11}$$

Relaxation of the assumption $E[\xi(t)] = 0$ means to introduce a constant value μ_ξ for the mean value: it must be added to Eq. (1.10) while in Eq. (1.11) one adds μ_ξ^2.

In the special case that $\psi_\xi(\omega)$ is differentiable, i.e., $d\psi_\xi(\omega) = G_\xi(\omega)d\omega$, Eq. (1.11) becomes:

$$R_\xi(\tau) = \int_{-\infty}^{\infty} e^{i\omega\tau} G_\xi(\omega) d\omega \tag{1.12}$$

This expression and its inverse:

$$G_\xi(\omega) = \frac{1}{2\pi} \int_{-\infty}^{\infty} e^{-i\omega\tau} R_\xi(\tau) d\tau \tag{1.13}$$

form a Fourier transform pair. For a real-valued process $\xi(t)$, both R_ξ and G_ξ must be even functions of their respective arguments. Since for $\tau = 0$, $R_\tau(0)$ is the mean-square value of $\xi(t)$, i.e., $E[\xi^2(t)]$, and Eq. (1.13) says it equates the integral of G_ξ over the entire domain ω, $G_\xi(\omega)$ is called the spectral density of $\xi(t)$. In the general case in which $\psi_\xi(\omega)$ is not differentiable, the mean-square value is just the value of $\psi_\xi(\infty)$ and $\psi_\xi(\omega)$ is the spectral distribution function of $\xi(t)$.

1.3.1 Time-Invariant Linear Systems

Provided the additive excitation $\xi(t)$ on a single-input-single-output (SISO) time-invariant linear system is correlation-stationary and damping is present, the response $x(t)$ of the system tends to a correlation-stationary process after the initial transient. Thence, $x(t)$ can be represented in the form of Eq. (1.10), by introducing a second orthogonal-increment process:

$$x(t) = \int_{-\infty}^{\infty} e^{i\omega t} dZ_x(\omega) \tag{1.14}$$

Indeed both $\xi(t)$ and $x(t)$ are composed of sinusoids with complex coefficients $dZ_\xi(\omega)$ and $dZ_x(\omega)$, respectively. They are related through the frequency response function $H(\omega)$ (see [6], among others, for its definition):

$$dZ_x(\omega) = H(\omega)dZ_\xi(\omega) \tag{1.15}$$

If one multiplies both sides by the respective complex-conjugate and takes ensemble averages, the basic result:

$$d\psi_x(\omega) = |H(\omega)|^2 d\psi_\xi(\omega) \tag{1.16}$$

is achieved. For ψ_ξ differentiable, also $\psi_x(\omega)$ is differentiable and one writes:

$$G_x(\omega) = |H(\omega)|^2 G_\xi(\omega) \tag{1.17}$$

For a multi-input-multi-output (MIMO) time-invariant linear system, one introduces a matrix of functions $\psi_{\xi_j\xi_k}(\omega)$: each entry is the cross spectral distribution of two jointly correlation-stationary processes $\xi_j(t)$ and $\xi_k(t)$ and becomes a spectral distribution function when $j = k$. Then Eq. (1.15) is written:

$$dZ_{x_l}(\omega) = \sum_j H_{lj}(\omega)dZ_{\xi_j}(\omega) \tag{1.18}$$

where $H_{lj}(\omega)$ is the frequency response function corresponding to a single input $\xi_j(t)$ and a single output $x_l(t)$. Eqs. (1.16) and (1.17) become in matrix notation:

$$\mathbf{d}\boldsymbol{\psi}_{\boldsymbol{x}}(\omega) = [\boldsymbol{H}(\omega)]\mathbf{d}\boldsymbol{\psi}_{\boldsymbol{\xi}}(\omega)[\boldsymbol{H}^*(\omega)]^T \tag{1.19}$$

and

$$\boldsymbol{G}_{\boldsymbol{x}}(\omega) = [\boldsymbol{H}(\omega)]\boldsymbol{G}_{\boldsymbol{\xi}}(\omega)[\boldsymbol{H}^*(\omega)]^T \tag{1.20}$$

If the parameter of the stochastic process is the time, ω is called the circular frequency. When the process is correlation-homogeneous (instead of stationary) in space, one introduces the associated variable κ known as the wave number.

1.3.2 Models of Common Use

Let underlined symbols denote a realization of the corresponding random variable. Models of common use are, among others, the idealizations listed below:

white noise, $W(t)$: the spectral density function is constant, $G_W = G_0$, and the spectral distribution is linear, $\psi_W(\omega) = G_0\omega$. The mean-square value of

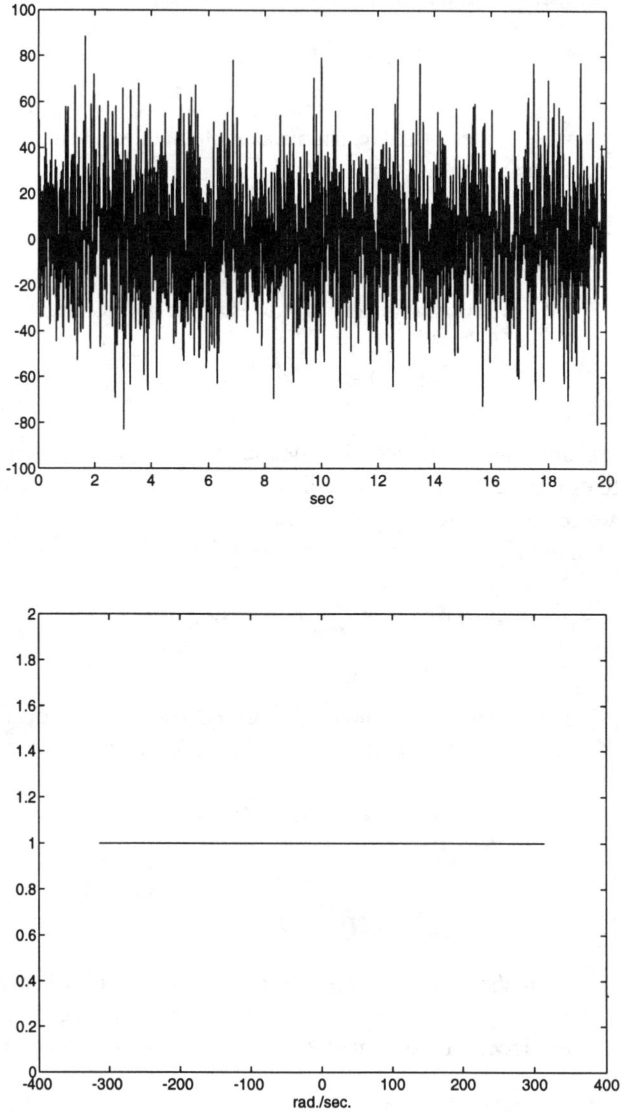

a)

b)

FIGURE 1.1
Realization $\underline{W}(t)$ (a) of the band limited (within (-50 Hz, 50 Hz)) white noise $W(t)$ of spectral density $G_W = 1$. (b).

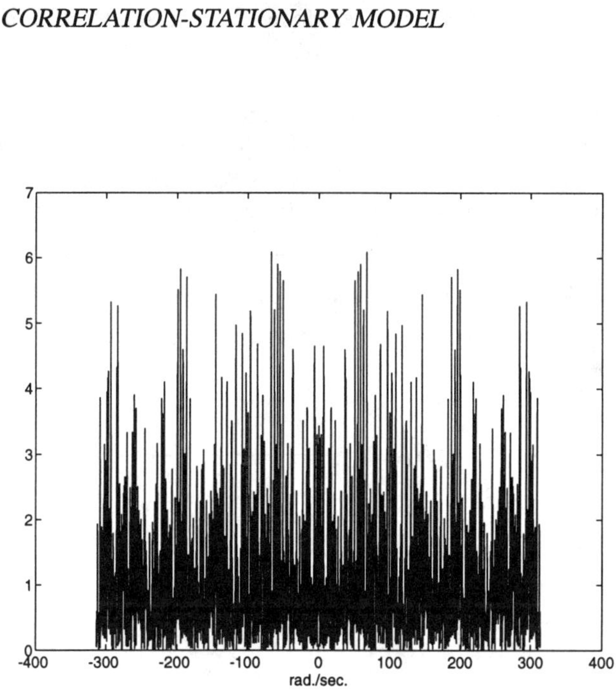

a)

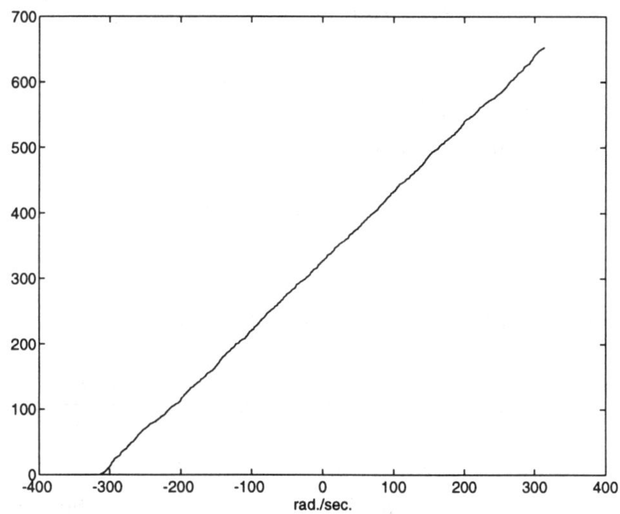

b)

FIGURE 1.2
With reference to the realization of Figure 1.1, (a) shows the estimated spectral density and (b) the estimated spectral distribution.

the process is therefore ∞, i.e., the process does not possess a physical meaning. In Figure 1.1 a realization $\underline{W}(t)$ of band limited white noise, covering the range (-50 Hz, 50 Hz), is represented. Spectral density and distribution estimated from the realization are given in Figure 1.2;

filtered white noise of type I $\xi_I(t)$: the output of a SISO governed by a second order differential equation of parameters ζ_g and ω_g (i.e., $\frac{d^2x(t)}{dt^2} + 2\zeta_g\omega_g\frac{dx(t)}{dt} + \omega_g^2 x(t) = \omega_g^2 \xi(t)$, with $x = \xi_I$ and $\xi(t) = W(t)$) shows the frequency response function:

$$H_I(\omega) = \frac{\omega_g^2}{\omega_g^2 - \omega^2 + 2i\zeta_g\omega_g\omega}$$

The model has spectral density

$$G_{\xi_I}(\omega) = G_0|H_I(\omega)|^2 = G_0\frac{\omega_g^4}{[\omega_g^2 - \omega^2]^2 + 4\zeta_g^2\omega_g^2\omega}$$

and [6] $E[\xi_I^2(t)] = \pi G_0\frac{\omega_g}{2\zeta_g}$.

filtered white noise of type II $\xi_{II}(t)$: the output of a SISO governed by a second order differential equation of parameters ζ_g and ω_g with basement motion (i.e., $\frac{d^2x(t)}{dt^2} + 2\zeta_g\omega_g\frac{dx(t)}{dt} + \omega_g^2 x(t) = \omega_g^2\xi(t) + 2\zeta_g\omega_g\frac{d\xi(t)}{dt}$, where $\xi(t)$ denotes the input displacement; in this case $x = \xi_{II}$ and $\xi(t) = W(t)$) shows the frequency response function:

$$H_{II}(\omega) = H_I(\omega)[1 + \frac{2i\zeta_g\omega}{\omega_g}]$$

The model has spectral density (see Figure 1.2)

$$G_{\xi_{II}}(\omega) = G_0|H_{II}(\omega)|^2 = G_0\frac{\omega_g^4 + 4\zeta_g^2\omega_g^2\omega^2}{[\omega_g^2 - \omega^2]^2 + 4\zeta_g^2\omega_g^2\omega^2}$$

and [6] $E[\xi_{II}^2(t)] = \pi G_0[\frac{\omega_g}{2\zeta_g} + 2\zeta_g\omega_g]$. Figure 1.3 shows a realization $\underline{\xi_{II}}(t)$ of the stochastic processes $\xi_{II}(t)$. The estimated spectral density $\underline{G_{\xi_{II}}}$ is shown together with the estimated spectral distribution in Figure 1.4. The former one results from the expression $|\underline{\xi_{II}}_F(\omega)|^2/(2\pi D)$, where $\underline{\xi_{II}}_F(\omega)$ denotes the Fourier transform of the realization and D its duration.

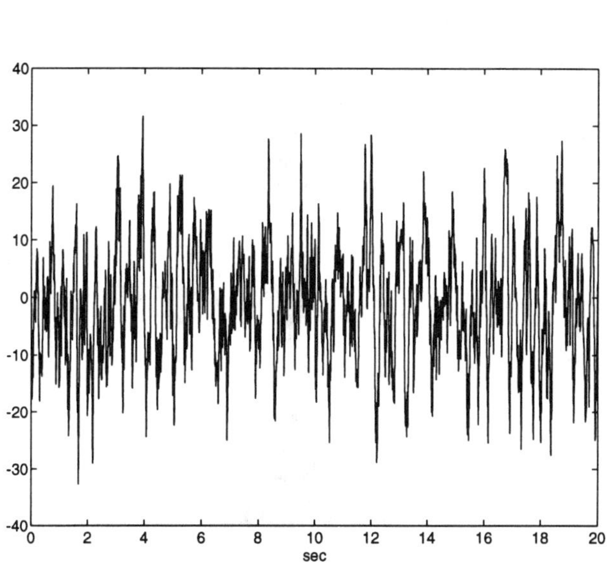

a)

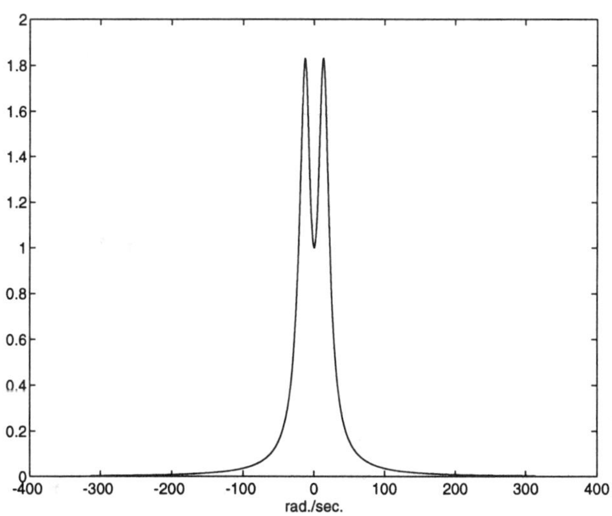

b)

FIGURE 1.3
Realization $\xi_{II}(t)$ (a) of the filtered white noise $\xi_{II}(t)$ of spectral density $G_{\xi_{II}}$ in (b).

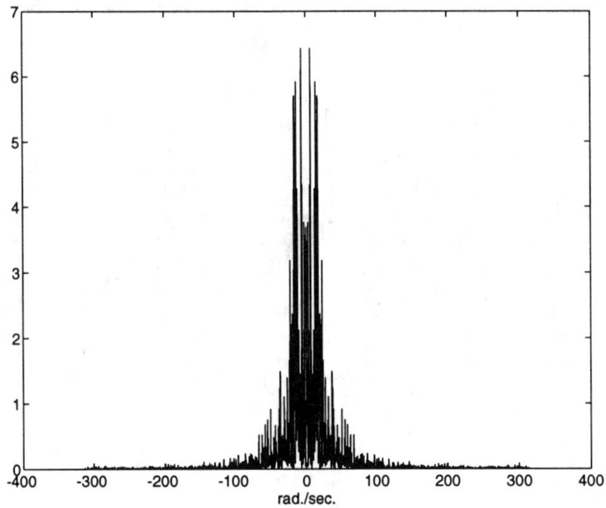

a)

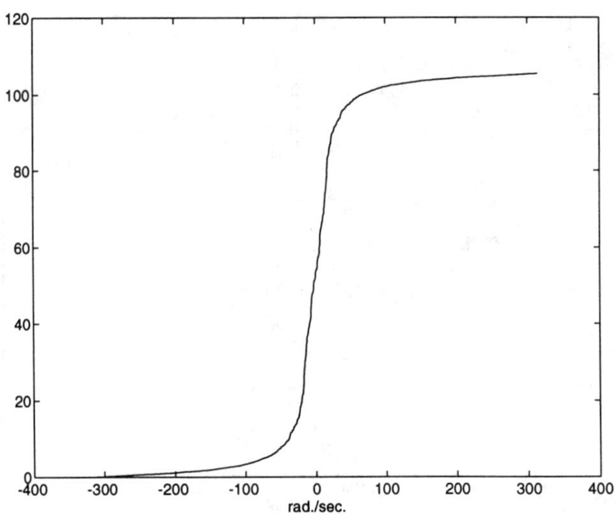

b)

FIGURE 1.4
With reference to the realization of Figure 1.3 (a) shows the estimated spectral density and (b) the estimated spectral distribution.

1.4 The Evolutionary Model

An *evolutionary* stochastic process $\xi(t)$ possesses the integral representation [18, 6]:

$$\xi(t) = \int_{-\infty}^{\infty} a(t,\omega)e^{i\omega t}dZ(\omega) \tag{1.21}$$

where $Z(\omega)$ is an orthogonal increment process and $a(t,\omega)$ is a deterministic (generally complex-valued) function of both t and ω. It has zero mean, but the correlation function depends on t_1 and t_2 separately:

$$E[\xi(t_1)\xi(t_2)] = \int_{-\infty}^{\infty} a^*(t_1,\omega)a(t_2,\omega)e^{i\omega(t_2-t_1)}d\psi(\omega) \tag{1.22}$$

and the mean square value of $\xi(t)$ is:

$$E[\xi^2(t)] = \int_{-\infty}^{\infty} |a(t,\omega)|d\psi(\omega) \tag{1.23}$$

In the special case that $\psi(\omega)$ is differentiable, i.e., $d\psi(\omega) = G(\omega)d\omega$, Eq. (1.23) becomes:

$$E[\xi^2(t)] = \int_{-\infty}^{\infty} |a(t,\omega)|G(\omega)d\omega \tag{1.24}$$

In the previous equations $\psi(\omega)$ and $G(\omega)$ are the spectral distribution and the spectral density, respectively, of some suitable stochastic process. The product $|a(t,\omega)|^2G(\omega)$ is the evolutionary spectral density but it has a physical meaning only if $a(t,\omega)$ is a slowly varying function of t (i.e., nearly independent of t within an interval much longer than $\frac{2\pi}{\omega}$ for any ω).

Figure 1.5 shows a realization of the evolutionary process with spectral density $G_{\xi_{II}}(\omega)$, but with ω_g and G_0 varying as the time functions in Figure 1.6. Figure 1.5 also shows the shape of the associated spectral density at different time instants. Figure 1.7 zooms the realization time history over initial and final ranges.

A sub-class of the evolutionary processes is that of the *uniformly modulated* processes [6]. It is obtained by restricting the term $a(t,\omega)$ to be dependent only on t. The simplest example is a nonstationary white-noise; it is called *shot-noise* and in its correlation function the term $2\pi G(t)$ multiplies the Dirac delta.

1.4.1 Time-Invariant Linear Systems

Provided the additive excitation $\xi(t)$ acting on a single-input-single-output (SISO) time-invariant linear system is an evolutionary process and damping is

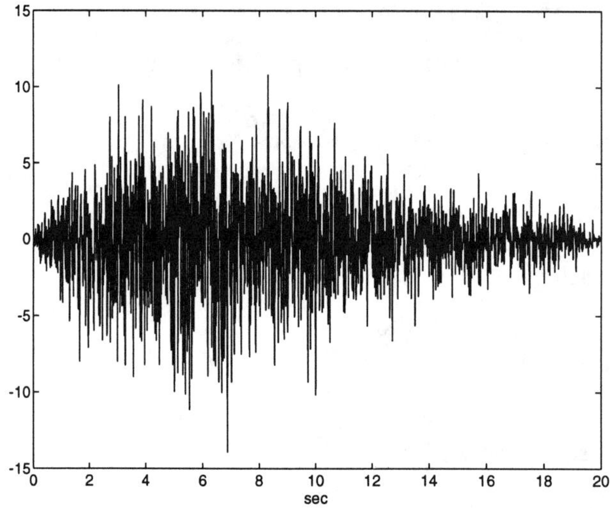

a)

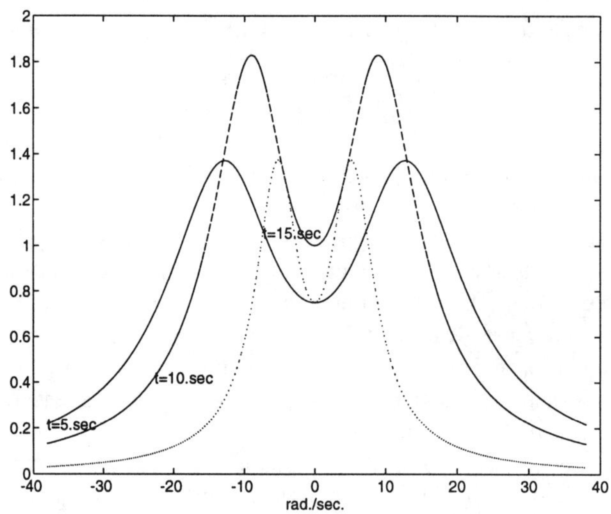

b)

FIGURE 1.5
Realization (a) of the evolutionary process which at each time t is modeled as ξ_{II} with intensity and ω_g depending on t (see Figure 1.6). The spectral density is shown in (b) for three different values of t.

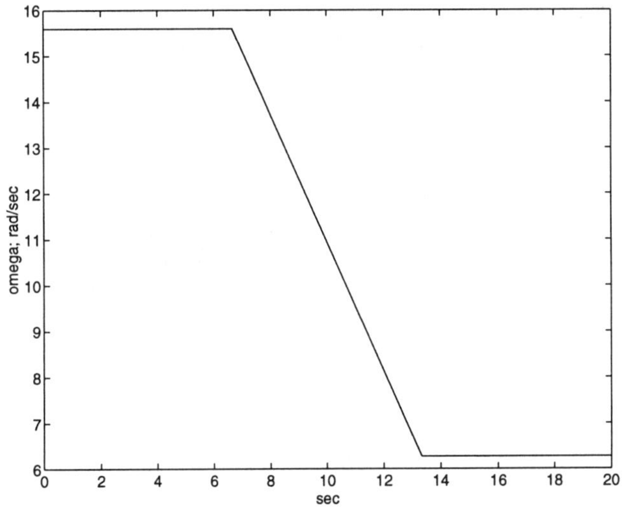

a)

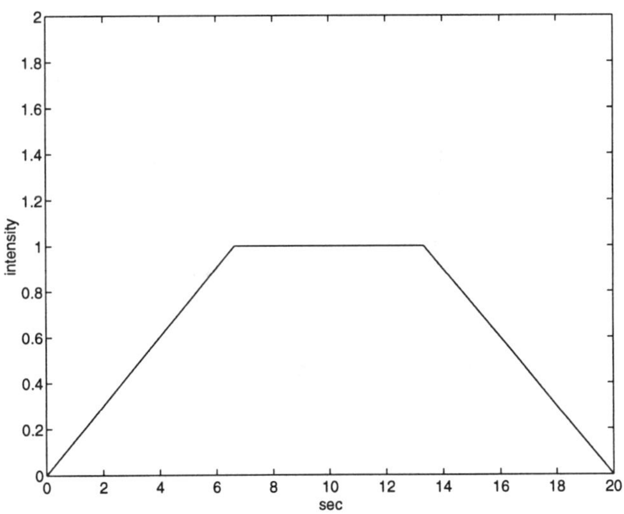

b)

FIGURE 1.6
Variability in time of the intensity (a) and of ω_g (b) for the evolutionary process in Figure 1.5.

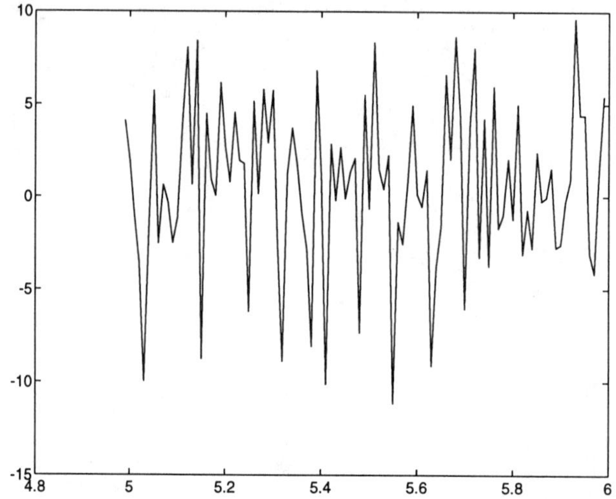

a)

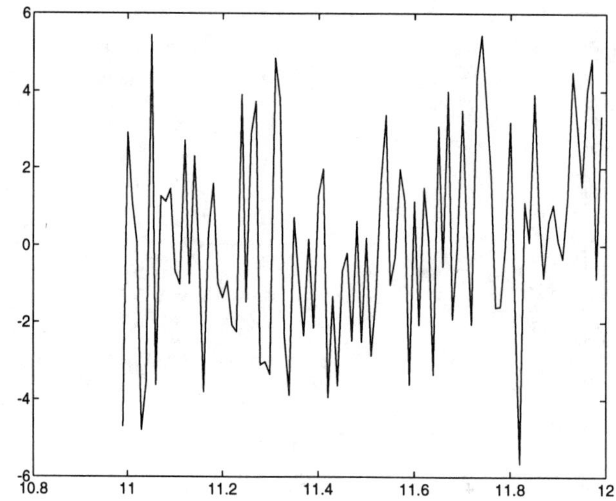

b)

FIGURE 1.7
Zoom over the ranges (5, 6 sec.) (a) and (11, 12 sec.) (b) for the realization of evolutionary process in Figure 1.5.

present, the response $x(t)$ of the system is also an evolutionary stochastic process. Indeed the response can be written:

$$x(t) = \int_0^t h(u)\xi(t-u)du \tag{1.25}$$

where $h(u)$ is the impulse response function, i.e., the Fourier pair of $H(\omega)$. After some algebra one can also write:

$$x(t) = \int_{-\infty}^{\infty} a_r(t,\omega)e^{i\omega t}dZ(\omega) \tag{1.26}$$

with:

$$a_r(t,\omega) = \int_0^t a(t-u,\omega)h(u)e^{-i\omega u}du \tag{1.27}$$

For ψ differentiable, one writes:

$$G_x(t,\omega) = |a_r(t,\omega)|^2 G(\omega) \tag{1.28}$$

and

$$E[x(t_1)x(t_2)] = \int_{-\infty}^{\infty} a_r^*(t_1,\omega)a_r(t_2,\omega)e^{i\omega(t_2-t_1)}G(\omega)d(\omega) \tag{1.29}$$

In the special case the input is covariance stationary, one obtains:

$$a_r(t,\omega) = \int_0^t h(u)e^{-i\omega u}du$$

which tends to the frequency response $H(\omega)$ as t increases, and the response $x(t)$ tends to the state of statistical stationarity.

1.5 Markov Processes

The models of the previous two sections are useful only if a dynamic system is linear and time-invariant and if random excitations are additive. For nonlinear systems or multiplicative random excitations (or both), the known exact solutions are quite limited. When a solution is available, there is generally the assumption that the system response is a Markov stochastic process in some sense.

The process $\xi(t)$ is a *Markov process* when the probability of visiting a state at time t_n, conditional to the values assumed at earlier time instants $t_1, t_2, ..., t_{n-1}$, equates the probability conditional to the only value at time t_{n-1}. Such a definition implies a sufficient condition for $\xi(t)$ being a Markov process: the increments in two non-overlapping intervals must be independent.

Such a model cannot exist physically. It is justified on the basis of how closely the sufficient condition of having independent increments is deemed to be satisfied.

The transition probability distribution function of a Markov vector process $\boldsymbol{\xi}(t)$, of size n, is the probability of reaching a state at time t, given the initial state at time t_0. When it is differentiable with respect to the state, one introduces the transition probability density $q_{\boldsymbol{\xi}}(\boldsymbol{y}, t|\boldsymbol{y}_0, t_0)$. The symbol $\boldsymbol{y}$ denotes the mathematical axes, in the space Ξ, along which the corresponding random variables $\boldsymbol{\xi}$ assume their values. If the Markov process is a Gaussian process, then $q_{\boldsymbol{\xi}}$ satisfies the Fokker-Planck equation [17]

$$\frac{\partial}{\partial t}q_{\boldsymbol{\xi}} + \frac{\partial}{\partial y_j}(A_j q_{\boldsymbol{\xi}}) - \frac{1}{2}\frac{\partial^2}{\partial y_j \partial y_k}(C_{jk} q_{\boldsymbol{\xi}}) = 0 \tag{1.30}$$

where a repeated index represents the summation over $1, \ldots, n$. The coefficients A_j and C_{jk} are the rates of first and second moment, respectively, on the condition that $\boldsymbol{\xi}(t) = \boldsymbol{y}$. Eq. (1.30) is not restricted to Gaussian processes: it also holds for some non-Gaussian processes generated by Gaussian processes. They are known as diffusive Markov processes or *diffusion processes.*

1.5.1 A Model of Common Use

The simplest example of Markov process is the Brownian motion process, also called the *Wiener process* $B(t)$. It is a zero-mean Gaussian process, starting at $t = t_0$ from a zero value with correlation function

$$E[B(t_1), B(t_2)] = \sigma^2 \min(t_1, t_2) \tag{1.31}$$

σ^2 being a positive constant. The moment in Eq. (1.31) is continuous for $t_1 = t_2$ and, hence, $B(t)$ is continuous. Derivation leads one to write

$$E[\dot{B}(t_1), \dot{B}(t_2)] = \sigma^2 \delta(t_2 - t_1)$$

which is unbounded for $t_1 = t_2$ (i.e., $B(t)$ is not differentiable). The Fourier transform, i.e., the associated spectral density, of the r.h.s. of the previous equation is constant and, hence, $\dot{B}(t)$ is formally a white-noise, denoted as $W(t)$ throughout this chapter. Nevertheless, its sample functions are almost surely continuous as well as not differentiable everywhere. This says that the Wiener process is just a mathematical idealization: a physical process can only be close, in some sense, to a Wiener process. Putting $t_1 = t_2$, the previous equation provides:

$$E[dB^2(t)] = \sigma^2 dt$$

from which one recognizes that dB is of the same order of $\sqrt{dt}$; higher order moments are infinitesimal of order higher than dt.

Writing Eq. (1.31) in terms of $E[dB(t_1), dB(t_2)]$, it shows that a Wiener process is a process with orthogonal increments. Due to its Gaussian nature, the increments of $B(t)$ in non-overlapping intervals must be independent and this is a sufficient condition for being a Markov process. The associated Fokker-Planck equation results

$$\frac{\partial q_B}{\partial t} - \frac{1}{2}\sigma^2 \frac{\partial^2 q_B}{\partial y_B^2} = 0$$

1.5.2 Itô Stochastic Differential Equation

An arbitrary scalar Markov process $x(t)$ may be generated by the following procedure. Starting from Eq. (1.1) simplified into

$$\dot{x}(t) = f(x,t) + F(x,t)W$$

one writes

$$dx(t) = f(x,t)dt + F(x,t)dB(t) \tag{1.32}$$

which is fully equivalent to

$$\int_{t_0}^{t} dx(\tau) = \int_{t_0}^{t} f(x,\tau)d\tau + \int_{t_0}^{t} F(x,\tau)dB(\tau)$$

The latter integral in the r.h.s. of the previous equation is neither a Riemann nor a Fourier-Stieljes integral because the corresponding limit as n tends to ∞ and $\Delta\tau_{\max}$ tends to 0 of the sum $\sum_{i=1}^{n} F(x,\tau_i)(B(t_i + \Delta\tau_i) - B(t_i))$, with $t_i \leq \tau_i \leq t_i + \Delta\tau_i$, does not attain a unique value but depends on the intermediate point τ_i one selects. The Itô interpretation (see Appendix) suggests to evaluate $F(x,t)$ at the beginning of the interval for which the increment dB is computed. This interpretation ensures that $dB(\tau)$ is independent of $x(\tau)$, i.e., $x(t)$ is a non-anticipating function.

A deterministic or random function $\varphi(t)$ is said to be a *non-anticipating* function if for any $t > t_j$ the value of the function at time t_j is independent of the increment of the Wiener process $B(t) - B(t_j)$. For it one can write

$$\int_a^b \varphi(t)(dB(t))^2 = \sigma^2 \int_a^b \varphi(t)dt$$

Examples of non-anticipating functions are the solution $x(t)$ of Eq. (1.32) or every function of this solution. Also, every deterministic function possesses this property. In this framework, Eq. (1.32) is called an *Itô stochastic differential equation.*

Let $x(t)$ be the solution of Eq. (1.32). Then the integral

$$\int_{t_0}^{t} \varphi(x(\tau), \tau) dB(\tau)$$

computed according to the Stratonovich scheme (see Appendix) is the sum of the same Itô integral plus the *Wong-Zakai correction term*:

$$\frac{1}{2} \int_{t_0}^{t} \varphi(x(\tau), \tau) \frac{\partial \varphi(x(\tau), \tau)}{\partial \mathrm{x}} \sigma^2 d\tau$$

The roman character x is used to distinguish the axis along which the stochastic process $x(t)$ assumes its values from the process itself. The derivative under the integral in the previous equation requires an explanation: one considers the function $\varphi(\mathrm{x})$, derives it with respect to x and evaluates the result for $\mathrm{x}(t) = x(t)$. In this sense sometimes one says that the function is derived with respect to the process.

For $\varphi(x(t), t) = F(x(t), t)$, the previous comparison can be done between the differential forms

$$F(x(t), t) dB(t)$$

corresponding to the Stratonovich and the Itô schemes. They differ for the quantity

$$\frac{1}{2} F(x(t), t) \frac{\partial F(x(t), t)}{\partial \mathrm{x}} \sigma^2 dt$$

Therefore, if Eq. (1.32) is regarded as a differential Stratonovich equation (see the difference from a computational point of view in Appendix), the Itô integral must be conducted over an equation whose first term in the r.h.s. (the *drift coefficient*) is the sum of the actual drift plus the difference between the two differential forms. The corresponding Itô differential equation writes:

$$dx(t) = (f(x, t) + \frac{1}{2} F(x(t), t) \frac{\partial F(x(t), t)}{\partial \mathrm{x}} \sigma^2) dt + F(x, t) dB(t) \tag{1.33}$$

Eq. (1.33) integrated in the Itô sense gives the same results in mean of those one obtains by integrating Eq. (1.32) in the Stratonovich sense. In the case of additive excitation, $F(x(t), t)$ is independent of $x(t)$, the correction term is zero and the Itô and Stratonovich integrals provide the same result.

In order to incorporate both the expressions in one relation, it is convenient to introduce the drift term $m(x, t)$: it is $f(x, t)$ in the Itô framework and $f(x, t) +$

$\frac{1}{2}F(x(t),t)\partial F(x(t),t)/\partial \mathrm{x}\sigma^2$ in the Stratonovich sense. Once Eq. (1.32) has been transformed in an Itô equation by modifying the drift coefficient, the Itô differential rule can be applied. The Itô differential rule for any scalar, real valued function $\varphi(x(t),t)$ with the suitable properties of differentiability is a Taylor expansion retaining all terms of order lower than or equal to dt

$$d\varphi = \frac{\partial\varphi}{\partial t}dt + \frac{\partial\varphi}{\partial \mathrm{x}}dx + \frac{1}{2}\frac{\partial^2\varphi}{\partial \mathrm{x}^2}(dx)^2 \tag{1.34}$$

For $x(t)$ being the solution of Eq. (1.32), one obtains (see [21], p. 137, for the proof)

$$d\varphi = [\frac{\partial\varphi}{\partial t} + m(x,t)\frac{\partial\varphi}{\partial \mathrm{x}} + \frac{1}{2}F^2(x,t)\frac{\partial^2\varphi}{\partial \mathrm{x}^2}]dt + F(x,t)\frac{\partial\varphi}{\partial \mathrm{x}}dB \tag{1.35}$$

The third term arises by the symbolic notation $dB^2 = dt$, but it disappears for φ a linear function of x.

A result of basic importance for this class of processes is that there are only two coefficients in the Fokker-Planck equation which are different from zero:

$$A_j(\mathrm{x},t) = m(\mathrm{x},t)$$

$$C_{jk}(\mathrm{x},t) = F^2(\mathrm{x},t)$$

where again the symbol x instead of x means that one is regarding the two terms in the r.h.s. as functions into the definition domain $\mathcal{X}$. Therefore, the Fokker-Planck equation for Eq. (1.32) writes

$$\frac{\partial}{\partial t}q_x(\mathrm{x},t) + \frac{\partial}{\partial \mathrm{x}}(m(\mathrm{x},t)q_x(\mathrm{x},t)) - \frac{1}{2}\frac{\partial^2}{\partial \mathrm{x}^2}(F^2(\mathrm{x},t)q_x(\mathrm{x},t) = 0$$

One selects then the function φ as x^k: after stochastic averaging and division by dt, one eventually obtains the rate of the moment $E[x^k]$ and these expressions are the basis for the moment approach. For this purpose the non-anticipating property in the Itô interpretation

$$E[\varphi(x,t)dB] = E[\varphi(x,t)]E[dB] = 0$$

$$E[\varphi(x,t)(dB)^2] = E[\varphi(x,t)]E[(dB)^2] = E[\varphi(x,t)]\sigma^2 dt$$

play a dominant role in writing:

$$\dot{E}[x^k] = kE[x^{k-1}m(x,t)] + \frac{k(k-1)}{2}E[x^{k-2}F^2(x,t)]\sigma^2 \tag{1.36}$$

The previous equation can be obtained by multiplying the Fokker-Planck equation by x^k and integrating the result in the range $\mathcal{X}$.

1.5.3 Some Examples

For the linear time invariant system through which an additive Markov excitation $\xi(t)$ is filtered, in particular, the classical Lyapunov matrix equation holds:

$$\frac{d}{dt}\boldsymbol{\Sigma}_{\boldsymbol{x}}(t) = \frac{d}{dt}[E[x_j(t)x_k(t)]] = \boldsymbol{f}\boldsymbol{\Sigma}_{\boldsymbol{x}}(t) + \boldsymbol{\Sigma}_{\boldsymbol{x}}(t)\boldsymbol{f}^T + [E[\xi x^T]] + [E[x\xi^T]] \tag{1.37}$$

where the last two terms simplify into a matrix with all zero except $2\pi G(t)$ for the entry corresponding to the equation driven by the Wiener process. An extension of Eq. (1.37) can also be written for a time varying $\boldsymbol{f}(t)$ and for a multivariate input of Wiener processes.

A figure in [5] (p. 164) shows for instance the evolution in time of the probability density function of the response of a Duffing oscillator, i.e., an oscillator with a cubic restoring force:

$$dx = ax^3 + F(t)dB$$

The Wong-Zakai term is zero since the excitation is additive and the Itô differential rule is:

$$dx^k = kx^{k-1}(ax^3dt + F(t)dB) + \frac{k(k-1)}{2}x^{k-2}F^2\sigma^2dt$$

The resulting moment equation is:

$$\dot{E}[x^k] = akE[x^{k+2}] + \frac{k(k-1)}{2}F^2\sigma^2E[x^{k-2}]$$

In the stationary case the l.h.s. of the previous equation becomes zero: the odd moments are all zero while the moments of order 4, 8, 12, ... are immediately computed; the moments of order 6, 10, 14 are obtained once the second moment is known. The response process does not present a Gaussian nature. One says that the equations of the moments form a hierarchy.

The simplest multiplicative case considers $f(x,t) = f(t)x(t) + \rho(t)$ and $F(x,t) = F(t)x(t) + \beta(t)$, ρ and β being two known functions of time. Then the Wong-Zakai term is written $\frac{1}{2}[F^2(t)x(t) + F(t)\beta(t)]\sigma^2$ and:

$$m(x,t) = f(t)x(t) + \rho(t) + \frac{1}{2}[F^2(t)x(t) + F(t)\beta(t)]\sigma^2$$

Adopting $\varphi(x,t) = x^k$ in Eq. (1.34), stochastic averaging and division by dt leads one to write:

$$\dot{E}[x^k] = kfE[x^k] + k\rho E[x^{k-1}] + \tfrac{1}{2}kF^2E[x^k]\sigma^2 + \tfrac{1}{2}kF\beta E[x^{k-1}]\sigma^2 +$$

$$\tfrac{1}{2}k(k-1)F^2E[x^k]\sigma^2 + \tfrac{1}{2}k(k-1)\beta^2E[x^{k-2}]\sigma^2 +$$

$$k(k-1)F\beta E[x^{k-1}]\sigma^2$$

which do not form a hierarchy.

1.5.4 Approximation of Mechanical Processes: Physical versus Itô Equations

Assume that $\xi(t)$ is not observed continuously but at discrete time instants $t_0 + k\Delta t$ with $k = 1, 2, \ldots$ The observed increments of the process realization will be $\Delta\xi_1, \Delta\xi_2, \ldots$. The sufficient condition for $\xi(t)$ being a Markov process requires that these increments are independent, regardless of how short the time intervals are. This property can never be found in a real physical process and, hence, the Markov process is a mathematical idealization.

Nevertheless, as long as there is true randomness in the mechanical process, one can regard the time steps as observation time laps: their length will be selected long enough that the observed increments show a Markov nature.

1.6 The Random Pulse Train Model

The *random pulse train* is a process $\xi(t)$ built as the sum of a random number $N(t)$ of pulses $w(t, \tau_j)$ of deterministic shape, arriving at time τ_j with random intensity Y_j:

$$\xi(t) = \sum_{j=1}^{N(t)} Y_j w(t, \tau_j) \tag{1.38}$$

In addition, one can often accept that the Y_j are independent but identically distributed random variables (which are also independent of the arrival time) and that $w(t, \tau_j)$ is a function of $t - \tau$:

$$\xi(t) = \sum_{j=1}^{N(t)} Y_j w(t - \tau_j) \tag{1.39}$$

One introduces:

- the random pulse-arrival rate $\Lambda(t)$. Its correlation properties are described by the cumulant functions. In the special case of independent (Poisson) arrivals all the cumulants $g_l(\tau_1, ..., \tau_l)$ are zero except the first cumulant g_1. This is the case of a filtered Poisson process.
- the moments of the random variable Y as $E[Y] = 0$ and $E[Y^2] \neq 0$ but bounded.

Under the latter assumptions the correlation structure of the problem becomes [18];

$$E[\xi(t_1)\xi(t_2)] = E[Y^2]\int_{-\infty}^{\infty} g_1(\tau)w(t_1-\tau)w(t_2-\tau)d\tau \tag{1.40}$$

Since g_1 is non-negative, one can put

$$a(t,\omega) = \int_{-\infty}^{\infty} w(u)\sqrt{g_1(t-u)}e^{-i\omega u}du$$

and after some algebra Eq. (1.40) is re-written

$$E[\xi(t_1)\xi(t_2)] = \frac{1}{2\pi}E[Y^2]\int_{-\infty}^{\infty} a^*(t_1,\omega)a(t_2,\omega)e^{i\omega(t_2-t_1)}d\omega \tag{1.41}$$

This shows that the correlation structure of the random pulse train of Eq. (1.38) is the same as the one of an evolutionary process of evolutionary spectral density:

$$G_\xi(t,\omega) = \frac{1}{2\pi}E[Y^2]|a(t,\omega)|^2 \tag{1.42}$$

1.6.1 The Delta-Correlated Model

Note that a discrete form of white noise can be simply expressed in the form (1.39) with $w(t-\tau_j) = \delta(t-\tau_j)$, $\delta(\ )$ being the Dirac's function. It is called a *delta-correlated process*:

$$W_\Lambda(t) = \sum_{j=1}^{N(t)} Y_j\delta(t-\tau_j) \tag{1.43}$$

with cumulants [22]

$$g_r(t_1,t_2,\ldots,t_r) = \Lambda(t)E[Y^r]\delta(t_1-t_2)\delta(t_1-t_3)\ldots\delta(t_1-t_r) \tag{1.44}$$

A discrete form of the Wiener process can be written introducing the Heaviside function $U(t-\tau_j)$. It is called the *homogeneous compound Poisson process*

$$B_{\Lambda}(t) = \sum_{j=1}^{N(t)} Y_j U(t-\tau_j) \tag{1.45}$$

The formal derivative of $B_{\Lambda}(t)$ is the Poisson delta correlated process $W_{\Lambda}(t)$. If $\Lambda(t)$ is constant in time then the process $W_{\Lambda}(t)$ is a strongly stationary one. Moreover, if, as $\Lambda \to \infty$, $\Lambda E[Y^2]$ keeps a constant value and $\Lambda E[Y^r] \to 0$ $\forall r > 2$ then the non–normal delta correlated process becomes a normal white noise process $W(t)$ and the homogeneous compound Poisson process becomes a Wiener process $B(t)$.

The cumulants of an increment of the process $B_{\Lambda}(t)$ can be written in the form

$$\begin{aligned} g_1(t) &= E[dB_{\Lambda}(t)] = \gamma_1(t)dt \\ g_2(t_1,t_2) &= E[dB_{\Lambda}(t_1)dB_{\Lambda}(t_2)] - E[dB_{\Lambda}(t_1)]E[dB_{\Lambda}(t_2)] \\ &= \gamma_2(t_1)\delta(t_1-t_2)dt_1dt_2 \\ &\dots \end{aligned} \tag{1.46}$$

in which $\gamma_j(t)$=$\Lambda(t)E[Y^j]$. Letting t_1=t_2=…=t_r=t, Eq. (1.46) gives

$$g_1(t) = \gamma_1(t)dt, \quad g_2(t,t) = \gamma_2(t)dt, \quad g_3(t,t,t) = \gamma_3(t)dt, \quad \dots \tag{1.47}$$

From Eq. (1.47) one recognizes that moments and cumulants of $dB_{\Lambda}(t)$ coincide and are all infinitesimal of first order. This is the substantial difference between the normal Wiener process and the compound Poisson process. In fact, in the former case, $dB(t)$ is of order $(dt)^{1/2}$ and this is the key of the Itô calculus, while for the $dB_{\Lambda}(t)$ most of the simplifications connected with the order of $dB_{\Lambda}(t)$ cannot be used.

1.6.2 Fokker Planck and Moment Equations for Parametric Delta Correlated Input

Let the equation of motion be given in the form

$$dx(t) = f(x,t)dt + F(x,t)dB_{\Lambda}(t) \tag{1.48}$$

In this case the evolution of the probability density function $q_x(\mathrm{x}, t)$ can be written in the form [8]

$$\frac{\partial}{\partial t} q_x(\mathrm{x}, t) = -\frac{\partial}{\partial \mathrm{x}}(q_x(\mathrm{x}, t) f(\mathrm{x}, t)) + \sum_{k=1}^{\infty} \frac{(-1)^k}{k!} \frac{\partial^k}{\partial \mathrm{x}^k}(q_x(\mathrm{x}, t)\Psi_k(\mathrm{x}, t)) \quad (1.49)$$

where

$$\Psi_1(\mathrm{x}, t) = \sum_{j=1}^{\infty} \frac{1}{j!} F^{(j)}(\mathrm{x}, t)\gamma_j(t) \quad (1.50)$$

$$\Psi_2(\mathrm{x}, t) = \sum_{j=1}^{\infty}\sum_{r=1}^{\infty} \frac{1}{j!r!} F^{(j)}(\mathrm{x}, t) F^{(r)}(\mathrm{x}, t)\gamma_{j+r}(t) \quad (1.51)$$

$$\ldots$$

The terms $F^{(j)}(\mathrm{x}, t)$ can be evaluated by the recursive form

$$F^{(1)}(\mathrm{x}, t) = F(\mathrm{x}, t); \qquad F^{(j)}(\mathrm{x}, t) = \frac{\partial F^{(j-1)}(\mathrm{x}, t)}{\partial \mathrm{x}} F(\mathrm{x}, t) \quad (1.52)$$

In the case of external excitation, that is $F(x, t)$=$F(t)$, then the Fokker Planck Eq. (1.49) writes

$$\frac{\partial}{\partial t} q_x(\mathrm{x}, t) = -\frac{\partial}{\partial \mathrm{x}}(q_x(\mathrm{x}, t) f(\mathrm{x}, t)) + \sum_{k=1}^{\infty} \frac{(-1)^k}{k!} \frac{\partial^k q_x(\mathrm{x}, t)}{\partial \mathrm{x}^k} F^k(t)\gamma_k(t) \quad (1.53)$$

For the case of normal white noise, that is $\Lambda E[Y^r]$=0 $\forall r > 2$ (since cumulants and moments coincide), Eq. (1.49) reduces to the well known Fokker Planck equation.

Extension of the Itô stochastic differential calculus to integro–differential equations subjected to a parametric–type discontinuous process with independent increments and arbitrary set of jumps can be found in [11]. However, no connection between the differential Eq. (1.48) and the integro–differential equation studied by Gihman and Skorohod in [11] has been found up to date, while in the case of external excitation, Eq. (1.53) proves to be equivalent to that proposed by Roberts in [20].

Some exact solutions of the generalized Fokker Planck equation (1.49) have been found in [24] and could serve as benchmark solutions for comparison with approximate and numerical techniques.

In order to find the differential equations governing the evolution of moments of every order of the response process $x(t)$, solution of Eq. (1.48), one can multiply

Eq. (1.49) by x^r and performing the integrations, one has:

$$\dot{\mu}_r[x] = rE[x^{r-1}f(x,t)] + \sum_{k=1}^{r} \frac{r!(-1)^{k+1}}{k!(r-k)!} E[\Psi_k(x,t)x^{r-k}] \tag{1.54}$$

In the case of external excitation, that is $F(x,t)$=$F(t)$, Eq. (1.54) simplifies into

$$\dot{\mu}_r[x] = rE[x^{r-1}f(x,t)] + \sum_{k=1}^{r} \frac{r!(-1)^{k+1}}{k!(r-k)!} \mu_{r-k}[x]F^k(t)\gamma_k(t) \tag{1.55}$$

At least if the system is linear, that is $f(x,t)$=$f(t)x + \rho(t)$, $F(x,t)$=$F(t)$, then Eq. (1.55) reduces to

$$\dot{\mu}_r[x] = r\mu_r[x]f(t)+r\rho(t)\mu_{r-1}[x]+\sum_{k=1}^{r} \frac{r!(-1)^{k+1}}{k!(r-k)!} \mu_{r-k}[x]F^k(t)\gamma_k(t) \tag{1.56}$$

From Eq. (1.56) one realizes that, for linear systems, the equation of moments of every order can be exactly formed starting from the first order moment. On the contrary, also in the case of nonlinear systems under external or parametric delta correlated process, the moments of every order cannot be computed exactly, in general, because the moment of a fixed order requires the knowledge of the probability density function that in turn depends on the moments. It follows that in order to solve them, some closure schemes will be adopted (see e.g. [26] and [15]).

1.6.3 Quasi–linear Systems

Another relevant example in which the solution, in terms of moments, can be exactly found is the so called quasi–linear (also called bilinear or simply linear) system, that is:

$$dx(t) = (f(t)x(t) + \rho(t))dt + (F(t)x(t) + \beta(t))dB_{\Lambda} \tag{1.57}$$

In this case, the functions $F^{(j)}$ simply write $F^{(1)}(x,t) = F(t)x(t) + \beta(t)$, $F^{(j)}(x,t) = F^{j-1}(x,t)(F(t)x(t) + \beta(t))$, and the differential equations of moments of every order of the response write (with argument omitted)

$$\begin{aligned}\dot{\mu}_r = \quad & r(f\mu_r + \beta\mu_{r-1}) + r(e_{1,1}\alpha_{1,1}\mu_r + e_{1,0}\alpha_{1,0}\mu_{r-1}) \\ & +\tfrac{r(r-1)}{2}(e_{2,2}\alpha_{2,2}\mu_r + e_{2,1}\alpha_{2,1}\mu_{r-1} + e_{2,0}\alpha_{2,0}\mu_{r-2}) + \cdots \\ & +\tfrac{r!}{r!}(e_{r,r}\alpha_{r,r}\mu_r + e_{r,r-1}\alpha_{r,r-1}\mu_{r-1} + \cdots + e_{r,0}\alpha_{r,0})\end{aligned} \tag{1.58}$$

In Eq. (1.58)

$$\alpha_{p,q} = \beta^{(p-q)} \sum_{j_1=1}^{\infty} \sum_{j_2=1}^{\infty} \cdots \sum_{j_p=1}^{\infty} \frac{F^{j_1+j_2+\ldots+j_p+q-p}}{j_1! j_2! \ldots j_p!} \gamma_{j_1+j_2+\ldots+j_p} \quad p \geq q$$

and $e_{p,q}$ are those derived from a Tartaglia triangle, i.e.,

$$e_{p,p} = e_{p,0} = 1 \quad ; \quad e_{p,q} = e_{p-1,q} + e_{p-1,q-1}$$

It will be noted that Eq. (1.58) does not constitute an infinity hierarchy in the sense that the moment equation of fixed order, say r, involves only moments of order lower than r and then they can be exactly solved. If $B_\Lambda(t)$ in Eq. (1.48) is the normal Wiener process $B(t)$, Eq. (1.58) reduces to

$$\dot{\mu}_r = r f_r(t)\mu_r - r\left(\rho(t) + \frac{1}{2}F(t)\beta(t)\gamma_2(t)\right)\mu_{r-1} \tag{1.59}$$

where

$$f_r(t) = f(t) + \frac{r}{2}F^2(t)\gamma_2(t)$$

If $f(t)$, $F(t)$, $\rho(t)$, $\beta(t)$, and $\gamma_2(t)$ are constant quantities and the system operates from t=$-\infty$, then the response $x(t)$ attains the stationary solution and the moment equations (1.59) are given in the simple form:

$$\mu_r = (-1)^r \frac{\left(\rho + \frac{1}{2}F\beta\gamma_2\right)^r}{\prod_{j=1}^r f_j} \qquad r = 1, 2, \cdots, \infty.$$

It will be noted that in the case of normal white noise input the response $x(t)$ of the quasi linear systems is not normal, in fact the cumulants of order greater than two prove to be different from zero. This is due to the coupling effect of the term $F(t)x(t)dB_\Lambda$.

1.6.4 Simulation of Delta Correlated Processes and Response

In Figure 1.8 a sample function of delta correlated process is plotted. From this figure one realizes that it consists of well spaced impulses. Therefore, the generation of a sample function requires the generation of a realization $\underline{Y}_j$ of the random variable Y_j with assigned distribution and a realization $\underline{\tau}_j$ of the random variable τ_j distributed according to the Poisson law. The integration of any sample function, then, requires the solution of the deterministic differential equation:

$$\dot{\underline{x}}(t) = f(\underline{x}, t) + F(\underline{x}, t)\underline{Y}_j\delta(t - \underline{\tau}_j) \quad ; \quad \underline{x}(\underline{\tau}_{j-1}) = \underline{x}_{j-1} \tag{1.60}$$

where $\underline{Y}_j$ is the amplitude of the spike and $\underline{x}_{j-1}$ is the initial condition for the time interval τ_{j-1}, τ_j. For $\underline{\tau}_{j-1} < t < \underline{\tau}_j$ and $\underline{\tau}_j < t < \underline{\tau}_{j+1}$ the solution is that of the differential equation $\dot{\underline{x}}(t) = f(\underline{x}, t)$. The only problem in the integration of Eq. (1.60) is that, immediately before and after the impulse, $\underline{x}(t)$ exhibits a jump $\Delta\underline{x}_j$ given as:

$$\Delta\underline{x}_j = \underline{x}(\underline{\tau}_j^+) - \underline{x}(\underline{\tau}_j^-)$$

where $\underline{\tau}_j^-$ and $\underline{\tau}_j^+$ indicate the time instants immediately before and after, respectively, the impulse arrival. Now the question is whether or not it is possible to obtain an exact prediction of the jump. Rewriting Eq. (1.60) in integral form, on the basis of elementary considerations, one recognizes that $\int_{\underline{\tau}_j-\Delta t}^{\underline{\tau}_j+\Delta t} F(\underline{x}, t)\underline{Y}\delta(t - \underline{\tau}_j)dt$ is no longer a Riemann integral nor a Fourier–Stieltjes integral since the limit of the Riemann sum does not attain an unique value but depends on the intermediate point selected. Very recently Caddemi and Di Paola [4] showed that the exact jump is provided by the series:

$$\Delta x_j = dx + \frac{1}{2}d^2x + \frac{1}{3!}d^3x + \cdots = \sum_{k=1}^{\infty}\frac{d^kx}{k!} \tag{1.61}$$

In words, as the response exhibits a jump in $t = \underline{\tau}_j$, higher order differentials give contribution of the same order as dx.

Using the rules of differentiation of composite function the jump can then be evaluated in the form:

$$\Delta\underline{x}_j = \sum_{k=1}^{\infty}\frac{\underline{Y}_j^k}{k!}\left[F^{(k)}(\underline{x}, t)\right]_{t=\underline{\tau}_j^-} \tag{1.62}$$

where $F^{(k)}(x, t)$ has been already defined in Eq. (1.52). With simple considerations, it can also be shown that the first term of the numerical series (1.62) coincides with the jump evaluated in the Itô sense: the jump only depends on the value obtained by the response immediately before the impulse. Moreover, truncation of the numerical series (1.62) after the second term leads one to the approximation obtained by performing the integral in the Stratonovich sense:

$$(\text{Itô}) \rightarrow \Delta x_j \qquad = Y_jF(x, \tau_j^-)$$

$$(\text{Stratonovich}) \rightarrow \Delta x_j = Y_jF(x, \tau_j^-) + \frac{Y_j^2}{2}F(x, \tau_j^-)\left[\frac{\partial F(x,t)}{\partial x}\right]_{t=\tau_j^-}$$

Note that the additional term $(Y_j^2/2)F(x, \tau_j^-)\,[\partial F(x, t)/\partial x]_{t=\tau_j^-}$ presents exactly the same form of the Wong Zakai correction term. The remaining terms of the

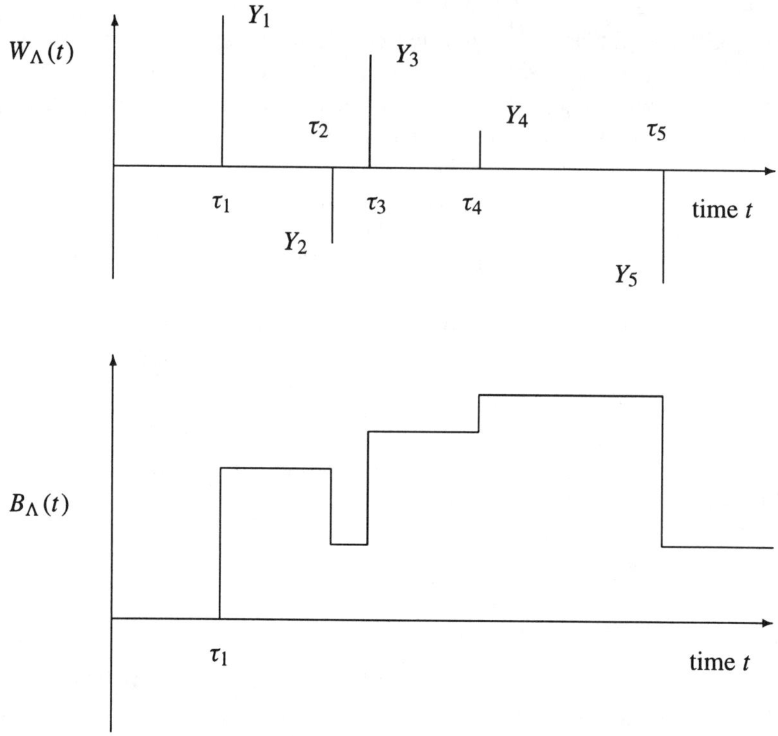

FIGURE 1.8
Sample of Poisson train of pulses $W_\Lambda(t)$ and corresponding homogeneous compound Poisson process.

series (1.62) possess the same mathematical form of the terms of the infinite series introduced by Di Paola and Falsone [9] to account for the non–normality of the input in the case of parametric Poisson white noise. From the previous considerations one can summarize that Itô and Stratonovich integrals are approximations of the jump obtained by the response. But the exact total jump can only be evaluated by means of Eq. (1.62). In the case of external excitation the series (1.62) reduces to the first term, so that Itô and Stratonovich integrals coincide.

In summary, the simulation of delta correlated processes and response can be done following, for each sample function, the following steps:

1. generate a time instant $\underline{\tau}_j$ from a Poisson law with mean arrival number Λ;

2. generate the corresponding value $\underline{Y}_j$ from its assigned probability distribution;

3. integrate, between two subsequent pulses occurring in (τ_{j-1}, τ_j), the differential equation $\dot{\underline{x}} = f(\underline{x}, t)$ with initial condition $\underline{x}(\underline{\tau}_{j-1}^+)$;

4. evaluate the jump $\Delta \underline{x}_j$ from Eq. (1.62) and compute $\underline{\tau}_j^+$, which acts as initial condition for the time interval $(\underline{\tau}_j, \underline{\tau}_{j+1})$.

1.6.5 Simulation of Normal White Noise Input and Response

Standard methods exist for the simulation of Gaussian white noise realizations. They are illustrated in classical text books (see [2, 6] among others) and were adopted for the simulations of Sections 1.3 and 1.4.

In this section the simulation of normal white noise input and response is framed in the context outlined in the previous subsection for reasons that will soon appear evident. This problem was also examined in [13] in the context of Itô stochastic differential calculus ([16, 1, 14, 25, 21]). How it was observed before a normal white noise can be seen as an impulsive Poisson process when $\Lambda \to \infty$ and $\Lambda E[Y^2] = \gamma_2$ attains a constant value.

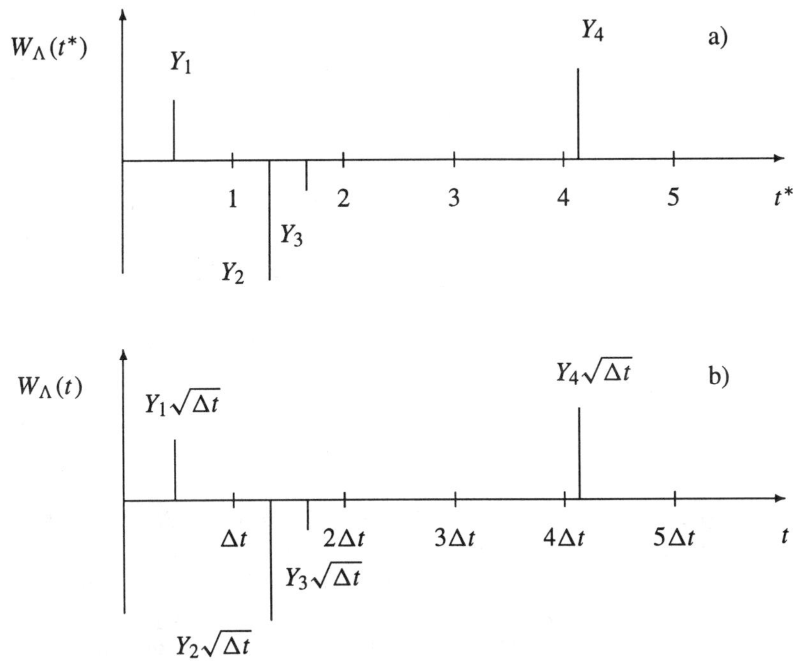

FIGURE 1.9
Poisson processes with $\Lambda = 1$ (a) and $1/\Delta t$ (b).

In order to simulate a normal white noise, let t^* be a fictitious reference time such that $\Lambda = 1$ that means one pulse per second is expected in mean. Moreover, let $\Lambda E[Y^2]$ be equal to γ_2. Select an integration time step Δt and map the original time t^* into a new time axis t in such a way that the unitary interval becomes Δt. From Figure 1.9 one recognizes that the location of the impulses along the axis t^* can be obtained by stretching the effective time t of $1/\Delta t$. Since in the new time axis t, Λ is $1/\Delta t$, one has to reduce each individual pulse Y_j along the time axis t multiplying by the factor $\sqrt{\Delta t}$ in order to have the same value of $\Lambda E[Y^2]$ along the two axes t^* and t. In this way the two processes have the same value of γ_2 and, as $\Delta t \to 0$, the process $W_\Lambda(t)$ tends to the normal white noise: it is independent of the original distribution of Y.

Because one is dealing with an impulsive process, extensive use of the relevant concepts can be done. For Δt very small one has

$$x(t_{j+1}) = x(t_j) + f(x, t_j)\Delta t + \sum_{k=1}^{\infty} \frac{Y_j^k (\Delta t)^{k/2}}{k!} F^{(k)}(x, t_j) \tag{1.63}$$

where Y_j is simply the sum of the amplitude of all pulses occurring in the time step from t_j to t_{j+1}. From Eq. (1.63) one realizes that the coefficients of the series decrease with law $(\Delta t)^{k/2}/k!$; therefore, if Δt is very small, the first two terms in the summation (1.63) give a contribution of the same order of $f(x, t_j)\Delta t$. As $\Delta t \to dt$, only the first two terms of the series in Eq. (1.63) give contributions of order dt and Eq. (1.63) reduces to the Itô equation.

If Δt is small but a finite quantity, the results are affected by errors: the first error is associated with the central difference scheme adopted; the second one with the terms of the series (1.63) included in the analysis. In order to minimize the latter error, assume for Y a normal distribution with zero mean and standard deviation $\sqrt{\gamma_2}$; in this way $E\left[Y^3\right] = 0$, $E\left[Y^4\right] = 3\gamma_2^2$, and hence the third term in the summation (1.63) is zero and the fourth term is of order $3\Delta t^2/4!$. Then, for a zero mean normal white noise, the central scheme difference simply writes

$$x\left(t_{j+1}\right) = x\left(t_j\right) + f(x, t_j)\Delta t + \frac{Y_j^2}{2!} F^{(2)}\left(x, t_j\right)\Delta t \tag{1.64}$$

Another source of error is associated with the fact that Eq. (1.63) does not account for the exact location of the impulse in the time step Δt, because $F^{(k)}(x, t)$ is evaluated at the beginning of the step τ_j independently of the location of the total impulse Y_j. However, this is a source of minor error because the jump in a very small interval is independent of the shape of the impulse but only depends on the total intensity of the impulses occurring in the time interval from t_j to t_{j+1} [4]. It follows that, for the generation of normal white noise realizations and the associated response, one can define the time step Δt and generate a random variable Y with assigned distribution law (say normal, with variance $\gamma_2\Delta t$): then one can conduct the integration using at each time step Eq. (1.63) and retaining only the first two terms in the summation.

As an example, consider the quasi linear oscillator studied in [9] and governed by the differential equation

$$\dot{x} = a + \gamma_0 x W_\Lambda(t) \quad ; \quad x(0) = 0 \tag{1.65}$$

$W_\Lambda(t)$ is the Poisson white noise defined in Eq. (1.43). The distribution of the random variable Y is uniform over the interval $(0, u)$, and the moments of Y are

$$E\left[Y^n\right] = \frac{u^{n+1}}{n+1} \quad ; \quad n = 1, 2, \ldots \tag{1.66}$$

The intensities $\gamma_j \quad j = 1, 2, \ldots$ are given as $\gamma_j = \Lambda E\left[Y^j\right]$, Λ being the mean number of arrivals per unit time. The parameters chosen for the analysis of Eq. (1.65) are: $a = 1$, $\gamma_0 = -1$, $\Lambda = 1$, $u = 1$. A Monte Carlo simulation has been performed by generating one thousand waves of input and response. The analysis for each sample and at each impulse occurrence has been performed by using one term in the summation of the series (1.62) (i.e., evaluating the Itô integral), two terms (i.e., evaluating the Stratonovich integral) and ten terms. The results in terms of first order and second order moments are reported in Figures 1.10 and 1.11, respectively. The results are compared with those evaluated by integrating Eq. (1.58) and retaining one term, two terms, and ten terms of the corrective coefficients $\alpha_{p,q}$ in the series.

In Figures 1.12 and 1.13 the results in terms of first and second order moments of Eq. (1.65) are reported for a larger value of Λ (Λ=100), and γ_1=0, γ_2=$\Lambda E[Y^2]$=1. In this way the non-normal white noise is approximately a normal one, with $\gamma_0 = -0.5$. The analysis has been performed by using a Monte Carlo simulation with two terms only for each impulses occurrence and by a second simulation allowing for ten terms: the two results cannot be distinguished.

1.6.6 Orthogonal-Increment Model for Delta Correlated Processes

Let $W_\Lambda(t)$ be a Poisson delta correlated process: it can be represented in the form (1.10), that is

$$W_\Lambda(t) = \int_{-\infty}^{+\infty} e^{i\omega t} dZ_{W_\Lambda}(\omega) \tag{1.67}$$

Because of the non–normality of the process $W_\Lambda(t)$, additional information on the orthogonal increments process $Z_{W_\Lambda}(\omega)$ is necessary in order to guarantee the strong stationarity of the non-normal process $W_\Lambda(t)$. These conditions are the so called "generalized orthogonality conditions" [8], that is the cumulants of $Z_{W_\Lambda}(\omega)$ obey the following relationships:

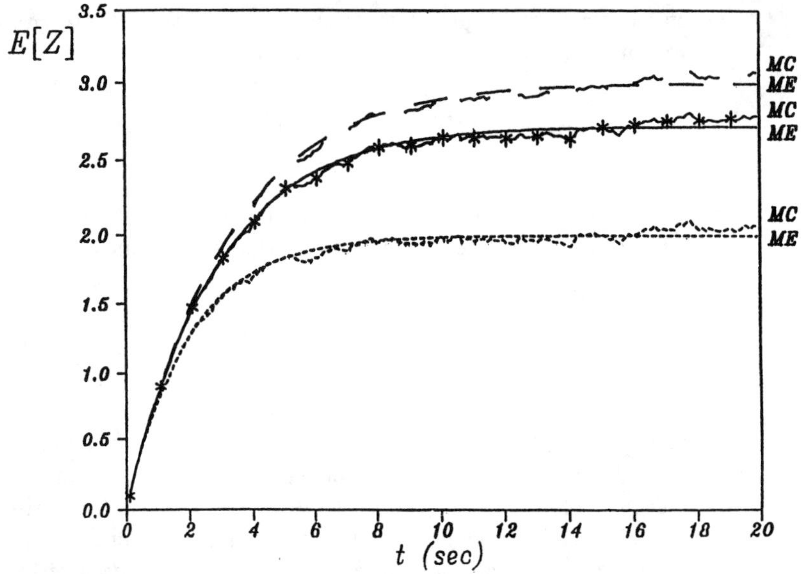

FIGURE 1.10
Non-normal delta correlated input. Evolution of the first order moment of the response of Eq. (1.65) by means of Monte Carlo simulation (MC) and moment equation (ME): short dashed line with one correction term; long dashed line with two correction terms; solid line with ten correction terms.

$$g_1(dZ_{W_\Lambda}(\omega_0)) = \Delta_K(\omega_0)d\psi^{(1)}_{W_\Lambda}(-\omega_0)$$

$$g_2(dZ_{W_\Lambda}(\omega_0)dZ_{W_\Lambda}(\omega_1)) = \Delta_K(\omega_0+\omega_1)d\psi^{(2)}_{W_\Lambda}(-\omega_0, \omega_1) \quad (1.68)$$

$$g_3(dZ_{W_\Lambda}(\omega_0)dZ_{W_\Lambda}(\omega_1)dZ_{W_\Lambda}(\omega_2)) = \Delta_K(\omega_0+\omega_1+\omega_2)d\psi^{(3)}_{W_\Lambda}(-\omega_0, \omega_1, \omega_2)$$

$$\ldots$$

where $\Delta_K(\cdot)$ is the Kronecker delta ($\Delta_K(\omega) = 1$, if $\omega = 0$, $\Delta_K(\omega) = 0$ otherwise) and for the delta correlated model:

$$d\psi^{(j+1)}_{W_\Lambda}(\omega_1+\omega_2+\ldots+\omega_j, \omega_1, \omega_2, \ldots, \omega_j) = \frac{\Lambda}{(2\pi)^j}E[Y^{j+1}]d\omega_1 d\omega_2 \ldots d\omega_j \quad (1.69)$$

The corresponding cumulants of the process W_Λ, in fact, write

$$g_{j+1}(t_0, t_1, \ldots, t_j) = \frac{\Lambda}{(2\pi)^j}E[Y^{j+1}] \times \quad (1.70)$$

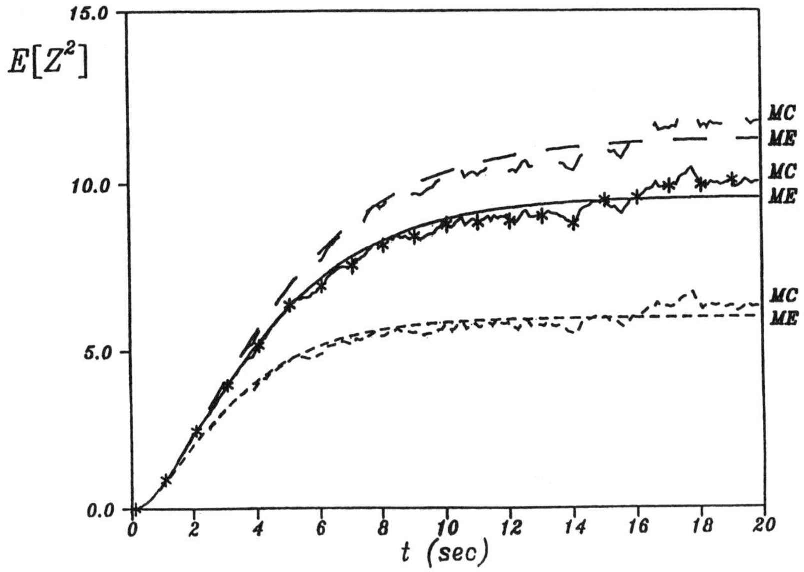

FIGURE 1.11
Non-normal delta correlated input. Evolution of the second order moment of the response of Eq. (1.65) by means of Monte Carlo simulation (MC) and moment equation (ME): short dashed line with one correction term; long dashed line with two correction terms; solid line with ten correction terms.

$$\int_{-\infty}^{\infty} \cdots \int_{-\infty}^{\infty} e^{i(\omega_0 t_0 + \omega_1 t_1 + \ldots + \omega_j t_j)} \Delta(\omega_0 + \omega_1 + \ldots + \omega_j) d\omega_0 d\omega_1 \ldots d\omega_j$$

Using the main properties of the Kronecker delta, one obtains

$$g_{j+1}(t_0, t_1, \ldots, t_j)$$

$$= \frac{\Lambda}{(2\pi)^j} E[Y^{j+1}] \int_{-\infty}^{\infty} \cdots \int_{-\infty}^{\infty} e^{i(\omega_1 \tau_1 + \omega_2 \tau_2 + \ldots + \omega_j \tau_j)} d\omega_1 d\omega_2 \ldots d\omega_j$$

$$= \Lambda E[Y^{j+1}] \delta(\tau_1) \delta(\tau_2) \ldots \delta(\tau_j) \tag{1.71}$$

in which $\tau_j = t_j - t_0$. The Poisson delta correlated process defined in Eq. (1.43) can be also represented in the integral form of Eq. (1.67) in which the orthogonal increment stochastic process $Z_{W_\Lambda}(\omega)$ obeys the generalized orthogonality condition (1.68).

As an example of response, consider the classical linear system of impulse

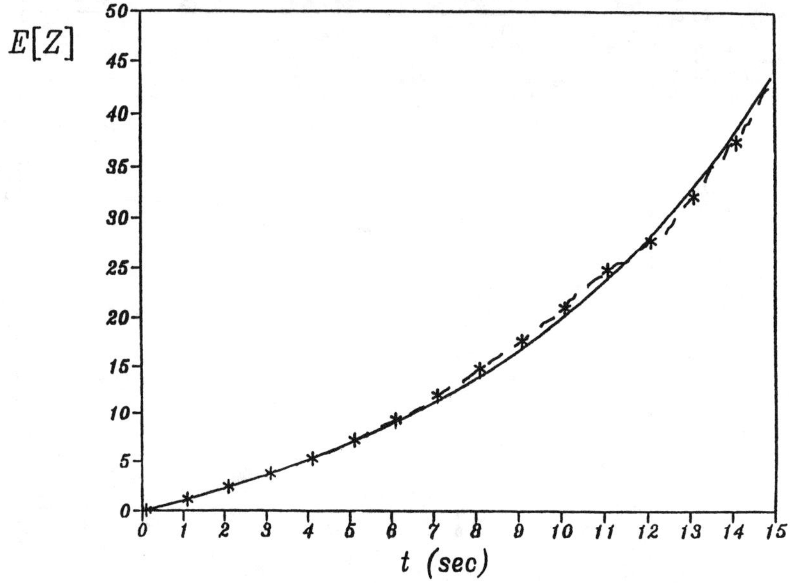

FIGURE 1.12
Normal white noise input. Evolution of the first order moment of the response of Eq. (1.65) (Λ=100) by means of Monte Carlo simulation with two (***) and ten (- - -) correction terms and by the moment equation.

response function (which is the Fourier pair of $H(\omega)$ in Eq. (1.15))

$$h(t) = \frac{1}{\omega_d} e^{\zeta \omega t} sin(\omega_d t)$$

with $\omega_d = \omega\sqrt{1-\zeta^2}$. Let the system be excited by a Poisson delta correlated process W_Λ with zero mean ($E[Y] = 0$). The response of the system is a process $x(t)$ with zero mean given by

$$x(t) = \int_{-\infty}^{t} h(t-\tau) W_\Lambda(\tau) d\tau$$

Thence one writes:

$$x(t_0) = \int_{-\infty}^{\infty} \int_{-\infty}^{t_0} h(t_0 - \tau_0) e^{i\omega_0 t_0} dZ_{W_\lambda}(\omega_0)$$

$$x(t_j) = \int_{-\infty}^{\infty} \int_{-\infty}^{t_j} h(t_j - \tau_j) e^{i\omega_j t_j} dZ_{W_\lambda}(\omega_j)$$

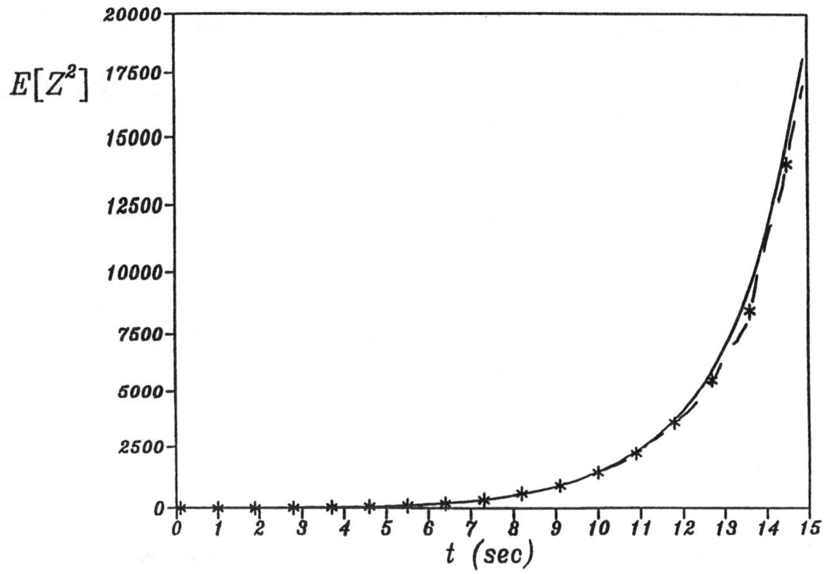

FIGURE 1.13
Normal white noise input. Evolution of the second order moment of the response of Eq. (1.65) (Λ=100) by means of Monte Carlo simulation with two (***) and ten (- - -) correction terms and by the moment equation.

and therefore, for $\tau_1 = t_1 - t_0$,

$$g_2(t_0, t_1) = E[x(t_0)x(t_1)] = \frac{\Lambda E[Y^2]}{2\pi} \int_{-\infty}^{\infty} |H(\omega_1)|^2 e^{i\omega_1\tau_1} d\omega_1$$

The variance of the process is easily obtained by putting $\tau_1 = 0$ in the previous equation. The general expression for g^{j+1} is:

$$g_{j+1}(t_0, t_1, \ldots, t_j) = \frac{\Lambda E[Y^{j+1}]}{2\pi}^j$$

$$\int_{-\infty} \infty \ldots \int_{-\infty} \infty H^*(\omega_1 + \omega_2 + \ldots \omega_j) H(\omega_1) H(\omega_2) \ldots H(\omega_j)$$

$$\times e^{i(\omega_1\tau_1 + \omega_2\tau_2 + \ldots + \omega_j\tau_j)} d\omega_1 d\omega_2 \ldots d\omega_j$$

1.7 Multidegree-of-Freedom Systems Under Parametric Delta Correlated Input

Let the equation of motion be given in the form

$$d\boldsymbol{x}(t) = \boldsymbol{f}(\boldsymbol{x}, t)dt + \boldsymbol{F}(\boldsymbol{x}, t)d\boldsymbol{B}_{\Lambda}(t) \tag{1.72}$$

where $\boldsymbol{f}\ (\boldsymbol{x}, t)$ is a vector of state variables of size n, $\boldsymbol{F}\ (\boldsymbol{x}, t)$ is an $n \times m$ matrix, and $\boldsymbol{B}_{\Lambda}(t)$ is a vector of size m of homogeneous compound Poisson processes, its s–th component being defined as

$$B_{\Lambda s}(t) = \sum_{j=1}^{N(t)} Y_{sj} U(t - \tau_{sj})$$

The cumulants of the process $W_{\Lambda s}(t)$ associated to $B_{\Lambda s}(t)$ are given in the form

$$g_{sr}(t_1, t_2, \ldots, t_r) = E[dW_{\Lambda s}(t_1), dW_{\Lambda s}(t_2), \ldots, dW_{\Lambda s}(t_r))$$

$$= \gamma_{sr}\delta(t_1 - t_2)\delta(t_1 - t_3) \cdots \delta(t_1 - t_r)$$

In this case, the evolution of the probability density function of the response write [8]

$$\frac{\partial q_{\boldsymbol{x}}(\mathbf{x}, t)}{\partial t} = -\tilde{\nabla}_{\mathbf{X}}\left(q_{\boldsymbol{x}}(\mathbf{x}, t)\boldsymbol{f}(\mathbf{x}, t)\right) - \sum_{k=1}^{\infty} \frac{(-1)^k}{k!} \tilde{\nabla}_{\mathbf{X}}^{[k]}\left(q_{\boldsymbol{x}}(\mathbf{x}, t)\boldsymbol{\Psi}_k(\mathbf{x}, t)\right) \tag{1.73}$$

where $\tilde{\nabla}_{\mathbf{X}} = \left[\frac{\partial}{\partial \mathrm{X}_1}, \frac{\partial}{\partial \mathrm{X}_2}, \cdots, \frac{\partial}{\partial \mathrm{X}_n}\right]$ (see appendix) and $\boldsymbol{\Psi}_k\ (\mathbf{x}, t)$ are vectors defined as

$$\boldsymbol{\Psi}_1(\mathbf{x}, t) = \sum_{j=1}^{\infty} \frac{1}{j!} \boldsymbol{F}^{(j)}(\mathbf{x}, t)\gamma_j(t)$$

$$\boldsymbol{\Psi}_2(\mathbf{x}, t) = \sum_{j=1}^{\infty}\sum_{r=1}^{\infty} \frac{1}{j!r!} \boldsymbol{F}^{(j)}(\mathbf{x}, t) \otimes \boldsymbol{F}^{(r)}(\mathbf{x}, t)\gamma_{j+r}(t) \tag{1.74}$$

$$\cdots$$

In Eq. (1.73) the exponent into square brackets means Kronecker power $\tilde{\nabla}_{\mathbf{X}}^{[k]} = \tilde{\nabla}_{\mathbf{X}} \otimes \tilde{\nabla}_{\mathbf{X}} \otimes \ldots \otimes \tilde{\nabla}_{\mathbf{X}}$ (k–fold) and the symbol $\otimes$ means Kronecker product. Some few fundamentals of Kronecker algebra are given in the appendix. The matrices $\boldsymbol{F}$ are defined in the recursive form

$$\boldsymbol{F}^{(1)}(\mathbf{x},t) = \boldsymbol{F}(\mathbf{x},t)$$

$$\boldsymbol{F}^{(j)}(\mathbf{x},t) = \left[\tilde{\nabla}_{\mathbf{X}} \otimes \boldsymbol{F}^{(j-1)}(\mathbf{x},t)\right]\left[\boldsymbol{F}^{(j)}(\mathbf{x},t) \otimes \boldsymbol{I}_m^{[r-1]}\right];\ j = 2, 3, \ldots \tag{1.75}$$

$\boldsymbol{I}_m$ being an identity matrix of order m. At last, the γ_j vectors appearing in equation (1.74), are vector of order m^j, whose components are the strengths of the delta correlated vector process $\boldsymbol{W}_{\Lambda}(t)$.

In the case of zero mean normal white noise input, that is γ_j=0 $\forall j > 2$, Eq. (1.73) reduces to the classical Fokker–Planck equation for multidimensional systems

$$\frac{\partial q_{\boldsymbol{x}}(\mathbf{x},t)}{\partial t} = -\tilde{\nabla}_{\mathbf{X}}\left(q_{\boldsymbol{x}}(\mathbf{x},t)\mathbf{f}(\mathbf{x},t)\right) - \tilde{\nabla}_{\mathbf{X}}\left(q_{\boldsymbol{x}}(\mathbf{x},t)\boldsymbol{F}^{(2)}(\mathbf{x},t)\right)\gamma_2(t)$$
$$+\tfrac{1}{2!}\tilde{\nabla}_{\mathbf{X}}^{[2]}\left(q_{\boldsymbol{x}}(\mathbf{x},t)\boldsymbol{F}^{(2)}(\mathbf{x},t)\right)\gamma_2(t) \tag{1.76}$$

In the case of purely external excitation, that is $\boldsymbol{F}(\mathbf{x},t)$ is a deterministic matrix depending on t only, ($\boldsymbol{F}(\mathbf{x},t) = \boldsymbol{F}(t)$), then Eq. (1.73) reduces to:

$$\frac{\partial q_{\boldsymbol{x}}(\mathbf{x},t)}{\partial t} = -\tilde{\nabla}_{\mathbf{X}}\left(q_{\boldsymbol{x}}(\mathbf{x},t)\boldsymbol{f}(\mathbf{x},t)\right) - \sum_{k=1}^{\infty}\frac{(-1)^k}{k!}\tilde{\nabla}_{\mathbf{X}}^{[k]}\left(q_{\boldsymbol{x}}(\mathbf{x},t)\right)\boldsymbol{F}^{[j]}(t)\gamma_j(t) \tag{1.77}$$

1.7.1 Moment Equation Approach for MDOF Systems

In order to find the differential equations of any order say r of the vector response $\boldsymbol{x}(t)$, one can multiply Eq. (1.77) by $\boldsymbol{x}^{[r]}$ and an integration yields

$$\dot{\mu}_r[\boldsymbol{x}] = \mathbf{Q}_r E[\boldsymbol{x}^{[r-1]} \otimes \boldsymbol{f}[\boldsymbol{x},t)] + \mathbf{Q}_r \sum_{j=1}^{\infty} E[\boldsymbol{x}^{[r-1]} \otimes \boldsymbol{F}^{(j)}(\boldsymbol{x},t)]\gamma_j(t) +$$
$$\tfrac{1}{2}\mathbf{Q}_r(\mathbf{Q}_{r-1} \otimes \boldsymbol{I}_n)\sum_{j=1}^{\infty}\sum_{k=1}^{\infty}\frac{1}{j!k!}E[\boldsymbol{x}^{[r-2]} \otimes \boldsymbol{F}^{(j)}(\boldsymbol{x},t) \otimes \boldsymbol{F}^{(k)}(\mathbf{x},t)]\gamma_{j+k}(t) \tag{1.78}$$

where $\boldsymbol{\mu}_r[\boldsymbol{x}]=E[\boldsymbol{x}^{[r]}]$ is an n^r vector collecting all the moments of order r of the vector process $\boldsymbol{x}(t)$, and $\mathbf{Q}_r$ are the permutation matrices described in the appendix.

If the system is driven by external delta correlated input

$$\boldsymbol{F}(\boldsymbol{x}, t) = \boldsymbol{F}(t),$$

then Eq. (1.78) simplifies into

$$\dot{\boldsymbol{\mu}}_r[\boldsymbol{x}] = \mathbf{Q}_r E[\boldsymbol{x}^{[r-1]} \otimes \boldsymbol{f}(\boldsymbol{x}, t)] + \sum_{j=1}^{r} \frac{1}{j!} \boldsymbol{F}_{jr}(t) \gamma_j(t) \tag{1.79}$$

where $\boldsymbol{F}_{jr}(t)$ is defined as

$$\boldsymbol{F}_{jr}(t) = \mathbf{Q}_r(\mathbf{Q}_{r-1} \otimes \boldsymbol{I}_n) \dots (\mathbf{Q}_{r-j+1} \otimes \boldsymbol{I}_n^{[j-1]})(\boldsymbol{\mu}_{r-j}[\boldsymbol{x}] \otimes \boldsymbol{I}_n^{[j]}) \otimes \boldsymbol{F}\boldsymbol{x}^{[j]}(t)$$

In the further case of linear systems, that is

$$\boldsymbol{f}(\boldsymbol{x}, t) = \mathbf{A}(t)\boldsymbol{x} + \boldsymbol{\rho}(t)$$

then the moment equations can be written in the form

$$\dot{\boldsymbol{\mu}}_r[\boldsymbol{x}] = \mathbf{A}_r(t)\boldsymbol{\mu}_r[\boldsymbol{x}] + \boldsymbol{\rho}_r(t)\boldsymbol{\mu}_{r-1}[\boldsymbol{x}] + \sum_{j=1}^{r} \frac{1}{j!} \boldsymbol{F}_{jr}(t) \gamma_j(t) \tag{1.80}$$

where

$$\mathbf{A}_r(t) = \mathbf{A} \oplus \mathbf{A} \oplus \dots \oplus \mathbf{A}, \qquad \boldsymbol{\rho}_r(t) = \boldsymbol{\rho} \otimes \boldsymbol{\rho} \oplus \dots \oplus \boldsymbol{\rho}$$

in which the symbol $\oplus$ means Kronecker sum (see appendix).

Due to the Lyapunov form of the matrix $\mathbf{A}_r(t)$, the eigenvalues $\boldsymbol{\Gamma}_r(t)$ and the eigenvectors $\boldsymbol{\Phi}_r(t)$ of $\mathbf{A}_r(t)$ are related to the eigenvalues and eigenvectors of $\mathbf{A}(t)$ by the following relationships

$$\boldsymbol{\Phi}_r(t) = \boldsymbol{\Phi}^{[r]}(t), \qquad \boldsymbol{\Gamma}_r(t) = \boldsymbol{\Gamma}(t) \oplus \boldsymbol{\Gamma}(t) \oplus \dots \oplus \boldsymbol{\Gamma}(t)$$

$\boldsymbol{\Phi}(t)$ being the eigenmatrix of $\mathbf{A}(t)$ and $\boldsymbol{\Gamma}(t)$ the corresponding eigenvalue diagonal matrix of $\mathbf{A}(t)$.

1.7.2 Simulation of Multivariate Delta Correlated Processes and Response

The simulation of a multivariate delta correlated process is very simple. It mainly consists in generating a multidimensional vector $\boldsymbol{Y}$ with assigned joint probability density function, an n–variate vector of instants distributed according to Poisson law, each of them with assigned mean number of occurrences per unit time. In this way the counterpart of Eq. (1.60) in the multidimensional case writes

$$\dot{\underline{x}}(t) = \boldsymbol{f}(\underline{x}, t) + \boldsymbol{F}(\underline{x}, t)\mathbf{R}_j\underline{\boldsymbol{Y}}\delta(t - \underline{\tau}_j); \qquad \underline{x}(\underline{\tau}_{j-1}) = \underline{x}_{j-1} \tag{1.81}$$

where $\mathbf{R}_j$ is a diagonal Boolean matrix describing the location of the active impulses in the time interval $(\underline{\tau}_{j-1}, \underline{\tau}_j)$, i.e., of the Dirac's deltas of the various components. Moreover, $\underline{\tau}_j$ is the time instant in which one or more components of the sample function vector exhibit a Dirac's delta. Keeping this in mind, Eq. (1.81) can be rewritten in the time intervals $(\underline{\tau}_{j-1}, \underline{\tau}_j)$, $(\underline{\tau}_j, \underline{\tau}_{j+1})$, $\underline{\tau}_{j+1}$ being a new time in which the sample function exhibits a Dirac's delta in the form

$$\dot{\underline{x}}(t) = \boldsymbol{f}(\underline{x}, t) + \chi_j(\underline{x}, t)\delta(t - \underline{\tau}_j); \qquad \forall t \in (\underline{\tau}_{j-1}^+, \underline{\tau}_{j+1}^-) \tag{1.82}$$

where $\chi_j(\underline{x}, t)$=$\boldsymbol{F}(\underline{x}, t)\mathbf{R}_j\underline{\boldsymbol{Y}}$.

For all time instants $\underline{\tau}_{j-1}^+ < t < \underline{\tau}_j^-$ and $\underline{\tau}_j^+ < t < \underline{\tau}_{j+1}^-$, the integration follows the usual schemes, because $\dot{\underline{x}}(t)$=$\boldsymbol{f}(\underline{x}, t)$, while in $t = \underline{\tau}_j$ the response exhibits a jump $\Delta\underline{x}$ given as

$$\Delta\underline{x} = \underline{x}(\underline{\tau}_j^+) - \underline{x}(\underline{\tau}_j^-)$$

$\underline{\tau}_j^-$ and $\underline{\tau}_j^+$ being the time instants immediately before and after, respectively, the Dirac's occurrence. The jump can be evaluated by means of the series (1.62) extended to the case of multidimensional vector, one obtains for the total jump

$$\Delta\underline{x} == \sum_{k=1}^{\infty} \frac{1}{k!}[\chi_j^{(k)}(\underline{x}(t), t)]_{t=\underline{\tau}_j^-} \tag{1.83}$$

where the vectors $\chi_j^{(k)}(\mathbf{x}(t), t)$ can be recursively evaluated in the form

$$\chi_j^{(1)}(\mathbf{x}(t), t) = \chi_j(\mathbf{x}(t), t)$$

$$\chi_j^{(k)}(\mathbf{x}(t), t) = \left(\tilde{\nabla}_{\mathbf{x}} \otimes \chi_j^{(k-1)}(\mathbf{x}(t), t)\right) \chi_j^{(1)}(\mathbf{x}(t), t);$$

In the previous expression $\tilde{\nabla}_{\mathbf{x}} \otimes \chi_j^{(k)}(\mathbf{x}(t), t)$ is the gradient operator of the vector

$\chi_j^{(k)}(\mathbf{x}(t), t)$, that is

$$\tilde{\nabla}_{\mathbf{x}} \otimes \chi_j^{(k)}(\mathbf{x}(t), t) = \begin{bmatrix} \dfrac{\partial \chi_{j1}^{(k)}}{\partial x_1} & \dfrac{\partial \chi_{j1}^{(k)}}{\partial x_2} & \cdots & \dfrac{\partial \chi_{j1}^{(k)}}{\partial x_n} \\ \dfrac{\partial \chi_{j2}^{(k)}}{\partial x_1} & \dfrac{\partial \chi_{j2}^{(k)}}{\partial x_2} & \cdots & \dfrac{\partial \chi_{j2}^{(k)}}{\partial x_n} \\ \vdots & \vdots & \ddots & \vdots \\ \dfrac{\partial \chi_{jn}^{(k)}}{\partial x_1} & \dfrac{\partial \chi_{jn}^{(k)}}{\partial x_2} & \cdots & \dfrac{\partial \chi_{jn}^{(k)}}{\partial x_n} \end{bmatrix} \tag{1.84}$$

$\chi_{js}^{(k)}$ being the s–th element of the vector $\chi_j^{(k)}$.

If all the elements $\Lambda_s \to \infty$ (s=1, 2, ..., m), and at the same time $\Lambda_j^{[2]} E[\mathbf{Y}^{[r]}]$ remains a constant vector, then the input process becomes a normal white noise vector. In this case, as stated before, if one selects a time step Δt, all components Y_k (k=1, 2, ..., n) of the input process have in mean one impulse of amplitude $\underline{Y}_{kj}\sqrt{\Delta t}$ in the time interval $(t_j, t_j + \Delta t)$. If Δt is small then only the first two terms of the series (1.83) give contributions of the same order of Δt, while higher order terms can be neglected so that the central difference scheme for zero mean normal white noise writes

$$\underline{\boldsymbol{x}}(t_{j+1}) = \underline{\boldsymbol{x}}(t_j) + \boldsymbol{f}(\underline{\boldsymbol{x}}(t_j), t_j)\Delta t + \frac{1}{2}\boldsymbol{F}^{(2)}(\underline{\boldsymbol{x}}(t_j), t_j)\boldsymbol{Y}_j^{[2]}\Delta t$$

where, according to Eq. (1.75)

$$\boldsymbol{F}^{(2)}(\mathbf{x}(t), t) = [\tilde{\nabla}_{\mathbf{x}} \otimes \boldsymbol{F}(\mathbf{x}(t), t)][\boldsymbol{F}(\mathbf{x}(t), t) \otimes \boldsymbol{I}_m]$$

1.8 Conclusions and References

This chapter deals, in the same context, with evolutionary and delta correlated models of stochastic processes. Such a bridge between different schemes of reasoning (see [19] and [8]) gives the chance of a unified understanding for the idealizations adopted in the following chapters.

An extension of the stochastic differential calculus to systems driven by Poisson delta correlated processes is also provided. In this context a Gaussian white noise can be regarded as a particularization of the Poisson delta correlated process.

References

[1] Arnold L., *Stochastic Differential Equations: Theory and Applications*, John Wiley and Sons, New York, N.Y., 1974.

[2] Augusti G., Baratta A., and Casciati F., *Probabilistic Methods in Structural Engineering*, Chapman & Hall, London, 1984.

[3] Brewer J.W., Kronecker Products and Matrix Calculus in System Theory, *IEEE Transactions on Circuits and Systems*, Vol.25, No. 9, 1978, pp. 772–781.

[4] Caddemi S. and Di Paola M., Itô and Stratonovich Integrals for Deterministic Parametric Input, to appear in the *Journal of Applied Mechanics*, 1995.

[5] Casciati F. (Ed.), *Dynamic Motion: Chaotic and Stochastic Behaviour*, Springer Verlag, Vienna, 1993.

[6] Casciati F. and Faravelli L., *Fragility Analysis of Complex Structural Systems*, Research Studies Press, Taunton, 1991.

[7] Casciati F. and Roberts J.B. (Eds.), *Reliability Problems: General Principles and Applications in Mechanics of Solids and Structures*, Springer Verlag, Vienna, 1991.

[8] Di Paola M., Stochastic Differential Calculus, in Casciati F. (Ed.), *Dynamic Motion: Chaotic and Stochastic Behaviour*, Springer Verlag, Vienna, 1993, pp. 29–92.

[9] Di Paola M. and Falsone G., Stochastic Dynamics of Nonlinear Systems Driven by Nonnormal Delta-correlated Processes, *Journal of Applied Mechanics*, 60, 1993, pp. 141–148.

[10] Gardiner C.W., *Handbook of Stochastic Methods*, Springer Verlag, Berlin, 1985.

[11] Gihman I.I. and Skorohod A.V., *Stochastic Differential Equations*, Springer–Verlag, Berlin, 1972.

[12] Grigoriu M., White Noise Processes, *Journal Engineering Mechanics, ASCE*, Vol. 113, 1987, pp. 757–765.

[13] Grigoriu M., Simulation of Diffusion Processes, *Journal of Engineering Mechanics, ASCE*, Vol. 116, 1990, pp. 1524–1542.

[14] Horstemke W. and Lefever R., *Noise-induced Transitions, Theory and Application in Physics, Chemistry and Biology*, Springer–Verlag, New York, N.Y., 1984.

[15] Ibrahim R. A. and Soundararajan A., Stochastic Response of Nonlinear Dynamics Systems Based on a Non-Gaussian Closure, *Journal of Applied Mechanics*, Vol. 52, 1985, pp. 965–970.

[16] Itô K., On a Formula Concerning Stochastic Differential, *Nagoya Mathematical Journal*, Vol. 3, 1951, pp. 55–65.

[17] Lin Y.K., *Probabilistic Theory of Structural Dynamics*, R.E. Krieger, Huntington, 1976.

[18] Lin Y.K. and Cai G.Q., *Probabilistic Structural Dynamics: Advanced Theory and Applications*, McGraw-Hill, New York, 1995.

[19] Priestley M.B., *Spectral Analysis and Time Series*, Academic Press, London, 1981.

[20] Roberts J.B., System Response to Random Impulses, *Journal of Sound and Vibration*, 24, 1972, pp. 23–34.

[21] Soong T.T. and Grigoriu M., *Random Vibration of Mechanical Structural Systems*, Prentice Hall, Englewood Cliffs, 1993.

[22] Stratonovich R.L., *Topics in the Theory of Random Noise*, Gordon and Breach, New York, 1963.

[23] Stratonovich R.L., A New Form of Representing Stochastic Integrals and Equations, *Journal SIAM Control*, 4, 1966, pp. 362–371.

[24] Vasta M., Exact Stationary Solution for a Class of Non–linear Systems Driven by a Non–normal Delta Correlated Process, *International Journal of Non Linear Mechanics*, Vol. 30, 1995, pp. 407–418.

[25] Wong E. and Zakai M., On the Relation between Ordinary and Stochastic Differential Equations, *International Journal of Engineering Science*, Vol. 3, 1965, pp. 213–229.

[26] Wu W.F. and Lin Y.K., Cumulant Closure for Non–linear Oscillator under Random Parametric and External Excitation, *International Journal of Non Linear Mechanics*, Vol. 19, 1984, pp. 349–362.

1.9 Appendix

An experienced researcher should already possess the information summarized in this Appendix. He may go through the Section just to fix the symbol list adopted in this Chapter. For the proofs of the results and for further details the reader is referred to [22, 17, 10, 6, 21].

1.9.1 Characterization of Random Variables

A continuous random variable x assumes values in a domain $\mathcal{X}$ where it is defined as the *probability density function* $p_x(\mathrm{x})$. It is a non-negative valued function whose integral over $(\mathrm{x}_1, \mathrm{x}_2)$ gives the probability that x belongs to that interval. The integral over $\mathcal{X}$ must be 1. Let θ be a real parameter, then the *characteristic function* is written:

$$M_x(\theta) = \int_{-\infty}^{\infty} \exp(-i\theta \mathrm{x}) p_x(\mathrm{x}) d\mathrm{x} \tag{1.85}$$

where i is the imaginary unit. Its real part is even and its imaginary part is odd. The inverse relation says that:

$$p_x(x) = \frac{1}{2\pi} \int_{-\infty}^{\infty} \exp(i\theta \mathrm{x}) M_x(\theta) d\theta \tag{1.86}$$

The Taylor expansion of M_x can be put in the form:

$$M_x(\theta) = \sum_{j=0}^{\infty} \frac{(-i)^j}{j!} E[x^j] \theta^j \tag{1.87}$$

where the moment of order j is defined as

$$E[x^j] = \int_{-\infty}^{\infty} \mathrm{x}^j p_x(\mathrm{x}) d\mathrm{x}$$

Analogously one can introduce the *log-characteristic function* $\ln(M_x)$ and its Taylor expansion

$$\ln(M_x(\theta)) = \sum_{j=1}^{\infty} \frac{(-i)^j}{j!} g_j[x]\theta^j \tag{1.88}$$

where the term g_j is called the cumulant of order j. Similarly, one introduces the symbol μ_j to denote the moment of order j (but the index 1 is often replaced by the variable x when reference is made to the mean value).

The relations between moments and cumulant can be written:

$$\begin{aligned}
\mu_1[x] &= g_1[x] \\
\mu_2[x] &= g_2[x] + g_1[x]\mu_1[x] \\
\mu_3[x] &= g_3[x] + 2g_2[x]\mu_1[x] + g_1[x]\mu_2[x] \\
&\vdots
\end{aligned} \tag{1.89}$$

where the coefficients of the recursive formulae are those of the Tartaglia triangle

$$\begin{array}{ccccccc}
 & & & 1 & & & \\
 & & 1 & & 1 & & \\
 & 1 & & 2 & & 1 & \\
1 & & 3 & & 3 & & 1 \\
 & & & \vdots & & &
\end{array}$$

Note that the second cumulant $g_2[x]$ coincides with the variance σ_x^2, i.e., the central second order moment (the second order moment minus the square of the first order moment).

When all the cumulants of order greater than 2 are zero, one says that the variable x follows a Gaussian (or normal) distribution:

$$\ln M_x(\theta) = -i\mu_x\theta + \frac{(-i)^2}{2}\sigma_x^2\theta^2$$

and, hence,

$$p_x(\mathrm{x}) = \frac{1}{\sqrt{2\pi}\sigma_x} \exp\{-\frac{(\mathrm{x} - \mu_x)^2}{2\sigma_x^2}\}$$

1.9.2 Joint Characterization of Random Variables

In the multivariate case one distinguishes finite from infinite dimension: random vectors, denoted by bold symbols, characterize the first case; functions of one or more parameters must be specified in the second case.

1.9.2.1 Random Vectors

A set of n continuous random variables $\boldsymbol{x}$ assumes values in a domain $\mathcal{X}^n$ where it is defined as the *joint probability density function* $p_{\boldsymbol{x}}(\mathbf{x})$. It is a non-negative valued function whose integral over a cell in $\mathcal{X}^n$ gives the probability that $\boldsymbol{x}$ belongs to that cell. The integral over $\mathcal{X}^n$ must be 1. The integral along a single variable (or a subset of them) provides the so called *marginal distribution.*

Let $\boldsymbol{\theta}$ be a real parameter vector, then the *characteristic function* is written:

$$M_{\boldsymbol{x}}(\boldsymbol{\theta}) = \int_{-\infty}^{\infty} \ldots \int_{-\infty}^{\infty} \exp(-i\boldsymbol{\theta}^T \mathbf{x}) p_{\boldsymbol{x}}(\mathbf{x}) d\mathbf{x} \tag{1.90}$$

The inverse relation says that:

$$p_{\boldsymbol{x}}(\boldsymbol{\theta}) = \frac{1}{2\pi} \int_{-\infty}^{\infty} \ldots \int_{-\infty}^{\infty} \exp(i\boldsymbol{\theta}^T \mathbf{x}) M_{\boldsymbol{x}}(\boldsymbol{\theta}) d\boldsymbol{\theta} \tag{1.91}$$

The Taylor expansions of $M_{\boldsymbol{x}}$ and $\ln(M_{\boldsymbol{x}})$ can be put in the form:

$$M_{\boldsymbol{x}}(\boldsymbol{\theta}) = 1 + (-i) \sum_{j=0}^{n} E[x_j]\theta_j + \frac{(-i)^2}{2!} \sum_{j=1}^{n} \sum_{k=1}^{n} E[x_j x_k]\theta_j \theta_k + \ldots \tag{1.92}$$

and

$$\ln(M_{\boldsymbol{x}}(\boldsymbol{\theta})) = (-i) \sum_{j=1}^{n} g_1[x_j]\theta_j + \frac{(-i)^2}{2!} \sum_{j=1}^{n} \sum_{k=1}^{n} g_2[x_j x_k]\theta_j \theta_k + \ldots \tag{1.93}$$

where the vector $\boldsymbol{g}_j$ is the vector of the cumulants of order j. Similarly, one introduces the vector $\boldsymbol{\mu}_j$ to collect the moments of order j (but the index 1 is often replaced by the variable $\boldsymbol{x}$ for the mean value).

The relations between moments and cumulant can be written:

$$\begin{aligned}
g_1[x_j] &= \mu_1[x_j] \\
g_2[x_j x_k] &= \mu_2[x_j x_k] - \mu_1[x_j]\mu_1[x_k] \\
g_3[x_j x_k x_l] &= \mu_3[x_j x_k x_l] - \mu_2[x_k x_l]\mu_1[x_j] - \mu_2[x_j x_l]\mu_1[x_k] - \\
&\quad \mu_2[x_j x_k]\mu_1[x_l] - \mu_1[x_j]\mu_1[x_k]\mu_1[x_l]
\end{aligned} \tag{1.94}$$

Note that the second cumulants $g_2[x]$ can also be arranged into a two-dimensional matrix $\boldsymbol{\Sigma_x}$ that is called the covariance matrix: it is symmetric and positive definite; the diagonal terms are the variances $\sigma_{x_j}^2$ and the out-of-diagonal terms are the covariances $\sigma_{x_j x_k}$.

When all the cumulants of order greater than 2 are zero, one says that the variables $\boldsymbol{x}$ follow a joint Gaussian (or normal) distribution:

$$\ln M_{\boldsymbol{x}}(\boldsymbol{\theta}) = -i \sum_{j=1}^{n} \mu_1[x_j]\theta_j + \frac{(-i)^2}{2} \sum_{j=1}^{n} \sum_{k=1}^{n} \sigma_{x_j x_k} \theta_j \theta_k$$

and, hence,

$$p_{\boldsymbol{x}}(\mathbf{x}) = \frac{1}{\sqrt{(2\pi)^n (\det \boldsymbol{\Sigma_x})}} \exp\{-\frac{1}{2}(\mathbf{x} - \boldsymbol{\mu_x})^T \boldsymbol{\Sigma_x}^{-1}(\mathbf{x} - \boldsymbol{\mu_x})\}$$

1.9.2.2 Random Processes

Given a parameter t (f.i. the time), the infinite set of continuous random variables $x(t)$ assumes values in a domain $\mathcal{X}^\infty$ where the following infinite hierarchy of probabilities is defined:

for a single value t_1 of the parameter t, the *probability density function* $p_{x(t_1)}(\mathrm{x})$;

for a couple of values t_1 and t_2 of the parameter t, the *joint probability density function* $p_{x(t_1),x(t_2)}(\mathrm{x}_1, \mathrm{x}_2)$;

...

for n values $\mathbf{t}$ of the parameter t, the *joint probability density function* $p_{x(\mathbf{t})}(\mathbf{x}; \mathbf{t})$.

...

When one continuously carries out values of $x(t)$ at increasing values of t, the resulting function $\underline{x}(t)$ is called a sample function or a *realization*. If one focuses attention on a couple of instants, one can introduce a covariance matrix $\boldsymbol{\Sigma}_x(t_1, t_2)$ whose diagonal terms coincide with the covariance function $\Sigma_x(t_1, t_2) = E[x(t_1), x(t_2)] - E[x(t_1)]E[x(t_2)]$. For zero mean processes, this diagonal term also coincides with the cumulant, which is called, in this case, correlation function and is indicated with $\Sigma'_x(t_1, t_2) = E[x(t_1), x(t_2)]$. For stationary processes these two functions only depend on the difference $\tau = t_2 - t_1$ and are denoted as $R_x(\tau)$ and $R'_x(\tau)$, respectively.

Gaussian processes are characterized by zero values of all the cumulants written for more than two instants of time.

1.9.3 Operation on Stochastic Processes

Consider a sequence of values $x(t_n) = x_n$ and let $E[x_n^2]$ be finite: one says that x is the limit, in a mean square sense, of x_n and writes:

$$\underset{n\to\infty}{\text{l.i.m.}}\, x_n = x \tag{1.95}$$

provided that

$$\lim_{n\to\infty} E[|x_n - x|^2] = 0 \tag{1.96}$$

This operator and the expectation commutes.

If $\Sigma_x(t_1, t_2)$ exists and is finite, then the process $x(t)$ is continuous and its derivative is defined as:

$$\dot{x}(t) = \underset{\tau\to 0}{\text{l.i.m.}} \frac{x(t+\tau) - x(t)}{\tau} \tag{1.97}$$

The following properties hold:

$$\frac{d}{dt} E[x(t)] = E[\dot{x}(t)]$$

$$\frac{d}{dt} E[x_1(t)x_2(t)] = E[\dot{x}_1(t)x_2(t)] + E[x_1(t)\dot{x}_2(t)]$$

$$\frac{\partial}{\partial t_1} \Sigma_x(t_1, t_2) = \Sigma_{\dot{x}x}(t_1, t_2)$$

Given a deterministic function $h(t, \tau)$, one introduces the mean square Riemann integral as the stochastic processes $y(t)$

$$y(t) = \int_a^b x(\tau)h(t,\tau)d\tau = \underset{n\to\infty\Delta n\to 0}{\text{l.i.m.}} \sum_{j=1}^{n} h(t, t_j^\circ)x(t_j^\circ)(t_j - t_{j-1}) \tag{1.98}$$

where $a = t_0 < t_1 < ... < t_n = b$, $\Delta n = \max_j(t_j - t_{j-1})$ and t_j° is a time instant in the interval (t_{j-1}, t_j). The concept of Stieltjes integral replaces $x(t_j^\circ)(t_j - t_{j-1})$ with $(x(t_j) - x(t_{j-1}))$.

Put

$$t_j^\circ = \alpha t_j + (1-\alpha)t_{j-1}$$

then one distinguishes:

- the *Itô integral* which corresponds to the adoption of $\alpha = 0$;
- the *Stratonovich integral* which corresponds to the adoption of $\alpha = \frac{1}{2}$.

1.9.4 Kronecker Algebra: Some Fundamentals

In this appendix some fundamentals of Kronecker algebra (see [3]) are listed. Let $\boldsymbol{A}$ and $\boldsymbol{B}$ be two given matrices of size $p \times q$ and $s \times t$, respectively. The Kronecker product of these two matrices, denoted by $\mathbf{A} \otimes \mathbf{B}$, is the matrix of size $(ps \times qt)$ obtained by multiplying each element a_{ij} of $\mathbf{A}$ by the whole matrix $\mathbf{B}$, that is

$$\mathbf{A} \otimes \mathbf{B} = \begin{bmatrix} a_{11}\mathbf{B} & a_{12}\mathbf{B} & \dots & a_{1q}\mathbf{B} \\ a_{21}\mathbf{B} & a_{22}\mathbf{B} & \dots & a_{2q}\mathbf{B} \\ \vdots & \vdots & \ddots & \vdots \\ a_{p1}\mathbf{B} & a_{p2}\mathbf{B} & \dots & a_{pq}\mathbf{B} \end{bmatrix}$$

The Kronecker product has the following properties:

$$\mathbf{A} \otimes (\mathbf{B} \otimes \mathbf{C}) = (\mathbf{A} \otimes \mathbf{B}) \otimes \mathbf{C}$$

$$(\mathbf{A} + \mathbf{B}) \otimes (\mathbf{C} + \mathbf{D}) = \mathbf{A} \otimes \mathbf{C} + \mathbf{A} \otimes \mathbf{D} + \mathbf{B} \otimes \mathbf{C} + \mathbf{B} \otimes \mathbf{D}$$

$$(\mathbf{A} \otimes \mathbf{B})(\mathbf{C} \otimes \mathbf{D}) = (\mathbf{AC}) \otimes (\mathbf{BD})$$

$$(\mathbf{A} \otimes \mathbf{B})^T = (\mathbf{A}^T \otimes \mathbf{B}^T)$$
$$(\mathbf{A} \otimes \mathbf{B})^{-1} = (\mathbf{A}^{-1} \otimes \mathbf{B}^{-1})$$

The Kronecker power of a matrix $\mathbf{A}$, denoted as $\mathbf{A}^{[k]}$, is defined as follows:

$$\mathbf{A}^{[1]} = \mathbf{A} \quad ; \quad \mathbf{A}^{[k+1]} = \mathbf{A} \otimes \mathbf{A}^{[k]} = \mathbf{A}^{[k]} \otimes \mathbf{A}$$

The Kronecker sum of two matrices $\mathbf{A}$ and $\mathbf{B}$ of order $p \times p$ and $s \times s$, respectively, is defined as:

$$\mathbf{A} \oplus \mathbf{B} = \mathbf{A} \otimes \mathbf{I}_s + \mathbf{I}_p \otimes \mathbf{B}$$

$\mathbf{I}_s$ and $\mathbf{I}_p$ being identity matrices of order s and p, respectively. Let $\mathbf{C}$ be a matrix of order $r \times r$, then

$$(\mathbf{A} \oplus \mathbf{B}) \oplus \mathbf{C} = \mathbf{A} \oplus (\mathbf{B} \oplus \mathbf{C})$$

Note that

$$\mathbf{A} \oplus \mathbf{A} = \mathbf{A} \otimes \mathbf{I}_p + \mathbf{I}_p \otimes \mathbf{A}$$

$$\mathbf{A} \oplus \mathbf{A} \oplus \mathbf{A} = \mathbf{A} \otimes \mathbf{I}_p^{[2]} + \mathbf{I}_p \otimes \mathbf{A} \otimes \mathbf{I}_p + \mathbf{I}_p^{[2]} \otimes \mathbf{A}$$

Let $\mathbf{E}_{k,j}$ denote the permutation matrix of order $(kj \times kj)$, consisting of $j \times k$ array of elementary submatrices $\mathbf{E}^{mn}$ of order $(k \times j)$

$$\mathbf{E}_{k,j} = \begin{bmatrix} \mathbf{E}^{11} & \mathbf{E}^{21} & \dots & \mathbf{E}^{k1} \\ \mathbf{E}^{12} & \mathbf{E}^{22} & \dots & \mathbf{E}^{k2} \\ \vdots & \vdots & \ddots & \vdots \\ \mathbf{E}^{1j} & \mathbf{E}^{2j} & \dots & \mathbf{E}^{kj} \end{bmatrix}$$

where each entry of the matrix $\mathbf{E}^{mn}$ has the value one in the (m, n)-th position and zero in all other positions. The permutation matrices have the following properties

$$\mathbf{E}_{k,1} = \mathbf{E}_{1,k} = \mathbf{I}_k$$

$$\mathbf{E}_{k,j}^T = \mathbf{E}_{k,j}^{-1} = \mathbf{E}_{j,k}$$

$\mathbf{I}_k$ being the identity matrix of order k.

Let $\mathbf{A}$ and $\mathbf{B}$ be two matrices of size $(p \times q)$ and $(s \times t)$, respectively, then

$$\mathbf{B} \otimes \mathbf{A} = \mathbf{E}_{p,s}(\mathbf{A} \otimes \mathbf{B})\mathbf{E}_{t,q}$$

Let ∇ be a differential operator of order $j \times k$, that is

$$\nabla = \begin{bmatrix} \partial/\partial x_{11} & \partial/\partial x_{12} & \dots & \partial/\partial x_{1k} \\ \partial/\partial x_{21} & \partial/\partial x_{22} & \dots & \partial/\partial x_{2k} \\ \vdots & \vdots & \ddots & \vdots \\ \partial/\partial x_{j1} & \partial/\partial x_{j2} & \dots & \partial/\partial x_{jk} \end{bmatrix}$$

and let the product $\mathbf{AB}$ exist. Then

$$\nabla \otimes (\mathbf{AB}) = (\nabla \otimes \mathbf{A})(\mathbf{I}_k \otimes \mathbf{B}) + (\mathbf{I}_j \otimes \mathbf{A})(\nabla \otimes \mathbf{B})$$

Moreover

$$\nabla \otimes (\mathbf{A} \otimes \mathbf{B}) = (\nabla \otimes \mathbf{A}) \otimes \mathbf{B} + (\mathbf{I}_j \otimes \mathbf{E}_{s,p})[(\nabla \otimes \mathbf{B}) \otimes \mathbf{A}](\mathbf{I}_k \otimes \mathbf{E}_{q,t})$$

For two vectors $\mathbf{a}$ and $\mathbf{b}$ having u and v components, respectively, the previous equation simplifies into

$$\nabla \otimes (\mathbf{a} \otimes \mathbf{b}) = (\nabla \otimes \mathbf{a}) \otimes \mathbf{b}_v + (\mathbf{I}_k \otimes \mathbf{E}_{v,u})[(\nabla \otimes \mathbf{b}) \otimes \mathbf{a}]$$

Let $\tilde{\nabla}_{\mathbf{X}}$ be the row differential vector

$$\tilde{\nabla}_{\mathbf{X}} = [\partial/\partial \mathrm{x}_1 \quad \partial/\partial \mathrm{x}_2 \cdots \partial/\partial \mathrm{x}_n]$$

Then

$$\tilde{\nabla}_{\mathbf{X}} \otimes (\mathbf{x}^{[k]}) = \mathbf{Q}_k(\mathbf{x}^{[k-1]} \otimes \mathbf{I}_n) = \mathbf{x}^{[k-1]} \otimes \mathbf{I}_n + \mathbf{x}^{[k-2]} \otimes \mathbf{I}_n \otimes \mathbf{x} + \cdots + \mathbf{I}_n \otimes \mathbf{x}^{[k-1]}$$

where $\mathbf{Q}_k$ is a matrix of order $(n^k \times n^k)$ given by

$$\mathbf{Q}_k = \sum_{j=0}^{k-1} \mathbf{E}_{n^{k-j},n^j}$$

The matrix $\mathbf{Q}_k$ has the following property

$$\mathbf{Q}_k \mathbf{e}^{[k]} = k\mathbf{e}^{[k]}$$

$\mathbf{e}$ being any vector of size n. Moreover

$$\tilde{\nabla}_{\mathbf{X}}^{[2]} \otimes (\mathbf{x}^{[k]}) = \mathbf{Q}_k(\mathbf{Q}_{k-1} \otimes \mathbf{I}_n)(\mathbf{x}^{[k-2]} \otimes \mathbf{I}_n^{[2]}).$$

2

Dimension Reduction and Discretization in Stochastic Problems by Regression Method

O. Ditlevsen
Technical University of Denmark, Lyngby

2.1 Introduction

It is not the intention in the following to use a rigorous mathematical style of presentation, but rather to stick to a heuristic style that makes the text possible to read for mathematically motivated engineers and scientists with an appreciation for applications of probabilistic concepts in their fields of work. A basic knowledge of elementary probability theory will be assumed including the definition of a vector of random variables and their joint distribution, expectations, variances, covariances, etc., the rules of operating with these concepts, and the generalizations to random processes and random fields. Neither will there be systematic references to the many brilliant mathematicians and statisticians that created these concepts and theories now belonging to the standard toolbox of probability and mathematical statistics. Predominantly the references will be to originators of applications that are related to structural reliability problems and stochastic mechanics problems.

Many different stochastic problems in engineering and in the sciences are defined in terms of random processes or random fields. In most of the problems these are non-countable infinite families of random variables: To each t in an ordered index set I (usually a subset of or the entire time axis or a subset of coordinates that define points in some space) a random variable (or random vector, or even a more general random entity) $X(t)$ is adjoined, and, in principle, (according to a theorem of Kolmogorov [1, 2]) this family of random variables is completely defined by the set of joint probability distributions that correspond to all finite subsets of I, given that the set of probability distributions satisfies some obvious consistency conditions. In the following the word "field" is used as a common short terminology for "random process" and "random field", except if otherwise is noted.

When it comes to the practical solution of the stochastic problems it is only in exceptional cases possible to proceed without introducing simplifications based on more or less approximate reasoning. Several different types of simplifications may be applicable on a given problem dependent on the type of the problem.

A frequently used simplification is to assume that the fields are related in some way to the class of Gaussian random variables, such that the fact that this class is closed with respect to linear operations can be utilized for obtaining the solution. Moreover, it can be utilized that the class of jointly distributed Gaussian random variables has the convenient property that the conditional expectation of any subset A of the Gaussian random variables given any subset B of the Gaussian random variables is coincident with the linear regression of the subset A on the subset B of random variables. The advantage is thus that the conditional expectations can be calculated solely by algebraic operations on the expectations and the covariances of the total set of Gaussian random variables. This is basic knowledge given in most elementary probability courses. Due to its importance for the present subject, the concept of linear regression will be introduced specifically in the following section.

Another similar type of simplification exists in the case where random point fields enter the problems. Then it is often used in practice to assume that Poissonian properties are present thus opening a catalogue of known results.

In order to be able to reach computational results, it is generally necessary to introduce simplifications by which the infinite dimensional set of random variables of the field is replaced by another infinite set of random variables defined completely in terms of a finite set of representative random variables. This replacement is denoted as discretization of the field. Problem independent automatic discretization procedures are generally based on direct approximation of the field (most often defined as input to the stochastic problem). Automatic procedures have obvious advantages in routine work. However, the finite set of random variables can be chosen such that it is sufficiently representative for the solution to the problem, noting that often it is less important whether or not the original field is well approximated when judged by direct comparison of sample functions. These aspects will be discussed in connection with the introduction of the mathematical tools of field discretization.

The opposite problem is about extending from a finite subset of the infinite set of random variables of a field about which nothing else is known except, perhaps, that it belongs to some given class of fields. The extension is intended to be to a field that is supposed to resemble the unknown field. This is the interpolation problem. It is obviously not solvable in general in the sense that it becomes possible to judge the goodness of the approximation by some well defined measure of error. A principle of simplicity may instead be taken as a way to choose between the possibly infinite number of fields that can be constructed as extensions.

The problem becomes still more ambiguous if only a single sample of values of the finite set of random variables is known. These values may even be given without any reference to a random variable model. They may be given as some measured almost error free values of a deterministic but otherwise unknown function. The interpolation problem is nevertheless solvable by reference to the same principle of simplicity as applied in the statistical theory of maximum likelihood estimation. This means that the measured values may be treated as if they were values from a realization of a suitably chosen type of field, the distribution parameters of which become estimated from the sample. The obtained estimate of the conditional mean value variation over the index set given that the measured values are reproduced by the conditional mean at the points of measurement may then be taken as the interpolation function. This method of interpolation is often called kriging after the South African mining engineer Krige [3]–[6], who first applied the principle of statistical-stochastic interpolation for estimating the size and properties of mineral deposits. The philosophy of statistical-stochastic interpolation is discussed and interpreted in terms of principles of deterministic interpolation in Section 2.10.

A further complication is added to the interpolation problem when the field values are uncertain due to superimposed measuring uncertainty. To deal with this problem it is necessary to make assumptions about the field properties of the measuring uncertainty. Two different measuring error models are considered in Section 2.12. The one is the standard model of independently added random errors and the other is a model where the error field has the same correlation structure as the unknown field itself. It is demonstrated for both models that the method of statistical-stochastic interpolation is well suited for the separation of the measuring uncertainty field from the object field. The importance of the principle of making independent double measurements of each field value is emphasized by this error analysis.

Certain types of problems in random vector processes can be analysed by discretization defined by linear regression on the derivatives at a single time point. In particular, this type of discretization is relevant in connection with the evaluation of the occurrence rate of events that may happen when a stationary Gaussian vector process crosses out of a given domain. Such investigations belong to the theory of so-called Slepian model processes. The linear regressions on derivatives are given in Section 2.13 and the problem of how to make unique conditioning on a crossing event of zero probability is discussed in Section 2.14. Being convinced that for the conditioning to be applicable on physical problems, the crossing events must be defined as "horizontal window" crossings, the Slepian process models for the sample function behavior at a level upcrossing follow directly. Several examples of interesting applications in stochastic mechanics are given in the last section.

Except for Section 2.9, this text deals solely with dimension reductions and field discretizations based directly on the concept of linear regression. There are alternative methods based on truncations of series expansions of the given random field with respect to some infinite orthogonal basis of functions. Like the linear regressions, these truncations are linear functions (or functionals) on the field, and they are therefore closely related to specific linear regressions. Most often the expansions are based on the so-called Karhunen-Loeve theorem [7] by which a field of mean zero can be represented as an infinite series where the ith term is the ith normalized eigenfunction to the eigenvalue problem with the covariance function of the field as kernel. The coefficients are then uncorrelated random variables, the ith of variance equal to the ith eigenvalue. Since exact solutions are known only for a few simple cases of correlation functions, the different methods apply various discretization approximations to reduce the integral equation eigenvalue problem to a matrix eigenvalue problem [8]–[11]. Also the random coefficients are usually taken to be Gaussian so that the expansion represents a Gaussian field. Some accuracy comparisons between different discretization methods are reported in the literature [12]. There can hardly be formulated a statement of superiority as to which discretization method that generally uses the smallest number of random variables to satisfy any given accuracy requirement.

2.2 Linear Regression

Consider a pair (X, Y) of random variables contained in some mathematical model of interest in a given engineering context. We might wish to simplify the random variable part of the model by reducing the dimension of the random vector (X, Y) from 2 to 1. Let us assume that Y is the least important of the two random variables for the considered problem. The simplest approximation next to replacing Y by a constant is to replace Y by an inhomogeneous linear function $a + bX$ of X.

The coefficients a and b should obviously be chosen such that some error measure becomes minimized. It is reasonable to require that the error measure definition is chosen so that it is directly related to the solution of the model. However, if a problem independent procedure is preferable from an operational point of view, the error measure must be related solely to X and Y. A reasonable procedure is to determine a and b such that the mean square deviation $E[(Y - a - bX)^2]$ becomes as small as possible. Minimum is obtained for the values of a and b that satisfy the equations

$$\frac{\partial}{\partial a} E[(Y - a - bX)^2] = -2E[Y - a - bX] = 0 \tag{2.1}$$

$$\frac{\partial}{\partial b} E[(Y - a - bX)^2] = -2E[(Y - a - bX)X] = 0 \tag{2.2}$$

from which it follows that

$$E[Y] = a + bE[X] \tag{2.3}$$

$$E[YX] = aE[X] + bE[X^2] \tag{2.4}$$

Since the variance Var$[X]$ and the covariance Cov$[X, Y]$ are

$$\text{Var}[X] = E[X^2] - E[X]^2 \tag{2.5}$$

$$\text{Cov}[X, Y] = E[XY] - E[X]E[Y] \tag{2.6}$$

respectively, (2.3) and (2.4) give

$$a = E[Y] - \frac{\text{Cov}[X, Y]}{\text{Var}[X]} E[X] \tag{2.7}$$

$$b = \frac{\text{Cov}[X, Y]}{\text{Var}[X]} \tag{2.8}$$

This linear approximation of Y in terms of X is called the linear regression of Y on X, and it is written as $\hat{E}[Y|X]$. The coefficient b given by (2.8) is called the regression coefficient. The result is

$$\hat{E}[Y|X] = E[Y] + \frac{\text{Cov}[X, Y]}{\text{Var}[X]}(X - E[X]) \tag{2.9}$$

The error, $Y - \hat{E}[Y|X]$, is called the residual, and it has the variance

$$\begin{aligned} \text{Var}[Y - \hat{E}[Y|X]] &= \text{Var}[Y] - \frac{\text{Cov}[X, Y]^2}{\text{Var}[X]} \\ &= \text{Var}[Y](1 - \rho[X, Y]^2) \end{aligned} \tag{2.10}$$

where

$$\rho[X, Y] = \frac{\text{Cov}[X, Y]}{D[X]D[Y]} \tag{2.11}$$

is the correlation coefficient. It is important to note that

$$\text{Cov}[X, Y - \hat{E}[Y|X]] = 0 \tag{2.12}$$

The conditions $E[Y - a - bX] = 0$ and Cov$[X, Y - a - bX] = 0$ can be chosen as an alternative basis for defining the linear regression.

Clearly it depends on the size of $\rho[X, Y]$ and of Var$[Y]$ how good the approximation is, if the residual is neglected. For example,

$$\text{Var}[X + Y] = \text{Var}[X] + 2\text{Cov}[X, Y] + \text{Var}[Y] \tag{2.13}$$

and

$$
\begin{aligned}
&\mathrm{Var}[X + \hat{E}[Y|X]] = (1 + \frac{\mathrm{Cov}[X,Y]}{\mathrm{Var}[X]})^2 \mathrm{Var}[X] = \\
&\mathrm{Var}[X] + 2\mathrm{Cov}[X,Y] + \rho[X,Y]^2 \mathrm{Var}[Y]
\end{aligned}
\tag{2.14}
$$

deviate by the residual variance.

The concept of linear regression of Y on X is directly generalized to the linear regression $\hat{E}[\mathbf{Y}|\mathbf{X}]$ of an m-dimensional random vector $\mathbf{Y} = \{Y_i\}$ on an n-dimensional random vector $\mathbf{X} = \{X_j\}$ as the best inhomogeneous linear approximation $\mathbf{a} + \mathbf{BX}$ to $\mathbf{Y}$ in terms of $\mathbf{X}$ in the sense that $\mathbf{a} = \{a_i\}$ and $\mathbf{B} = \{b_{ij}\}$ minimizes the mean square deviation

$$
\begin{aligned}
&E[(\mathbf{Y} - \mathbf{a} - \mathbf{BX})'(\mathbf{Y} - \mathbf{a} - \mathbf{BX})] = \\
&\sum_{i=1}^{m} E[(Y_i - a_i - \sum_{j=1}^{n} b_{ij} X_j)^2]
\end{aligned}
\tag{2.15}
$$

where prime $'$ attached to a matrix indicates transposition of the matrix. By minimizing each term on the right side, we directly get

$$
E[Y_i] = a_i - \sum_{j=1}^{n} b_{ij} E[X_j] \tag{2.16}
$$

in the same way as (2.3) follows from (2.1). Thus a_i can be eliminated from the ith term of (2.15) so that it becomes

$$
E[(Y_i - \sum_{j=1}^{n} b_{ij} X_j)^2] \tag{2.17}
$$

after renaming $X_i - E[X_i]$ and $Y_i - E[Y_i]$ to X_i and Y_i, respectively. These random variables now have zero mean. Partial differentiation of (2.17) with respect to b_{ik} (using that $\partial b_{ij}/\partial b_{ik} = \delta_{jk}$ where δ_{jk} is Kronecker's delta) and setting to zero gives the equation

$$
\begin{aligned}
&E[(Y_i - \sum_{j=1}^{n} b_{ij}X_j)\sum_{j=1}^{n}\delta_{jk}X_j] = \\
&E[Y_iX_k] - \sum_{j=1}^{n} b_{ij}E[X_jX_k] = \\
&\mathrm{Cov}[Y_i, X_k] - \sum_{j=1}^{n} b_{ij}\mathrm{Cov}[X_j, X_k] = 0
\end{aligned}
\tag{2.18}
$$

In matrix notation this equation reads

$$\mathrm{Cov}[\mathbf{Y}, \mathbf{X}'] = \mathbf{B}\mathrm{Cov}[\mathbf{X}, \mathbf{X}'] \tag{2.19}$$

from which it follows that

$$\mathbf{B} = \mathrm{Cov}[\mathbf{Y}, \mathbf{X}']\mathrm{Cov}[\mathbf{X}, \mathbf{X}']^{-1} \tag{2.20}$$

given that the covariance matrix of $\mathbf{X}$ is regular. This is the generalization of the regression coefficient b in (2.8) to the regression coefficient matrix $\mathbf{B}$ of type (m, n) for the linear regression of $\mathbf{Y}$ on $\mathbf{X}$.

The residual vector $\mathbf{Y} - \hat{E}[\mathbf{Y}|\mathbf{X}] = \mathbf{Y} - \mathbf{BX}$ has the covariance matrix

$$
\begin{aligned}
&\mathrm{Cov}[\mathbf{Y} - \mathbf{BX}, \mathbf{Y}' - \mathbf{X}'\mathbf{B}'] = \\
&\mathrm{Cov}[\mathbf{Y}, \mathbf{Y}'] - \mathrm{Cov}[\mathbf{Y}, \mathbf{X}']\mathbf{B}' - \mathbf{B}\mathrm{Cov}[\mathbf{X}, \mathbf{Y}'] + \mathbf{B}\mathrm{Cov}[\mathbf{X}, \mathbf{X}']\mathbf{B}' = \\
&\mathrm{Cov}[\mathbf{Y}, \mathbf{Y}'] - \mathrm{Cov}[\mathbf{Y}, \mathbf{X}']\mathrm{Cov}[\mathbf{X}, \mathbf{X}']^{-1}\mathrm{Cov}[\mathbf{X}, \mathbf{Y}']
\end{aligned}
\tag{2.21}
$$

called the residual covariance matrix. Moreover,

$$\mathrm{Cov}[\mathbf{Y} - \mathbf{BX}, \mathbf{X}'] = \mathrm{Cov}[\mathbf{Y}, \mathbf{X}'] - \mathbf{B}\mathrm{Cov}[\mathbf{X}, \mathbf{X}'] = \mathbf{0} \tag{2.22}$$

that is, the residual vector $\mathbf{Y} - \mathbf{BX}$ and $\mathbf{X}$ is uncorrelated, and

$$\mathrm{Cov}[\mathbf{Y} - \mathbf{BX}, \mathbf{Y}'] = \mathrm{Cov}[\mathbf{Y}, \mathbf{Y}'] - \mathbf{B}\mathrm{Cov}[\mathbf{X}, \mathbf{Y}'] \tag{2.23}$$

is the residual covariance matrix.

It is seen from (2.22) that if instead of requiring minimum of the mean square deviation (2.15) we may equivalently require that the residual $\mathbf{Y}-\mathbf{BX}$ is uncorrelated with $\mathbf{X}$ and that

$$E[\hat{E}[\mathbf{Y} \mid \mathbf{X}]] = E[\mathbf{Y}] \tag{2.24}$$

The linear regression

$$\hat{E}[\mathbf{Y}|\mathbf{X}] = E[\mathbf{Y}] + \mathbf{B}(\mathbf{X} - E[\mathbf{X}]) \tag{2.25}$$

has the important property of being linear in $\mathbf{Y}$. In fact, let $L_y(\mathbf{Y}) = \mathbf{KY} + \mathbf{k}$ be any inhomogeneous linear mapping of $\mathbf{Y}$ into $L_y(\mathbf{Y})$. Then it follows directly by substitution that

$$\hat{E}[L_y(\mathbf{Y})|\mathbf{X}] = L_y(\hat{E}[\mathbf{Y}|\mathbf{X}]) \tag{2.26}$$

a property that the linear regression has in common with the conditional expectation $E[\mathbf{Y}|\mathbf{X}]$ with which it coincides on the constant vectors.

Another important property shared with the conditional expectation $E[\mathbf{Y}|\mathbf{X}]$ is that the linear regression $\hat{E}[\mathbf{Y}|\mathbf{X}]$ is invariant to a one-to-one inhomogeneous linear mapping $\mathbf{X} \leftrightarrow L_x(\mathbf{X})$ of the conditioning vector $\mathbf{X}$:

$$\hat{E}[\mathbf{Y}|L_x(\mathbf{X})] = \hat{E}[\mathbf{Y}|\mathbf{X}] \tag{2.27}$$

If $\mathbf{X}$ consists of two subvectors $\mathbf{X}_1$ and $\mathbf{X}_2$ that are mutually uncorrelated, that is, $\text{Cov}[\mathbf{X}_1, \mathbf{X}_2'] = \mathbf{0}$, then obviously,

$$\hat{E}[\mathbf{Y} \mid \mathbf{X}] - E[\mathbf{Y}] = (\hat{E}[\mathbf{Y} \mid \mathbf{X}_1] - E[\mathbf{Y}]) + (\hat{E}[\mathbf{Y} \mid \mathbf{X}_2] - E[\mathbf{Y}]) \tag{2.28}$$

This means that the linear regression of $\mathbf{Y}$ on $\mathbf{X}_1$ (or $\mathbf{X}_2$) can be obtained from the linear regression of $\mathbf{Y}$ on $\mathbf{X}$ by removing all terms that contain elements of $\mathbf{X}_1$ (or $\mathbf{X}_2$). In particular, if the one-to-one inhomogeneous linear mapping $\mathbf{X} \leftrightarrow \mathbf{Z} = L_x(\mathbf{X})$ in (2.27) is chosen such that $\mathbf{Z}$ has zero mean vector and the unit matrix as covariance matrix, then (2.28) applies on any division of $\mathbf{Z}$ into subvectors. Thus the relative importance of the terms of $\hat{E}[\mathbf{Y} \mid \mathbf{Z}] = \mathbf{BZ}$, $\mathbf{B} = \text{Cov}[\mathbf{Y}, \mathbf{Z}']$, can be directly studied by comparing the residual covariance matrices $\text{Cov}[\mathbf{Y}, \mathbf{Y}'] - \mathbf{BB}'$ and $\text{Cov}[\mathbf{Y}, \mathbf{Y}'] - \mathbf{B}_1\mathbf{B}_1'$, where $\mathbf{B}_1$ is the matrix obtained from $\mathbf{B}$ by removing those columns that correspond to the elements of $\mathbf{Z}$ whose importance are investigated. Such investigations are particularly useful for the purpose of reduction of the dimension of randomness in reliability and stochastic finite element calculations (Section 2.10).

In this introduction to linear regression let us generalize further. Consider a pair of random processes $(X(t), Y(t))$. By direct generalization the linear regression of process Y on process X over the interval $[\alpha, \beta]$ takes the form

$$\hat{E}[Y(t)|X] = a(t) + \int_{\alpha}^{\beta} B(t,\tau)X(\tau)d\tau \tag{2.29}$$

where $a(t)$ and $B(t, s)$ are functions that are determined from the condition

$$E[\hat{E}[Y(t)|X]] = E[Y(t)] \tag{2.30}$$

that corresponds to (2.16), and the condition

$$\text{Cov}[Y(t) - \hat{E}[Y(t)|X], X(s)] = 0 \tag{2.31}$$

that corresponds to (2.22). From (2.29) and (2.30) it follows that

$$E[Y(t)] = a(t) + \int_{\alpha}^{\beta} B(t,\tau)E[X(\tau)]d\tau \tag{2.32}$$

and (2.29) and (2.31) give the equation

$$\text{Cov}[Y(t), X(s)] = \int_{\alpha}^{\beta} B(t,\tau)\text{Cov}[X(\tau), X(s)]d\tau \tag{2.33}$$

In particular, if the process pair is stationary we have covariance functions c_{YX} and c_X of one variable such that (2.33) in the case $\alpha = -\infty$, $\beta = \infty$ reduces to

$$c_{YX}(s-t) = \int_{-\infty}^{\infty} B(t,\tau)c_X(s-\tau)d\tau \tag{2.34}$$

or, by substituting $u = s - t$, $v = s - \tau$:

$$c_{YX}(u) = \int_{-\infty}^{\infty} B(s-u, s-v)c_X(v)dv \tag{2.35}$$

Since the left side is independent of s, we may put $s = u$. Thus

$$c_{YX}(u) = \int_{-\infty}^{\infty} b(u-v)c_X(v)dv \tag{2.36}$$

where $b(x) = B(0, x)$.

It follows from (2.33) or in the particular case of (2.36) that the determination of the linear regression of process Y on process X amounts to solving an integral equation, knowing the cross-covariance function $\text{Cov}[X(s), X(t)]$ of the process X.

Example As an example consider a stationary process pair that is periodic with period 2π. Then (2.36) can be written as

$$c_{YX}(u) = \int_0^{2\pi} h(u-v)c_X(v)dv \tag{2.37}$$

where

$$h(x) = \sum_{i=-\infty}^{\infty} b(x + 2\pi i) \tag{2.38}$$

By a substitution test the reader may show that the regression coefficient function is

$$h(x) = \frac{1}{\pi}\sum_{n=1}^{\infty} \frac{b_n}{c_n} \sin(nx) \tag{2.39}$$

and that the residual covariance function is

$$c_Y(u|X) = \frac{1}{2}a_0 + \sum_{n=1}^{\infty}[1 - \frac{b_n^2}{a_n c_n}]a_n \cos(nu) \tag{2.40}$$

where

$$a_n = \frac{1}{\pi}\int_0^{2\pi} c_X(x)\cos(nx)dx, \qquad n = 0, 1, ... \tag{2.41}$$

$$b_n = \frac{1}{\pi}\int_0^{2\pi} c_{YX}(x)\sin(nx)dx, \qquad n = 1, ... \tag{2.42}$$

$$c_n = \frac{1}{\pi}\int_0^{2\pi} c_Y(x)\cos(nx)dx, \qquad n = 0, 1, ... \tag{2.43}$$

are the Fourier coefficients of c_X, c_{YX}, and c_Y, respectively.

Considered as a function of n, the ratio $|b_n|/\sqrt{a_n c_n}$ is the so-called coherence function for the pair (X, Y) of random processes, and it is bounded in value between 0 and 1. The coherence function can for each n be interpreted as the absolute value of the correlation coefficient between the nth Fourier components of X and Y. This can be seen by comparing (2.10) and (2.40).

A technical application of the linear regression in this example concerns the modeling of random silo load fields in vertical circular cylindrical silos. Let $X(u)$ and $Y(u)$ be the horizontal wall shear stress and the wall normal stress, respectively, at the angular position u in a given horizontal plane. Assume that the random field $(X(u), Y(u))$ is homogeneous with respect to u with given covariance functions formulated such that the entire wall stress field is in global equilibrium. Then the linear regression $\hat{E}[Y(u)|X]$ defines a normal stress field that is in global equilibrium with the shear stresses set to zero everywhere. Thus the residual covariance function corresponds to a wall stress field that acts normal to the wall and is in global equilibrium. Such a wall stress field may act on a horizontally ideally smooth silo wall [13].

2.3 Normal Distribution

The standardized normal (or Gaussian) density

$$\varphi(x) = \frac{1}{\sqrt{2\pi}} e^{-\frac{1}{2}x^2}, \quad x \in \Re \tag{2.44}$$

is directly generalized to the n-dimensional standardized normal (or Gaussian) density

$$f_{U_1,\dots,U_n}(u_1, \dots, u_n) = \prod_{i=1}^{n} \varphi(u_i) = (\frac{1}{\sqrt{2\pi}})^n e^{-\frac{1}{2}r^2}, \quad (u_1, \dots, u_n) \in \Re^n \tag{2.45}$$

where $r^2 = u_1^2 + \dots + u_n^2$. This density is rotational symmetric with respect to origo and the covariance matrix of the random vector $\mathbf{U} = (U_1, \dots, U_n)$ is the unit matrix. According to the definition the random variables $U_1, \dots, U_n$ are mutually independent. The conditional density of any subvector of dimension $m < n$ given the complementary subvector of dimension $n - m$ is obviously the standardized m-dimensional normal distribution. Moreover, any random vector $\mathbf{V} = (V_1, \dots, V_n)$ obtained by an orthogonal mapping of $(U_1, \dots, U_n)$ has the n-dimensional standardized normal density.

For any regular linear mapping $\mathbf{X} = \mathbf{AU}$ the random vector $\mathbf{X}$ (or the distribution of $\mathbf{X}$) is said to be n-dimensional normal (or Gaussian) with expectation vector zero and regular covariance matrix $\text{Cov}[\mathbf{X}, \mathbf{X}'] = \mathbf{AA}'$. The density function is

$$f_{\mathbf{X}}(\mathbf{x}) = (\frac{1}{\sqrt{2\pi}})^n \frac{1}{|\det(\mathbf{A})|} \exp\{-\frac{1}{2}\mathbf{x}'\text{Cov}[\mathbf{X}, \mathbf{X}']^{-1}\mathbf{x}\} \tag{2.46}$$

where $|\det(\mathbf{A})| = \sqrt{\det(\mathrm{Cov}[\mathbf{X}, \mathbf{X}'])}$, ($\det(\mathbf{A})$ is the determinant of the square matrix $\mathbf{A}$). Conversely, any $\mathbf{X}$ with the density function (2.46) can in an infinity of ways be written as a regular linear mapping $\mathbf{X} = \mathbf{AU}$ of an n-dimensional standardized normal vector $\mathbf{U}$. All that is needed is to determine $\mathbf{A}$ such that

$$\mathbf{AA}' = \mathrm{Cov}[\mathbf{X}, \mathbf{X}'] \tag{2.47}$$

It is required for the definition (2.46) of the n-dimensional normal density to make sense that $\det\mathbf{A} \neq 0$, that is, that $\mathrm{Cov}[\mathbf{X}, \mathbf{X}']$ is a regular matrix. In that case the n-dimensional normal distribution is characterized as being regular. However, a probability distribution can be obtained in $\Re^n$ by a limit passage where $\mathrm{Cov}[\mathbf{X}, \mathbf{X}']$ approaches a singular matrix. Then the entire probability mass in the limit becomes situated on a subspace of $\Re^n$ the dimension of which is equal to the rank r of $\mathrm{Cov}[\mathbf{X}, \mathbf{X}']$ (as obtained in the limit). Then with probability 1 exactly $n - r$ of the random variables in $\mathbf{X}$ can be expressed as linear functions of the remaining r random variables. These r random variables jointly have a regular r-dimensional normal distribution. In any higher dimension than r the distribution is called singular normal.

Mathematical models of physical phenomena of engineering interest do not rarely contain nonlinear functions of random vectors. A way to make such models accessible to analytical solution methods is to replace them by linear models that in some sense are approximations to the nonlinear models. Phenomena that strongly depend on the nonlinear nature of the models are lost in this way, of course, but other properties such as the behavior of robust averages may be sufficiently well represented for engineering purposes by the approximating linear model.

In the sense of least mean square deviation the nonlinear function $F(\mathbf{X})$ is best approximated by the linear regression

$$\hat{E}[F(\mathbf{X}) \mid \mathbf{X}] = E[F(\mathbf{X})] + \mathrm{Cov}[F(\mathbf{X}), \mathbf{X}']\mathrm{Cov}[\mathbf{X}, \underline{f}X']^{-1}(\mathbf{X} - E[\mathbf{X}]) \tag{2.48}$$

It is seen that the calculation of the coefficients in this linear regression requires knowledge of the distribution of $\mathbf{X}$. If $\mathbf{X}$ is Gaussian, it is convenient to use the representation $\mathbf{X} = \mathbf{A}\,\mathbf{U} + E[\mathbf{X}]$ with $\mathbf{U}$ standardized Gaussian and $\mathbf{A}$ satisfying (2.47). Then $F(\mathbf{X}) = F(\mathbf{A}\,\mathbf{U} + E[\mathbf{X}])$ may be written as $G(\mathbf{U})$ and according to (2.27) we have

$$\hat{E}[F(\mathbf{X}) \mid \mathbf{X}] = \hat{E}[G(\mathbf{U}) \mid \mathbf{U}] = E[G(\mathbf{U})] + E[G(\mathbf{U})\mathbf{U}']\mathbf{U} \tag{2.49}$$

The ith element of $E[G(\mathbf{U})\mathbf{U}']$ becomes

$$E[G(\mathbf{U})U_i] = E[E[G(\mathbf{U}) \mid U_i]U_i] = \int_{-\infty}^{\infty} E[G(\mathbf{U}) \mid u_i]u_i\varphi(u_i)du_i =$$

$$[-E[G(\mathbf{U}) \mid u_i]\varphi(u_i)]_{-\infty}^{\infty} + \int_{-\infty}^{\infty} \varphi(u_i)\frac{d}{du_i}E[G(\mathbf{U}) \mid u_i]du_i = \tag{2.50}$$

$$\int_{-\infty}^{\infty} E\left[\frac{\partial G(\mathbf{U})}{\partial u_i} \mid u_i\right]\varphi(u_i)du_i = E\left[\frac{\partial G(\mathbf{U})}{\partial u_i}\right]$$

assuming that $E[G(U) \mid u_i]\varphi(u_i) \to 0$ for $u_i \to \pm\infty$, and that $\partial G(\mathbf{u})/\partial u_i$ exists everywhere except for a set of probability zero. Thus we have the result

$$E[G(\mathbf{U})\mathbf{U}] = E[\operatorname{grad} G(\mathbf{U})] \tag{2.51}$$

where grad $G(\mathbf{u})$ is the gradient of the scalar field $G(\mathbf{u})$. By use of the chain rule of partial differentiation it is easily seen that

$$\operatorname{grad} G(\mathbf{U}) = \mathbf{A}'\operatorname{grad} F(\mathbf{X}) \tag{2.52}$$

such that (2.49) becomes

$$\hat{E}[F(\mathbf{X}) \mid \mathbf{X}] = E[F(\mathbf{X})] + E[\operatorname{grad} F(\mathbf{X})]'(\mathbf{X} - E[\mathbf{X}]) \tag{2.53}$$

By comparison with (2.48) it is seen that

$$\operatorname{Cov}[F(\mathbf{X}), \mathbf{X}] = \operatorname{Cov}[\mathbf{X}, \mathbf{X}']E[\operatorname{grad} F(\mathbf{X})] \tag{2.54}$$

is valid for any nonsingular Gaussian vector **X**.

Example In random vibration engineering it is often relevant to study an n degree of freedom damped mass system with nonlinear restoring forces and subjected to Gaussian force process excitation. The matrix equation of motion given in terms of the response **X** then reads

$$\mathbf{M}\ddot{\mathbf{X}} + \mathbf{D}\dot{\mathbf{X}} + \mathbf{F}(\mathbf{X}) = \mathbf{Y} \tag{2.55}$$

where **M** and **D** are mass and damping matrices, respectively, **F**(**X**) is the vector of nonlinear restoring forces, and **Y** is the Gaussian vector process of force excitation. In general, it is difficult to solve (2.55) to obtain the probabilistic description of the response vector process $\mathbf{X}(t)$. However, if (2.55) is replaced by the linear differential equation

$$\mathbf{M}\ddot{\mathbf{X}} + \mathbf{D}\dot{\mathbf{X}} + \mathbf{K}\mathbf{X} = \mathbf{Y} \tag{2.56}$$

where $\mathbf{K}$ is a suitably chosen stiffness matrix (that may be time dependent), then $\mathbf{X}$ becomes Gaussian. One type of so-called ("equivalent") stochastic linearization then assumes that $\mathbf{X}$ is Gaussian and on this basis replaces the nonlinear restoring force $\mathbf{F}(\mathbf{X})$ (of zero mean) by the linear regression of $\mathbf{F}(\mathbf{X})$ on $\mathbf{X}$. Thus according to (2.53), $\mathbf{K}$ is defined as

$$\mathbf{K} = E[\text{grad}\,\mathbf{F}(\mathbf{X})] = \left\{E\left[\frac{\partial F_i(\mathbf{X})}{\partial x_j}\right]\right\}_{i,j=1,\ldots,n} \tag{2.57}$$

(i = row number, j = column number). The stiffness matrix is then determined iteratively by finding the parameters of the Gaussian distribution of $\mathbf{X}$ from equations obtained from (2.56) with some initial guess on $\mathbf{K}$ substituted, next using these parameters to calculate a new $\mathbf{K}$ from (2.57), and thereafter proceed iteratively in the same way until a level of sufficiently small changes is reached. This particular procedure is called stochastic linearization using Gaussian closure. How much truth there is in the word "equivalent" often used in connection with this technique can only be investigated for cases where exact solutions are known, or by comparisons with empirical results obtained by simulation [14].

Now let us define a random vector $\mathbf{X} = (X_1, \ldots, X_n)$ recursively by the linear equations

$$\begin{aligned} X_1 &= \delta_1 U_1 \\ X_2 &= b_{21} X_1 + \delta_2 U_2 \\ &\vdots \\ X_n &= b_{n1} X_1 + \ldots + b_{n,n-1} X_{n-1} + \delta_n U_n \end{aligned} \tag{2.58}$$

where $\delta_1, \ldots, \delta_n > 0$ and $b_{21}, \ldots, b_{n,n-1}$ are constant coefficients. By solution we get

$$\mathbf{X} = (\mathbf{I} - \mathbf{B})^{-1} \Delta \mathbf{U} \tag{2.59}$$

where $\mathbf{I}$ is the unit matrix, $\Delta = \lceil \delta_1 \ldots \delta_n \rfloor$ is a diagonal matrix, and

$$\mathbf{B} = \begin{bmatrix} 0 & 0 & \cdots & 0 & 0 \\ b_{21} & 0 & \cdots & 0 & 0 \\ \cdots \\ b_{n1} & b_{n2} & \cdots & b_{n,n-1} & 0 \end{bmatrix} \tag{2.60}$$

Clearly $\mathbf{I} - \mathbf{B}$ is regular so that the solution (2.59) exists. Thus $\mathbf{X}$ has a normal distribution with expectation vector zero and covariance matrix

$$\text{Cov}[\mathbf{X}, \mathbf{X}'] = (\mathbf{I} - \mathbf{B})^{-1} \Delta [(\mathbf{I} - \mathbf{B})^{-1} \Delta]' \tag{2.61}$$

for which the inverse is

$$\mathrm{Cov}[\mathbf{X}, \mathbf{X}']^{-1} = [\Delta^{-1}(\mathbf{I} - \mathbf{B})]'\Delta^{-1}(\mathbf{I} - \mathbf{B}) \tag{2.62}$$

where $\Delta^{-1}(\mathbf{I} - \mathbf{B})$ as well as $(\mathbf{I} - \mathbf{B})^{-1}\Delta$ are lower triangular matrices, that is, all elements above the diagonal are zero.

If the covariance matrix $\mathrm{Cov}[\mathbf{X}, \mathbf{X}']$ is given and is regular, a lower triangular matrix $\mathbf{A}$ can be uniquely determined by Cholesky decomposition of $\mathrm{Cov}[\mathbf{X}, \mathbf{X}']$ such that (2.47) is satisfied. Thus $(\mathbf{I} - \mathbf{B})^{-1}\Delta$ and its inverse $\Delta^{-1}(\mathbf{I} - \mathbf{B})$ can be determined uniquely by Cholesky decomposition of $\mathrm{Cov}[\mathbf{X}, \mathbf{X}']$ and $\mathrm{Cov}[\mathbf{X}, \mathbf{X}']^{-1}$, respectively, implying that any Gaussian random vector $\mathbf{X}$ of zero expectation and regular covariance matrix can be written uniquely as in (2.58).

Obviously the linear regression

$$\hat{E}[X_i|X_1, ..., X_{i-1}] = b_{i1}X_1 + ... + b_{i,i-1}X_{i-1} \tag{2.63}$$

coincides with the conditional expectation $E[X_i|X_1, ..., X_{i-1}]$, and the residual variance

$$\mathrm{Var}[X_i - \hat{E}[X_i|X_1, ..., X_{i-1}]] = \delta_i^2 \tag{2.64}$$

coincides with the conditional variance $\mathrm{Var}[X_i|X_1, ..., X_{i-1}]$. However, writing $\mathbf{X}_k = (X_1, ..., X_k)$ we more generally have that

$$E[X_i|\mathbf{X}_k] = \hat{E}[X_i|\mathbf{X}_k] \tag{2.65}$$

and

$$\mathrm{Cov}[X_i, X_j|\mathbf{X}_k] = \mathrm{Cov}[X_i - \hat{E}[X_i|\mathbf{X}_k], X_j - \hat{E}[X_j|\mathbf{X}_k]] \tag{2.66}$$

for any $i, j, k \in \{1, ..., n\}$. The proofs of (2.65) and (2.66) are as follows.

The coincidence of the conditional expectation of X_i given $\mathbf{X}_k$ and the linear regression of X_i on $\mathbf{X}_k$ follows from the equations (2.58) using the one-to-one correspondence between $(X_1, ..., X_k)$ and $(U_1, ..., U_k)$. Thus we may replace the conditioning on $\mathbf{X}_k$ by conditioning on $\mathbf{U}_k = (U_1, ..., U_k)$. Since

$$E[U_i|\mathbf{U}_k] = \hat{E}[U_i|\mathbf{U}_k] = \begin{cases} U_i & \text{for } i \leq k \\ 0 & \text{for } i > k \end{cases} \tag{2.67}$$

it follows that (2.58) gives the same equations for the conditional expectations and the linear regressions. The unique solution is given by (2.59) replacing $\mathbf{U}$ by $E[\mathbf{U}|\mathbf{U}_k]$.

Similarly the conditional covariances $\text{Cov}[X_i, X_j|\mathbf{X}_k]$ can be obtained from (2.58) replacing X_i and U_i for $i \in \{1, ..., n\}$ by $\text{Cov}[X_i, X_j|\mathbf{X}_k]$ and $\text{Cov}[U_i, X_j|\mathbf{U}_k]$, respectively, for each $j \in \{1, ..., n\}$. The solution is obtained from (2.59) as

$$\text{Cov}[\mathbf{X}, \mathbf{X}'|\mathbf{X}_k] = (\mathbf{I} - \mathbf{B})^{-1}\Delta\text{Cov}[\mathbf{U}, \mathbf{X}'|\mathbf{U}_k] \tag{2.68}$$

where, according to (2.59),

$$\text{Cov}[\mathbf{U}, \mathbf{X}'|\mathbf{U}_k] = \text{Cov}[\mathbf{U}, \mathbf{U}'|\mathbf{U}_k]\Delta[(\mathbf{I} - \mathbf{B})^{-1}]' \tag{2.69}$$

and

$$\text{Cov}[U_i, U_j|\mathbf{U}_k] = \begin{cases} 1 \text{ for } i = j \notin \{1, ..., k\} \\ 0 \text{ otherwise} \end{cases} \tag{2.70}$$

It is seen that the conditional covariance does not depend on the value of $\mathbf{X}_k$ implying that

$$E[\text{Cov}[\mathbf{X}, \mathbf{X}'|\mathbf{X}_k]] = \text{Cov}[\mathbf{X}, \mathbf{X}'|\mathbf{X}_k] \tag{2.71}$$

The residual covariance matrix is

$$\begin{aligned}
&\text{Cov}[\mathbf{X} - \hat{E}[\mathbf{X}|\mathbf{X}_k],\ \mathbf{X}' - \hat{E}[\mathbf{X}'|\mathbf{X}_k]] = \\
&\text{Cov}[\mathbf{X} - E[\mathbf{X}, \mathbf{X}_k],\ \mathbf{X}' - E[\mathbf{X}', \mathbf{X}_k]] = \\
&E[\text{Cov}[\mathbf{X}, \mathbf{X}'|\mathbf{X}_k]] + \\
&\text{Cov}[E[\mathbf{X} - E[\mathbf{X}|\mathbf{X}_k]|\mathbf{X}_k],\ E[\mathbf{X}' - E[\mathbf{X}'|\mathbf{X}_k]|\mathbf{X}_k]] \\
&= \text{Cov}[\mathbf{X}, \mathbf{X}'|\mathbf{X}_k]
\end{aligned} \tag{2.72}$$

which proves (2.66). Thus we have the important result that the conditional covariance matrix is identical to the residual covariance matrix.

For $n = 2$ and $\text{Var}[X_1] = \text{Var}[X_2] = 1$, we have

$$\text{Cov}[\mathbf{X}, \mathbf{X}'] = \begin{bmatrix} 1 & \rho \\ \rho & 1 \end{bmatrix} = \begin{bmatrix} 1 & 0 \\ \rho & \sqrt{1-\rho^2} \end{bmatrix} \begin{bmatrix} 1 & \rho \\ 0 & \sqrt{1-\rho^2} \end{bmatrix} \tag{2.73}$$

so that $X_1 = U_1$, $X_2 = \rho X_1 + \delta_2 U_2$ where $\delta_2 = \sqrt{1-\rho^2}$. The joint density of

(X_1, X_2) is

$$
\begin{aligned}
&f_{X_1,X_2}(x_1,x_2) = f_{X_2}(x_2|X_1 = x_1) f_{X_1}(x_1) = \\
&\frac{1}{\delta_2}\varphi\left(\frac{x_2-\rho x_1}{\delta_2}\right)\varphi(x_1) = \frac{1}{2\pi\delta_2}\exp\left\{-\frac{1}{2\delta_2^2}[(x_2-\rho x_1)^2 + (1-\rho^2)x_1^2]\right\} = \\
&\frac{1}{2\pi\sqrt{1-\rho^2}}\exp\left[-\frac{1}{2(1-\rho^2)}(x_1^2 - 2\rho x_1 x_2 + x_2^2)\right]
\end{aligned}
\tag{2.74}
$$

This density function is a standard denoted as $\varphi_2(x_1, x_2; \rho)$.

For $n = 3$ and $\mathrm{Var}[X_1] = \mathrm{Var}[X_2] = \mathrm{Var}[X_3] = 1$, we have

$$
\mathrm{Cov}[\mathbf{X}, \mathbf{X}'] = \begin{bmatrix} 1 & \rho_{12} & \rho_{13} \\ \rho_{12} & 1 & \rho_{23} \\ \rho_{13} & \rho_{23} & 1 \end{bmatrix} = \mathbf{A}\,\mathbf{A}' \tag{2.75}
$$

with

$$
\mathbf{A} = \begin{bmatrix} 1 & 0 & 0 \\ \rho_{12} & \delta_2 & 0 \\ \rho_{13} & \dfrac{\rho_{23} - \rho_{12}\rho_{13}}{\sqrt{1-\rho_{12}^2}} & \delta_3 \end{bmatrix} \tag{2.76}
$$

where

$$
\delta_2 = \sqrt{1-\rho_{12}^2} \tag{2.77}
$$

$$
\delta_3 = \sqrt{\frac{1-\rho_{12}^2-\rho_{13}^2-\rho_{23}^2+2\rho_{12}\rho_{13}\rho_{23}}{1-\rho_{12}^2}} \tag{2.78}
$$

so that

$$
\begin{aligned}
X_1 &= U_1 \\
X_2 &= \rho_{12}U_1 + \delta_2 U_2 = \rho_{12}X_1 + \delta_2 U_2 \\
X_3 &= \rho_{13}U_1 + \frac{\rho_{23}-\rho_{12}\rho_{13}}{\sqrt{1-\rho_{12}^2}}U_2 + \delta_3 U_3 = \\
&= \frac{\rho_{13}-\rho_{12}\rho_{23}}{\sqrt{1-\rho_{12}^2}}X_1 + \frac{\rho_{23}-\rho_{12}\rho_{13}}{\sqrt{1-\rho_{12}^2}}X_2 + \delta_3 U_3
\end{aligned}
\tag{2.79}
$$

The joint density of (X_1, X_2, X_3) is

$$f_{X_1,X_2,X_3}(x_1, x_2, x_3) =$$

$$f_{X_3}(x_3|X_1 = x_1, X_2 = x_2) f_{X_2}(x_2|X_1 = x_1) f_{X_1}(x_1) =$$

$$\frac{1}{\delta_3}\varphi\left(\frac{x_3 - b_{31}x_1 - b_{32}x_2}{\delta_3}\right)\frac{1}{\delta_2}\varphi\left(\frac{x_2 - b_{21}x_1}{\delta_2}\right)\varphi(x_1) =$$

$$\left(\frac{1}{2\pi}\right)^{3/2}\frac{1}{\delta_2\delta_3}\exp\{-\frac{1}{2(\delta_2\delta_3)^2}[(1-\rho_{23}^2)x_1^2 + (1-\rho_{13}^2)x_2^2 + (1-\rho_{23}^2)x_3^2$$

$$-2(\rho_{12} - \rho_{13}\rho_{23})x_1x_2 - 2(\rho_{13} - \rho_{12}\rho_{23})x_1x_3 - 2(\rho_{23} - \rho_{12}\rho_{13})x_2x_3]\}$$

$$(2.80)$$

Finally, an n-dimensional vector $\mathbf{X}$ with expectation $E[\mathbf{X}] = \boldsymbol{\mu}$ is said to be normal or Gaussian, if $\mathbf{Y} = \mathbf{X} - \boldsymbol{\mu}$ is Gaussian.

2.4 Non-Gaussian Distributions and Linear Regression

In the previous section it is shown that the multi-dimensional normal distribution has the property that the conditional expectation of any normal vector $\mathbf{Y}$ given any normal vector $\mathbf{X}$ is coincident with the linear regression of $\mathbf{Y}$ on $\mathbf{X}$ provided the joint distribution of $(\mathbf{X}, \mathbf{Y})$ is normal. Non-Gaussian distributions generally do not have this property. It is easy to see that if the conditional expectation

$$E[\mathbf{Y}|\mathbf{X}] = \int_{\Re^n} \mathbf{y} f_{\mathbf{Y}}(\mathbf{y}|\mathbf{X})\mathbf{dy} \tag{2.81}$$

is linear in $\mathbf{X}$, then $E[\mathbf{Y}|\mathbf{X}] = \hat{E}[\mathbf{Y}|\mathbf{X}]$. This follows from the fact that the expectation of any random variable is equal to the value relative to which the mean square deviation of the random variable is smallest. However, the conditional covariance matrix $\text{Cov}[\mathbf{Y}, \mathbf{Y}'|\mathbf{X}]$ may vary with $\mathbf{X}$ and thus be different from the residual covariance matrix $\text{Cov}[\mathbf{Y} - \hat{E}[\mathbf{Y}|\mathbf{X}], \mathbf{Y}' - \hat{E}[\mathbf{Y}'|\mathbf{X}]]$.

Important classes of m-dimensional non-Gaussian distributions can be defined by suitable non-linear transformations of a normal vector $\mathbf{X}$. The simplest and most often used type of non-linear transformation maps each element X_i of $\mathbf{X}$ into $Y_i = g_i(X_i)$, where $g_1, ..., g_m$ are non-linear increasing functions of one variable. This type of m-dimensional transformation may conveniently be denoted as an increasing marginal transformation.

Let $\mathbf{g} = (\mathbf{g}_1, \mathbf{g}_2)$ be an increasing marginal transformation of the normal vector $(\mathbf{X}, \mathbf{Y})$. Obviously the linear regression $\hat{E}[\mathbf{g}_2(\mathbf{Y})|\mathbf{g}_1(\mathbf{X})]$ is generally not simply related to the conditional expectation $E[\mathbf{Y}|\mathbf{X}] = \hat{E}[\mathbf{Y}|\mathbf{X}]$ or to the conditional expectation $E[\mathbf{g}_2(\mathbf{Y})|\mathbf{g}_1(\mathbf{X})]$ (except, of course, if $\mathbf{g}$ is linear). However, it can be generally stated that $g_2(\hat{E}[\mathbf{Y}|\mathbf{X}])$ is the marginal median point of the conditional density of $\mathbf{g}_2(\mathbf{Y})$ given $\mathbf{g}_1(\mathbf{X})$ (or given $\mathbf{X} = \mathbf{g}_1^{-1}[\mathbf{g}_1(\mathbf{X})]$), that is, any given element of $\mathbf{g}_2(\mathbf{Y})$ takes a value below or above the corresponding element value of $g_2(\hat{E}[\mathbf{Y}|\mathbf{X}])$ with probability 1/2. Generally the point $\mathbf{g}_2(\hat{E}[\mathbf{Y}|\mathbf{X}])$ is simpler to calculate than the conditional expectation point $E[\mathbf{g}_2(\mathbf{Y})|\mathbf{g}_1(\mathbf{X})]$.

Example In practice the most frequently used increasing marginal transformation is the exponential transformation $g_i(x) = e^x$, $i = 1, ..., n$: Define $\log\mathbf{X} \equiv (\log X_1, ..., \log X_n)$ as a normal vector. Then $\mathbf{X}$ is said to have a lognormal distribution. The relations between $E[\log\mathbf{X}]$, $\mathrm{Cov}[\log\mathbf{X}, \log\mathbf{X}']$ and $E[\mathbf{X}]$, $\mathrm{Cov}[\mathbf{X}, \mathbf{X}']$ are

$$E[\mathbf{X}] = \left\{\exp(E[\log X_i] + \frac{1}{2}\mathrm{Var}[\log X_i])\right\}_{i=1,...,n} \tag{2.82}$$

$$E[\log\mathbf{X}] = \left\{\log E[X_i] + \frac{1}{2}\log(1 + V_{X_i}^2)\right\}_{i=1,...,n} \tag{2.83}$$

$$\mathrm{Cov}[\mathbf{X}, \mathbf{X}'] = \left\{[\exp(\mathrm{Cov}[\log X_i, \log X_j]) - 1]E[X_i]E[X_j]\right\}_{i,j=1,...,n} \tag{2.84}$$

$$\mathrm{Cov}[\log\mathbf{X}, \log\mathbf{X}'] = \left\{\log[1 + \frac{\mathrm{Cov}[X_i, X_j]}{E[X_i]E[X_j]}]\right\}_{i,j=1,...,n} \tag{2.85}$$

Let $(\mathbf{X}, \mathbf{Y})$ be a lognormal vector, $\mathbf{X}$ of dimension m, $\mathbf{Y}$ of dimension n. Then the conditional distribution of $\mathbf{Y}$ given $\mathbf{X}$ is n-dimensional lognormal. According to (2.82) to (2.84) the conditional expectation of Y_i given $\mathbf{X}$ becomes

$$\begin{aligned}
&E[Y_i|\mathbf{X}] = E[Y_i|\log\mathbf{X}] = \\
&\exp(E[\log Y_i|\log\mathbf{X}] + \tfrac{1}{2}\mathrm{Var}[\log Y_i|\log\mathbf{X}]) = \\
&\exp(\hat{E}[\log Y_i|\log\mathbf{X}] + \tfrac{1}{2}\mathrm{Var}[\log Y_i - \hat{E}[\log Y_i|\log\mathbf{X}]]) = \\
&\exp\{E[\log Y_i] + \mathrm{Cov}[\log Y_i, \log\mathbf{X}']\mathrm{Cov}[\log\mathbf{X}, \log\mathbf{X}']^{-1}(\log\mathbf{X} - E[\log\mathbf{X}]) \\
&+\tfrac{1}{2}(\mathrm{Var}[\log Y_i] - \mathrm{Cov}[\log Y_i, \log\mathbf{X}']\mathrm{Cov}[\log\mathbf{X}, \log\mathbf{X}']^{-1} \\
&\quad \mathrm{Cov}[\log\mathbf{X}, \log Y_i])\}
\end{aligned} \tag{2.86}$$

Thus $E[Y_i|\mathbf{X}]$ depends nonlinearly on $\mathbf{X}$ and is therefore different from $\hat{E}[Y_i|\mathbf{X}]$. If the variance term is neglected we get the marginal median point $\exp(\hat{E}[\log Y_i|\log\mathbf{X}])$.

The conditional covariance between Y_i and Y_j given $\mathbf{X}$ becomes

$$\begin{aligned}&\text{Cov}[Y_i, Y_j|\mathbf{X}] = \text{Cov}[Y_i, Y_j|\log\mathbf{X}] = \\ &\left\{\exp(\text{Cov}[\log Y_i, \log X_j|\log\mathbf{X}]) - 1\right\} E[Y_i|\mathbf{X}]E[Y_j|\mathbf{X}]\end{aligned} \tag{2.87}$$

where $\text{Cov}[\log Y_i, \log X_j|\log\mathbf{X}]$ is equal to the covariance between the linear regression residuals $\log Y_i - \hat{E}[\log Y_i|\log\mathbf{X}]$ and $\log Y_j - \hat{E}[\log Y_j|\log\mathbf{X}]$, and therefore does not depend on $\mathbf{X}$. However, the conditional expectation factors in (2.87) depend on $\mathbf{X}$ as shown in (2.86).

The linear regression $\hat{E}[\mathbf{Y}|\mathbf{X}]$ plays no particular interesting role in the lognormal distribution except that $\hat{E}[\mathbf{Y}|\mathbf{X}]$ is that linear function of $\mathbf{X}$ that approximates the conditional expectation $E[\mathbf{Y}|\mathbf{X}]$ best in the sense of minimizing the expected squared difference $E[(E[\mathbf{Y}|\mathbf{X}] - \hat{E}[\mathbf{Y}|\mathbf{X}])^2]$.

2.5 Marginally Transformed Gaussian Processes and Fields

As stated in the introduction section, the word "field" will in the following be used as short for "random process" or "random field". A field $X(t)$ is said to be Gaussian if the random vector corresponding to any finite subset $\{t_1, ..., t_n\}$ of the index set I is a Gaussian vector. A Gaussian field is completely defined by the expectation or mean value function $\mu(t) \equiv E[X(t)]$ and the covariance function $c(s,t) \equiv \text{Cov}[X(s), X(t)]$. The last function must be nonnegative definite:

$$\begin{aligned}&\forall t_1, ..., t_n \in I : \{c(t_i, t_j) = c(t_j, t_i) \\ &\text{and } \forall x_1, ..., x_n \in \Re : \sum_{i=1}^{n}\sum_{j=1}^{n} c(t_i, t_j)x_i x_j \geq 0\}\end{aligned} \tag{2.88}$$

Given that $I \subset \Re^q$ for some q, a field is said to be homogeneous (or stationary, if the word "field" stands for "random process") within I if the joint distribution of $(X(t_1), ..., X(t_n))$ is identical to the joint distribution of $(X(t_1+\tau), ..., X(t_n+\tau))$ for any $\{t_1, ..., t_n\} \in I$ and any τ such that $\{t_1+\tau, ..., t_n+\tau\} \in I$.

A Gaussian field is homogeneous if and only if $\mu(t)$ is a constant and the covariance function $c(s,t)$ is a function solely of the difference $t - s$. If this condition is satisfied for a non-Gaussian field, the field is not necessarily homogeneous but it is then said to be weakly homogeneous or homogeneous up to the second order moments.

Let $Y(t)$ be a Gaussian field with zero mean value function $E[Y(t)] \equiv 0$, unit variance function $\text{Var}[Y(t)] \equiv 1$, and correlation function $\rho(s,t) = \rho[Y(s), Y(t)]$. Moreover, let $g(x,t)$ be some function of $x \in \Re$ and $t \in I$ for which

$$\int_{-\infty}^{\infty} g(x,t)\varphi(x)dx \equiv 0, \quad \int_{-\infty}^{\infty} g(x,t)^2\varphi(x)dx \equiv 1 \tag{2.89}$$

and let $\mu(t)$ and $\sigma(t) > 0$ be given functions of $t \in I$. Then the field

$$X(t) = \mu(t) + g[Y(t), t]\sigma(t) \tag{2.90}$$

has mean value function $\mu(t)$, variance function $\sigma(t)^2$, and correlation function

$$\rho[X(s), X(t)] = \int_{-\infty}^{\infty}\int_{-\infty}^{\infty} g(x,s)g(y,t)\varphi_2[x,y;\rho(s,t)]dx\,dy \tag{2.91}$$

The field $X(t)$ is said to be obtained by a marginal transformation of the field $Y(t)$, and it is Gaussian if $g(x,t)$ is a linear function in x. According to (2.89) this linear function then must be $g(x,t) = x$, and (2.91) gives $\rho[X(s), X(t)] = \rho[Y(s), Y(t)] \equiv \rho(s,t)$. If $\mu(t), \sigma(t)$ as well as the marginal transformation is independent of $t \in I$, and $Y(t)$ is homogeneous, then $X(t)$ is also homogeneous, and (2.91) simplifies to

$$r(t) = \int_{-\infty}^{\infty}\int_{-\infty}^{\infty} g(x)g(y)\varphi_2[x,y;\rho(t)]dx\,dy \tag{2.92}$$

where $r(t-s) = \rho[X(s), X(t)]$ and $\rho(t-s) = \rho[Y(s), Y(t)]$.

We may now quite naturally ask the question of whether it is always possible to determine the correlation function $\rho(t)$ of the Gaussian process such that a given nonnegative definite function $r(t)$ is the correlation function for the homogeneous field $X(t) = g[Y(t)]$. The answer is negative. By the right side of (2.92) the set of nonnegative definite functions is in general mapped into a genuine subset of the set of nonnegative definite functions. In other words, it is not granted that the integral equation (2.92) for a given nonnegative definite function $r(t)$ has a solution $\rho(t)$ in the set of nonnegative definite functions. If a nonnegative definite solution exists the field $X(t) = g[Y(t)]$ is well defined with zero mean value, unit variance, and given correlation function $r(t)$. This type of homogeneous non-Gaussian field is called a zero mean, unit variance homogeneous Nataf field [15, 16].

Example Let $I = \Re$ and

$$X(t) = \exp[a + bY(t)] \tag{2.93}$$

such that for any given $t \in \Re$, $X(t)$ has a lognormal distribution. The mean μ and the variance σ^2 are given by

$$\mu = \int_{-\infty}^{\infty} e^{a+bx} \varphi(x) dx = e^{a+b^2/2} \tag{2.94}$$

$$(\frac{\sigma}{\mu})^2 = \frac{1}{\mu^2} \int_{-\infty}^{\infty} e^{2(a+bx)} \varphi(x) dx - 1 = e^{b^2} - 1 \tag{2.95}$$

consistent with (2.82) and (2.84). Thus

$$a = \log(\mu/\sqrt{1+X^2}) \tag{2.96}$$

$$b = \sqrt{\log(1+V^2)} \tag{2.97}$$

where $V = \sigma/\mu$ is the coefficient of variation of $X(t)$. Then

$$g(x) = \frac{1}{V} \left(\frac{e^{x\sqrt{\log(1+V^2)}}}{\sqrt{1+V^2}} - 1 \right) \tag{2.98}$$

Substitution into (2.92) leads to a solvable integral. We get

$$r(t) = \frac{1}{V^2} \left[(1+V^2)^{\rho(t)} - 1 \right] \tag{2.99}$$

which by solution with respect to $\rho(t)$ gives

$$\rho(t) = \frac{\log[1+V^2 r(t)]}{\log(1+V^2)} \tag{2.100}$$

consistent with (2.84) and (2.85). It can by examples be shown that there exist positive definite functions $r(t)$ for which (2.100) gives functions $\rho(t)$ that are not positive definite.

In scientific or engineering applications it is usually so that the field $X(t)$ has a physical interpretation and that $X(t)$ must satisfy some physical or geometrical conditions. For example, it may be so that $X(t)$ of physical reasons should be positive for all t. Often a model with $X(t)$ being Gaussian is applicable in spite of the physical condition of nonnegativity, simply because $X(t)$ may have so small a coefficient of variation that the probability of getting negative values of $X(t)$ is small as compared to the calculated probability of any event that is relevant for the engineering application. However, for larger coefficients of variation of $X(t)$ the

Gaussian assumption may be in too gross conflict with the nonnegativity condition to be applicable. Then often the lognormal field or some other nonnegative marginal transformation of a Gaussian field is adopted as $X(t)$.

If the model is obtained solely by fitting to data, there will be no inconsistency in the covariance function modeling if all data are transformed inversely to data that are assumed to comply with a Gaussian field model. However, in some cases it may be so that $X(t)$ satisfies some physical equation. For example, if $X(t)$ models the normal pressure on the wall of a vertical cylindrical silo with horizontally ideally smooth wall, the horizontal equilibrium of the silo medium requires that $X(t)$ satisfies three global equilibrium equations that are linear in $X(t)$. These equations put restrictions on the choice of the covariance function of $X(t)$ among the nonnegative definite functions. Therefore, starting the modeling by obeying these equilibrium conditions and thereafter assuming that $X(t)$ is a homogeneous lognormal field, say, requires careful consideration of the nonnegative definiteness of $\rho(t)$ in (2.100) when a modeling candidate for $r(t)$ has been chosen. The nonlinear relation between $r(t)$ and $\rho(t)$ usually requires some corrective steps to be taken [17].

2.6 Discretized Fields Defined by Linear Regression on a Finite Set of Field Values

The linear regression of a field $X(t)$ on a finite set $\mathbf{X}_n = (X(t_1), ..., X(t_n))$ of random variables of the field with regular covariance matrix $\mathrm{Cov}[\mathbf{X}_n, \mathbf{X}_n']$ is

$$\hat{E}[X(t)|\mathbf{X}_n] = \mu(t) + \mathrm{Cov}[X(t), \mathbf{X}_n']\mathrm{Cov}[\mathbf{X}_n, \mathbf{X}_n']^{-1}(\mathbf{X}_n - \boldsymbol{\mu}_n) \tag{2.101}$$

where $\mu(t) = E[X(t)]$ and $\boldsymbol{\mu}_n = (\mu(t_1), ..., \mu(t_n))$. The linear regression defines a field $\hat{X}(t|t_1, ..., t_n) \equiv \hat{E}[X(t)|\mathbf{X}_n]$ that may be said to have a dimension of randomness equal to n. Since

$$\hat{X}(t_i|t_1, ..., t_n) = X(t_i) \tag{2.102}$$

the field $\hat{X}(t_i|t_1, ..., t_n)$ interpolates between the values $X(t_1), ..., X(t_n)$ of the field $X(t)$. The covariance functions of the row matrix $\mathrm{Cov}[X(t), \mathbf{X}_n']$ play the role as deterministic interpolation functions (shape functions). The mean value function is identical for the two fields. The covariance function is

$$\begin{aligned}&\mathrm{Cov}[\hat{E}[X(s)|t_1, ..., t_n],\ \hat{E}[X(t)|t_1, ..., t_n]] =\\ &\mathrm{Cov}[X(s), \mathbf{X}_n']\mathrm{Cov}[\mathbf{X}_n, \mathbf{X}_n']^{-1}\mathrm{Cov}[\mathbf{X}_n, X(t)]\end{aligned} \tag{2.103}$$

which added to the residual covariance function, see (2.21), gives the covariance function $\text{Cov}[X(s), X(t)]$ of the field $X(t)$.

In numerical calculations with fields it is most often necessary to discretize the fields in the sense of replacing fields of infinite dimension of randomness by approximating fields of finite dimension of randomness. This replacement is called random field discretization. It depends on the considered problem and the related error measure, which type of random field discretization is most effective and operationally convenient.

As mentioned in the introduction section, the replacement of $X(t)$ by $\hat{E}[X(t)|\mathbf{X}_n]$ is sometimes called kriging. The error of the calculation output comes from neglecting the residual field $X(t) - \hat{E}[X(t)|\mathbf{X}_n]$, an error that in some problems can be crudely evaluated at the output level by repeated calculations using different dimensions of randomness of the discretized field. Field discretization different from the kriging method, but all based on linear regression in one or the other form, will be treated in several of the following sections.

The linearity of $\hat{E}[X(t)|\mathbf{X}_n]$ with respect to $X(t)$ directly shows that if t is a parameter for which we can talk about differentiability or integrability of $X(t)$, and $X(t)$ has such differentiability or integrability properties, then

$$\hat{E}[dX(t)|\mathbf{X}_n] = d\hat{E}[X(t)|\mathbf{X}_n] \tag{2.104}$$

and

$$\hat{E}[\int_\Omega X(t)dt|\mathbf{X}_n] = \int_\Omega \hat{E}[X(t)|\mathbf{X}_n]dt \tag{2.105}$$

where the integration is over any suitably regular set $\Omega \subset I$.

2.7 Discretization Defined by Linear Regression on a Finite Set of Linear Functionals

In stochastic mechanics applications of random fields, the fields often appear as integrands in the solutions to the relevant equations. For example, several types of load intensities acting on a structure can conveniently be modeled as random fields. The internal stresses and the displacements caused by the acting load are functions of weighted integrals (i.e., linear functionals) of the load intensity over the structure. Another example concerns constitutive relations that contain spatially varying parameters modeled as outcomes of random fields. Macroscopic effects of such variation generally are determined by weighted integrals of the constitutive parameters over the structure.

Numerical approximations to the solutions may therefore in both examples be improved with respect to accuracy if the field discretizations are made such that a selected set of relevant linear functionals are not affected by the discretization. Linear regression on the set of linear functionals serves this purpose [18].

Let $F_1, \ldots, F_n$ be n different linear functionals defined on the field $X(t)$, and let $\hat{X}(t)$ be the field defined by the linear regression

$$\hat{X}(t) \equiv \hat{E}[X(t)|\mathbf{F}\{X\}] \tag{2.106}$$

where $\mathbf{F}\{X\} = (F_1\{X\}, \ldots, F_n\{X\})$. The linearity of the linear regression ensures that the linear functionals are invariant to the replacement of $X(t)$ by $\hat{X}(t)$:

$$\mathbf{F}\{\hat{X}\} = \hat{E}[\mathbf{F}\{X\}|\mathbf{F}\{X\}] = \mathbf{F}\{X\} \tag{2.107}$$

The linear regression reads:

$$\begin{aligned} &\hat{E}[X(t)|\mathbf{F}\{X\}] = E[X(t)]+ \\ &\mathrm{Cov}[X(t), \mathbf{F}\{X\}']\mathrm{Cov}[\mathbf{F}\{X\}, \mathbf{F}\{X\}']^{-1}(\mathbf{F}\{X\} - E[\mathbf{F}\{X\}]) = \\ &E[X(t)] + \mathbf{F}_v\{\mathrm{Cov}[X(t), X(v)]\}' \\ &(\mathbf{F}_u\mathbf{F}_v'\{\mathrm{Cov}[X(u), X(v)]\})^{-1}\mathbf{F}\{X - E[X]\} \end{aligned} \tag{2.108}$$

where the indices u and v indicate that $\mathbf{F}$ operates on the functions of u and v, respectively.

Next consider any linear functional G defined both on $X(t)$ and $\hat{X}(t)$. Then the discretization error with respect to G is

$$G\{X\} - G\{\hat{X}\} = G\{X\} - \hat{E}[G\{X\}|\mathbf{F}\{X\}] \tag{2.109}$$

that is, the discretization error is the residual corresponding to the linear regression of $G\{X\}$ on $\mathbf{F}\{X\}$. Thus the variance of the discretization error is

$$\begin{aligned} &\mathrm{Var}[G\{X\} - G\{\hat{X}\}] = \mathrm{Var}[G\{X\}]- \\ &\mathrm{Cov}[G\{X\}, \mathbf{F}\{X\}'](\mathbf{F}_u\mathbf{F}_v'\{\mathrm{Cov}[X(u), X(v)]\})^{-1}\mathrm{Cov}[G\{X\}, \mathbf{F}\{X\}] = \\ &G_sG_t\Big\{\mathrm{Cov}[X(s), X(t)] - \mathbf{F}_u\{\mathrm{Cov}[X(s), X(u)\}' \\ &\left(\mathbf{F}_u\mathbf{F}_v'\{\mathrm{Cov}[X(u), X(v)]\}\right)^{-1}\mathbf{F}_v\{\mathrm{Cov}[X(t), X(v)]\}\Big\} \end{aligned} \tag{2.110}$$

where the indices s and t indicate that G operates on the functions of s and t, respectively. Thus the residual variance is obtained by application of the bilinear functional $G_s G_t$ to the residual covariance function corresponding to the linear regression of the field X on the vector of linear functionals $\mathbf{F}\{X\}$.

Example Consider a beam resting on a linear elastic bed of stiffness $S(x)$ at the point x, and subdivide the length of the beam into intervals $[x_0, x_1], ...,]x_{m-1}, x_m]$. A displacement function $u(x)$ of the form as a polynomial of nth degree generates a reaction load intensity field that over the interval $]x_{i-1}, x_i]$ has the following resulting vertical reaction and moment with respect to $x = 0$:

$$\int_{x_{i-1}}^{x_i} u(\xi)S(\xi)d\xi = \sum_{j=0}^{n} a_j \int_{x_{i-1}}^{x_i} \xi^j S(\xi)d\xi \tag{2.111}$$

$$\int_{x_{i-1}}^{x_i} \xi u(\xi)S(\xi)d\xi = \sum_{j=0}^{n} a_j \int_{x_{i-1}}^{x_i} \xi^{j+1} S(\xi)d\xi \tag{2.112}$$

respectively. With $S(x)$ being a field, these reactions and moments are not affected by replacing $S(x)$ by the discretized field $\hat{S}(x)$ defined as the linear regression of $S(x)$ on the $m(n+2)$ linear functionals

$$F_{ij}\{S\} = \int_{x_{i-1}}^{x_i} \xi^j S(\xi)d\xi; \quad i = 1, ..., m; \; j = 0, ..., n+1 \tag{2.113}$$

for all displacements that vary as a polynomial of, at most, nth degree (or just as a function that within each of the subintervals vary as a polynomial of, at most, nth degree).

Example Assume that the flexibility (compliance) of a linear elastic Euler-Bernoulli beam is modeled as a field $C(x)$, and consider a straight beam element over the interval $[-L, L]$ loaded externally from the neighbor elements at the end points and by a distributed load varying as a polynomial of $(n-2)$th degree along the beam axis and acting orthogonal to the axis. Thus the bending moment $M(x)$ in the beam varies as an nth degree polynomial. It then follows from the principle of virtual work that the total angular rotation over $[-L, L]$ and the displacement of the one end point orthogonal to the tangent at the other end point have the form

$$\int_{-L}^{L} C(x)M(x)dx = \sum_{j=0}^{n} a_j \int_{-L}^{L} x^j C(x)dx \tag{2.114}$$

$$\int_{-L}^{L} xC(x)M(x)dx = \sum_{j=0}^{n} a_j \int_{-L}^{L} x^{j+1} C(x)dx \tag{2.115}$$

respectively. Thus the macro flexibility properties of the beam element ranging over $[-L, L]$ are invariant to the replacement of $C(x)$ by $\hat{C}(x)$ defined as the linear regression of $C(x)$ on the $n+2$ linear functionals

$$F_j\{C\} = \int_{-L}^{L} x^j C(x); \; j = 0, ..., n+1 \tag{2.116}$$

Example Consider a plane straight or curved beam with the beam axis completely defined in terms of the natural equation representation $\{s, \varphi(s)\}$, where s is the arch length along the beam axis and $\varphi(s)$ is the angle between the tangent at s and the tangent at the origin of s. The beam is subjected to the random load intensity fields $p(s)$ and $q(s)$ acting in the plane of the beam orthogonal and tangential to the beam axis, respectively, Figure 2.1. First let us assume that the beam structure is statically determinate. If we let the load fields act indirectly through a row of beams that are simply supported on the given beam, the internal forces in the beam are the same at the supports of the simply supported beams as in the directly loaded beam. Therefore, a representation of the random load fields by the corresponding orthogonal and tangential random support forces of the simple beams is sufficient for the stochastic representation of the internal forces at these discretization points. However, if the beam is statically indeterminate, the redundants are functionals of the directly acting load fields causing that the internal forces at the discretization points do not become error free if the directly acting load fields are replaced by the field of the concentrated support forces. This effect becomes more dominant for coarser discretization. An illustrative example is an Euler-Bernoulli beam of shape as a circular ring with uniform load field acting orthogonal to the beam axis. This directly acting load field causes a normal force but no bending moments in the beam. However, any system of indirect loading will generate non-zero bending moments that become at their extreme level, if the discretization is chosen, so coarse as to be defined by only two diametrically opposite points, Figure 2.2. Obviously this undesirable discretization effect is counteracted by reintroducing a directly acting load field that has some similarity with the given load field. This is achieved by using the linear regression of the given load field on the set of statically equivalent support forces as the replacement load field. Thus the method of regression on linear functionals in combination with the principle of indirect loading is a rational tool for stochastic finite element discretization of random load fields on beam structures [19].

At a discretization point, let $X = X_+ + X_-$ and $Y = Y_+ + Y_-$ be the orthogonal and tangential support forces, respectively, from the two adjacent simply supported imaginary beams assumed to carry the direct load fields over to the given beam structure. Let the origin of the arch length parameter s be at the discretization point. The support forces coming from the imaginary simply supported beam of span L_+ on the positive side of the origin and from the imaginary sim-

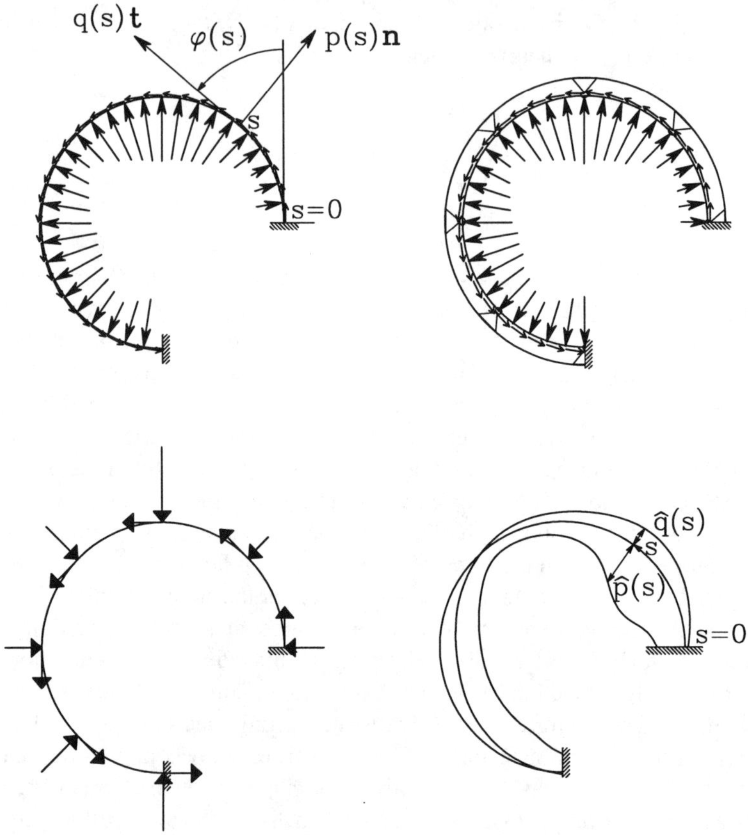

FIGURE 2.1
Top left: Load fields $p(s)$ and $q(s)$. Top right: Indirect loading on simply supported beams. Bottom left: Support forces from simple beams. Bottom right: Discretized replacement load fields $\hat{p}(s)$ and $\hat{q}(s)$ defined as the linear regressions of $p(s)$ and $q(s)$ on the support forces.

ply supported beam of span L_- on the negative side of the origin are X_+, Y_+, and X_-, Y_-, respectively. The 4 support forces are linear functionals of the load fields $p(s)$ and $q(s)$, and they depend solely on the natural equation representation $\{s, \varphi(s)\}$ in the interval from $-L_-$ to L_+. The corresponding 8 influence functions $I^+_{np}(s), I^+_{nq}(s), I^+_{tp}(s), I^+_{tq}(s), I^-_{np}(s), I^-_{nq}(s), I^-_{tp}(s), I^-_{tq}(s)$ are derived from elementary static analysis.

The principle of indirect loading suggests that it may be sufficient to let the replacement load fields $\hat{p}(s)$ and $\hat{q}(s)$ be the linear regressions on the resulting support forces $X = X_+ + X_-$ and $Y = Y_+ + Y_-$ at all discretization points instead

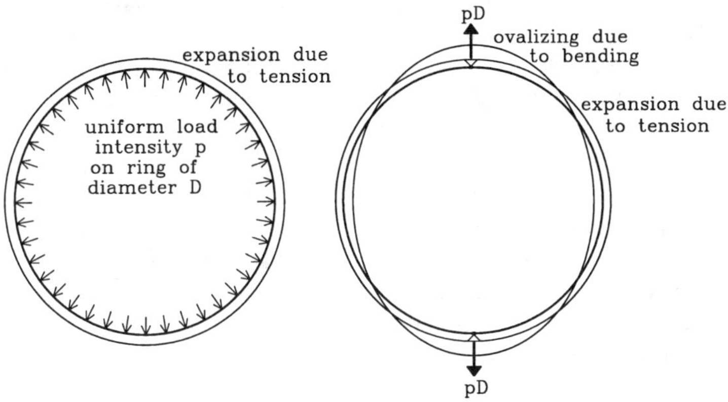

FIGURE 2.2
Left: Uniformly loaded circular ring carrying the load solely by tension. Right: The most coarse discretization of the load field into the support forces of two simple beams. The ring ovalizes due to bending. The linear regression $\hat{p} = \hat{E}[p \mid pD] = p$ **reestablishes the uniform load intensity.**

of being the linear regressions on the individual forces X_+, X_-, Y_+, Y_- at these points. Thereby the number of discretization variables are reduced to the half at the expense of an increased residual variance. Clearly, in the limit where the load fields on the positive side of the origin are stochastically independent of the load fields on the negative side of the origin, $\hat{p}(s)$ and $\hat{q}(s)$ for $s > 0$ should depend only on X_+, Y_+ and not on X_-, Y_-, that is, if $\hat{p}(s)$ and $\hat{q}(s)$ are defined as the linear regressions on X and Y, then $X_-$and Y_- add irrelevant random contributions to the replacement fields to be applied for $s > 0$.

2.8 Poisson Load Field Example

The material of this section is mathematically very technical and it is not used subsequently. The section serves to illustrate that the method of linear regression on linear functionals is applicable even to replace a load field with "sample curves" of highly singular nature by a discretized almost statically equivalent load field with continuous sample curves [18]. The considered example is of a class for which the direct kriging method is not applicable. However, the example is perfectly relevant for investigations concerning load effects from load fields of intermittent type as, for example, traffic load fields. The field is a homogeneous Poisson stream of single forces acting orthogonal to a straight line axis, Figure 2.3. For simplicity the load field is taken to be extended along the entire axis from $-\infty$ to ∞, even though the actual loaded beam has finite length. The mean number of Poisson points per length unit is λ, and the sequence of random forces F_i assigned to the sequence of Poisson points are assumed to be mutually independent, independent of the Poisson process, and all distributed as a given random variable F. According to the discretization principle of indirect loading explained in the last example of the previous section, we let the load field be applied to a row of imaginary simply supported beams of span L. For convenience we take L as length unit and choose the origin of the axis at the end point of an interval. Thus the Poisson process on a dimensionless s-axis has the intensity λL.

First we will consider the particular situation where a single force F is placed at random within the interval of the first three units. The nodal forces X_1 and X_2 at the abscissas 1 and 2 caused by the force F are $X_1 = FI_1(U)$ and $X_2 = FI_2(U)$, respectively, where $I_1(s) = s\mathbf{1}_{s\in[0,1]}+(2-s)\mathbf{1}_{s\in[1,2]}$ and $I_2(s) = (s-1)\mathbf{1}_{s\in[1,2]}+(3-s)\mathbf{1}_{s\in[2,3]}$ are the influence functions, and where U is a random variable which is uniformly distributed between 0 and 3. For a force placed outside the interval from 0 to 3, the nodal forces at 1 and 2 are zero. Clearly X_1 and X_2 are identically distributed with nth order moment

$$E[X_i^n] = E[F^n(U^n\mathbf{1}_{U\in[0,1]} + (2-U)^n\mathbf{1}_{U\in[1,2]})] =$$
$$2E[F^n]E[U^n\mathbf{1}_{x\in[0,1]}] = \frac{2}{3(n+1)}E[F^n] \tag{2.117}$$

giving $E[X_i] = \frac{1}{3}E[F]$, $\text{Var}[X_i] = \frac{2}{9}E[F^2] - \frac{1}{9}E[F^2] = \frac{2}{9}\text{Var}[F] + \frac{1}{9}E[F]^2$ and

$$E[X_1X_2] = E\left[F^2(2-U)(U-1)\mathbf{1}_{U\in[1,2]}\right] =$$
$$E\left[F^2(1-U)U\mathbf{1}_{U\in[0,1]}\right] = \frac{1}{18}E[F^2] \tag{2.118}$$

such that the covariance between X_1 and X_2 becomes

$$\text{Cov}[X_1, X_2] = \frac{1}{18}E[F^2] - \frac{1}{9}E[F]^2 = \frac{1}{18}(\text{Var}[F] - E[F]^2) \tag{2.119}$$

If N forces are placed independently and at random in the interval from 0 to 3, the points will be distributed exactly as the points in a realization of a homogeneous Poisson process given that there are N points within the interval. Conditional on N, the mean, variance, and covariance are obtained by applying the factor N on the above results. Using that $E[N] = \text{Var}[N]$ for a Poisson distribution, unconditioning gives

$$E[X_i] = \frac{1}{3}E[F]E[N] \tag{2.120}$$

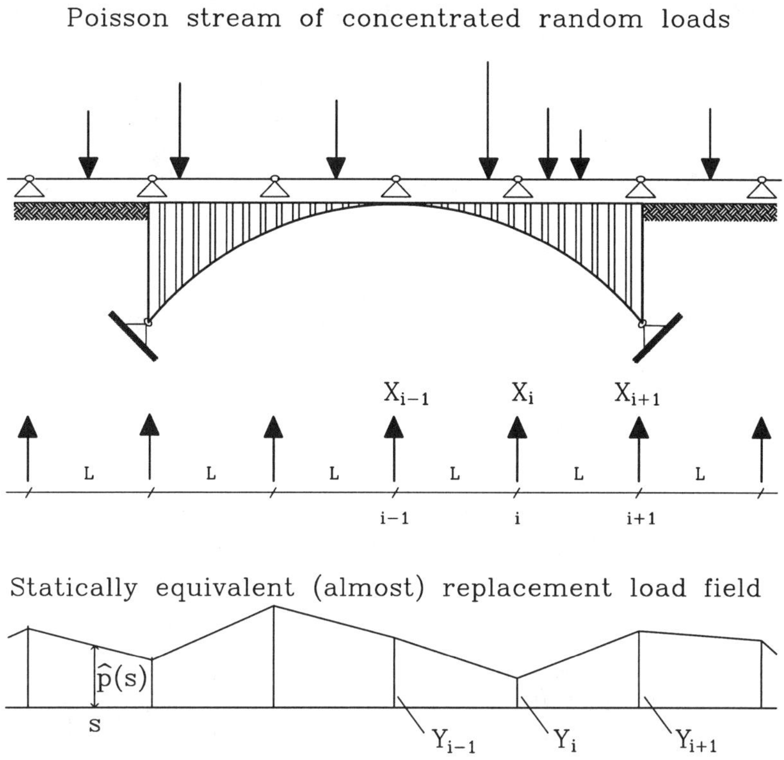

FIGURE 2.3
Poisson traffic load field discretized into an almost statically equivalent continuous and piecewise linear load field defined within any finite interval by a finite non-random number of random variables.

$$\text{Var}[X_i] = E[\text{Var}[X_i|N]] + \text{Var}[E[X_i|N]] = \left(\frac{2}{9}\text{Var}[F] + \frac{1}{9}E[F]^2\right)E[N] + \left(\frac{1}{3}E[F]\right)^2\text{Var}[N] = \frac{2}{9}E[F^2]E[N] \tag{2.121}$$

$$\text{Cov}[X_1, X_2] = E[\text{Cov}[X_1, X_2|N]] + \text{Cov}[E[X_1|N], E[X_2|N]] = \left(\frac{1}{18}E[F^2] - \frac{1}{9}E[F]^2\right)E[N] + \left(\frac{1}{3}E[F]\right)^2\text{Var}[N] = \frac{1}{18}E[F^2]E[N] \tag{2.122}$$

where $E[N] = 3\lambda L$. It follows from this that the homogeneous sequence of random nodal forces $X_i, i = ..., -2, -1, 0, 1, 2, ...$ has the mean and the covariances

$$E[X_i] = E[F]\lambda L \tag{2.123}$$

$$\text{Cov}[X_i, X_j] = \frac{1}{6}(\delta_{i(j-1)} + 4\delta_{ij} + \delta_{i(j+1)})\lambda L E[F^2] \tag{2.124}$$

respectively, where δ_{ij} is Kronecker's delta.

To obtain the linear regression $\hat{p}(s)$ of the Poisson load field $p(s)$ on the sequence of nodal forces X_i, we need to invert the covariance matrix of infinite order, and to calculate the covariances $\text{Cov}[X_i, p(s)]$. Noting that

$$\sum_{j=-\infty}^{\infty} (\delta_{i(j-1)} + 4\delta_{ij} + \delta_{i(j+1)})a^{|k-j|} = \begin{cases} a^{|k-i-1|}(a^2 + 4a + 1) \\ 2a + 4 \end{cases} \text{for} \begin{cases} k \neq i \\ k = i \end{cases} \tag{2.125}$$

it follows that this expression is proportional to δ_{ik} only for $a = -(\sqrt{3} + 2)$ or $a = \sqrt{3} - 2$. Since $a^{|i-j|}$ is bounded as $| i - j | \to \infty$ only for the last value of a for which the constant of proportionality is $2\sqrt{3}$, it follows that the inverse covariance matrix is the matrix of infinite order with the element in ith row and jth column equal to

$$\sqrt{3}(\lambda L E[F^2])^{-1} a^{-|i-j|} \tag{2.126}$$

with $a = (\sqrt{3} - 2)^{-1} = -(\sqrt{3} + 2)$.

The mean load intensity obviously is

$$E[p(s)] = E[X_i] \tag{2.127}$$

For the covariances $\text{Cov}[X_i, p(s)]$ we first consider the case $N = 1$. Then for $h > 0$ and as $h \to 0$:

$$\begin{aligned} &E[X_i p(s)] = \\ &\lim_{h\downarrow 0} E\left[F I_i(U)\mathbf{1}_{U\in[s,s+h]}\frac{F}{h}\right] = E[F^2]\lim_{h\downarrow 0}\left\{\frac{1}{h}E[I_i(U)\mathbf{1}_{U\in[s,s+h]}]\right\} = \\ &E[F^2]\lim_{h\downarrow 0}\left\{\frac{1}{h}E[I_i(s)\mathbf{1}_{U\in[s,s+h]}]\right\} = \frac{1}{3}E[F^2]I_i(s) \end{aligned} \tag{2.128}$$

so that $\text{Cov}[X_i, p(s)] = \frac{1}{3}E[F^2]I_i(s) - \frac{1}{9}E[F]^2$. Then

$$\text{Cov}[X_i, p(s)|N] = \{3E[F^2]I_i(s) - E[F]^2\}N/9 \tag{2.129}$$

Unconditioning with respect to N finally gives

$$\begin{aligned} &E[p(s)] = E[E[p(s)|N]] = E\left[E\left[\tfrac{F}{h}\mathbf{1}_{U\in[s,s+h]}\right]N\right] = \\ &\tfrac{1}{3}E[F]E[N] = \lambda L\, E[F] \end{aligned} \tag{2.130}$$

$$\begin{aligned} &\text{Cov}[X_i, p(s)] = \frac{1}{9}\left\{3E[F^2]I_i(s) - E[F]^2\right\}E[N]+ \\ &\frac{1}{3}E[X_i]E[F]\text{Var}[N] = \frac{1}{3}E[N]E[F^2]I_i(s) = \lambda L\, E[F^2]I_i(s) \end{aligned} \tag{2.131}$$

where

$$I_i(s) = (s - i + 1)\mathbf{1}_{s\in[i-1,i]} + (i + 1 - s)\mathbf{1}_{s\in[i,i+1]} \tag{2.132}$$

The linear regression $\hat{p}(s)$ of $p(s)$ on the sequence of nodal forces X_i is calculated as follows:

$$\begin{aligned} &\frac{1}{\sqrt{3}}\hat{p}(s) = \frac{1}{\sqrt{3}}\lambda L\, E[F] + \sum_{i=-\infty}^{\infty} I_i(s) \sum_{j=-\infty}^{\infty} a^{-|i-j|}[X_j - \lambda L\, E[F]] = \\ &\sum_{j=-\infty}^{\infty} X_j \sum_{i=-\infty}^{infty} I_i(s)a^{-|i-j|} = \sum_{j=-\infty}^{\infty} X_j \sum_{k=-\infty}^{infty} I_{k+j}(s)a^{-|k|} = \\ &\sum_{j=-\infty}^{\infty} X_j \sum_{k=[s]-j}^{[s]-j+1} I_0(s-k-j)a^{-|k|} = \\ &\sum_{j=-\infty}^{\infty} X_j \left(I_0(s-[s])a^{-|[s]-j|} + I_0(s-[s]-1)a^{-|[s]-j+1|}\right) = \\ &\sum_{k=-\infty}^{\infty} \left(X_{k+[s]}I_0(s-[s]) + X_{k+[s]+1}I_0(s-[s]-1)\right)a^{-|k|} \end{aligned} \tag{2.133}$$

in which $[s]$ is the integer part of s. It is seen that $\hat{p}(s)$ is obtained by linear interpolation between the adjacent random variables $Y_{[s]}, Y_{[s]+1}$ in the sequence

$$\{Y_i\} = \{\sqrt{3} \sum_{k=-\infty}^{\infty} X_{i+k} a^{-|k|}\} \tag{2.134}$$

This is illustrated in Figure 2.3. The mean and the covariances are

$$E[Y_i] = \sqrt{3} \sum_{k=-\infty}^{\infty} E[X_{i+k}] a^{-|k|} = E[X_i] = E[p(s)] \tag{2.135}$$

$$\begin{aligned} \mathrm{Cov}[Y_i, Y_j] = 3 \sum_{k=-\infty}^{\infty} \sum_{l=-\infty}^{\infty} \mathrm{Cov}[X_{i+k}, X_{j+l}] a^{-|k|-|l|} = \frac{1}{2}\lambda L\, E[F^2] \cdot \\ \sum_{k=-\infty}^{\infty} \sum_{l=-\infty}^{\infty} (\delta_{(i+k)(j+l-1)} + 4\delta_{(i+k)(j+l)} + \delta_{(i+k)(j+l+1)}) a^{-|k|-|l|} = \\ \frac{1}{2}\lambda L\, E[F^2] \sum_{l=-\infty}^{\infty} \left(a^{-|l-i-l|-|l-j|} + 4a^{-|l-i|-|l-j|} + a^{-|l-i+l|-|l-j|} \right) = \\ \sqrt{3}\lambda L\, E[F^2] a^{-|i-j|} \end{aligned} \tag{2.136}$$

respectively. The exponential decay with $| \ i - j \ |$ shows that the correlation structure is Markovian. The sequence of correlation coefficients $a^{-|i-j|}$ has the specific values 1, $\sqrt{3} - 2 \approx -0.268$, $7 - 4\sqrt{3} \approx 0.072$, $15\sqrt{3} - 26 \simeq -0.019$, $97 - 56\sqrt{3} \approx 0.005$, ...

It follows from (2.136) that the intensity of the Poisson process, the properties of the load sequence, and even the subdivision into intervals have no influence on the correlation coefficients of the random sequence $\{Y_i\}$ that determines the approximately statically equivalent load intensity $\hat{p}(s)$ to the given Poisson load field $p(s)$ by simple linear interpolation. It is noted that changing from the dimensionless abscissa s to the real geometrical abscissa implies that the dimensionless load intensity $\hat{p}(s)$ transforms to $\hat{p}(s)/L$. Thus the mean load intensity becomes $\lambda E[F]$ and the variance of Y_i/L becomes $\sqrt{3}\lambda E[F^2]/L$. Surprisingly, perhaps, the correlation coefficients remain invariant with respect to L.

The covariance function of $p(s)$ follows from

$$E[p(s)p(t)|N=1] = \lim_{h\downarrow 0}\left\{E\left[\left(\frac{F}{h}\right)^2 \mathbf{1}_{U\in[s,s+h]\cap[t,t+h]}|N=1\right]\right\} =$$
$$E[F^2]\lim_{h\downarrow 0}\left\{\frac{1}{h^2}E\left[\mathbf{1}_{U\in[s,s+h]\cap[t,t+h]}|N=1\right]\right\} =$$
$$E[F^2]\lim_{h\downarrow 0}\left\{\frac{1}{h^2}\frac{h}{3}I_0\left(\frac{s-t}{h}\right)\right\} = \frac{1}{3}E[F^2]\delta(s-t) \tag{2.137}$$

so that

$$\begin{aligned}&\mathrm{Cov}[p(s),p(t)|N] =\\ &N(E[p(s)p(t)|N=1]-E[p(s)|N=1]E[p(t)|N=1]) =\\ &N\left(\frac{1}{3}E[F^2]\delta(s-t)-\frac{1}{9}E[F]^2\right)\end{aligned} \tag{2.138}$$

and

$$\mathrm{Cov}[p(s),p(t)] = E\left[N\left(\frac{1}{3}E[F^2]\delta(s-t)-\frac{1}{9}E[F]^2\right)\right] + \mathrm{Var}[NE[F]] =$$
$$\lambda L\, E[F^2]\delta(s-t) \tag{2.139}$$

The residual covariance function is the difference between (2.139) and the function obtained from the linear regression by substituting the covariance $\mathrm{Cov}[X_i, p(t)]$ for X_i, that is, by replacing Y_i by

$$\begin{aligned}&\sum_{k=-\infty}^{\infty}\mathrm{Cov}[X_{i+k},p(t)]a^{-|k|} = \sqrt{3}\lambda L\, E[F^2]\sum_{k=-\infty}^{\infty} I_{i+k}(t)a^{-|k|} =\\ &\sqrt{3}\lambda L\, E[F^2]\sum_{k=[t]-i}^{[t]-i+1} I_0(t-i-k)a^{-|k|} =\\ &\sqrt{3}\lambda L\, E[F^2]\left[I_0(t-[t])a^{-|[t]-i|}+I_0(t-[t]-1)a^{-|[t]-i+1|}\right]\end{aligned} \tag{2.140}$$

Linear interpolation with respect to s between the two values corresponding to $i=[s]$ and $i=[s]+1$, then gives that, except for the factor $\lambda L\, E[F^2]$, the residual covariance function is

$$C(s,t) = \delta(s-t) - \Bigg\{\sqrt{3}(1 - 3\sigma - 3\tau + 6\sigma\tau + 3\Bigg[(\tau - \sigma)\mathbf{1}_{[s]<[t]} + \\ (\sigma + \tau - 2\sigma\tau)\mathbf{1}_{[s]=[t]} + (\sigma - \tau)\mathbf{1}_{[s]>[t]}\Bigg]\Bigg\} a^{-|[s]-[t]|} \tag{2.141}$$

where $\sigma = s - [s]$ and $\tau = t - [t]$ are the fractional parts of s and t, respectively.

Clearly the linear regression $\hat{p}(s)$ is not a good direct approximation to the Poisson load field $p(s)$. However, the accuracy should only be judged after application of the linear functional that maps the load field into the relevant load effect. In fact, since

$$G_s G_t\{\delta(s-t)\} = G\{g(s)\} \tag{2.142}$$

where $g(s)$ is the influence function of the considered load effect, the delta function singularity becomes eliminated. Let $g(s)$ be a function that can be defined by linear interpolation between its values on the integers. Obviously $g(s)$ can be expressed as a linear combination of the functions $I_i(s)$ for which the residual variance is zero. Thus the residual variance is zero also for $g(s)$ showing that the static equivalence is exact for any influence function of this particular type. To get a sufficiently general evaluation of the order of size of the residual variance, consider a piece-wise linear influence function of the form $g(s-l)$ for some $l \in [0,1]$. Then it is sufficient to consider the difference between $g(s-l)$ and the piece-wise linear function defined by the values of $g(s-l)$ on the integers. This difference function is zero on the integers. In the interval $[i-1, i]$ corresponding to $i = [s]$ it is proportional to the function

$$\frac{\sigma}{l}\mathbf{1}_{\sigma\in[0,l]} + \frac{1-\sigma}{1-l}\mathbf{1}_{\sigma\in[l,1]} \tag{2.143}$$

by a factor c_i. Applying (2.110), the residual variance becomes

$$\text{Var}[G(s) - \hat{G}(s)] = \lambda L\, E[F^2] R(l; \{c_i\}) \tag{2.144}$$

where

$$R(l;\{c_i\}) = \int_{-\infty}^{\infty}\int_{-\infty}^{\infty} C(s,t)\left(\mathbf{1}_{\sigma\in[0,l]}\frac{\sigma}{l} + \mathbf{1}_{\sigma\in[l,1]}\frac{1-\sigma}{1-l}\right)$$
$$\left(\mathbf{1}_{\tau\in[0,l[}\frac{\tau}{l} + \mathbf{1}_{\tau\in[l,1]}\frac{1-\tau}{1-l}\right) c_{[s]}c_{[t]}ds\,dt =$$
$$\left(\left[\sqrt{3}+2(\sqrt{3}-1)l(1-l)\right]/12\right)\sum_{i=-\infty}^{\infty} c_i^2 + \tag{2.145}$$
$$\left(\left[\sqrt{3}+2\sqrt{3}\,l(1-l)\right]/6\right)\sum_{i=-\infty}^{\infty} c_i \sum_{j=1}^{\infty} c_{i+j}a^{-j}$$

For $\{c_i\}$ given, the residual variance is maximal for $l = 0.5$ and minimal for $l = 0$ or 1 with the two coefficient values in the last factor equal to 0.175, 0.433 and 0.144, 0.289, respectively. Thus the dominating influence on the residual variance comes from the sequence $\{c_i\}$. By and large the best accuracy is obtained by choosing the equidistant stochastic finite element division points as closely as possible to the points of non-differentiability of the actual influence function. Of course, this attempt tends to counteract the goal of having few discretization random variables.

2.9 Stochastic Finite Element Methods and Reliability Calculations

The main features of finite element methods applied in structural engineering are assumed known to the reader. When random fields are included in the structural modeling and finite element methods are applied, the fields must be given a suitable representation on the finite element level. This leads to the notion of stochastic finite elements. Essentially the stochastic part of the finite element modeling consists of replacing the field of infinite dimensionality of randomness by a field defined by a finite set of random variables. Referring to the previous sections there are several possibilities of such field discretizations giving different levels of approximation errors, of course. Seeking for better approximations lead to increasing computational efforts even though it may not necessarily lead to an increasing conceptual complexity of the discretization method.

Based on the presentation in the previous sections the following field discretization methods can be listed [12]. The first 4 to 6 methods are listed by and large in the order of increasing output approximation accuracy for the same dimensionality of randomness.

1. Midpoint method
The field is replaced by a field of random values that are constant within each element. These values are taken as the field values of the original field at a central point of each element [20]. Thus the replacement field depends on the element mesh and is discontinuous over the element boundaries. The joint probability distribution of the finite set of random variables is directly given by the definition of the original field.

2. Integral average method
Same as 1 except that the random values are taken as the average field value over each element [21]. The joint probability distribution of the finite set of random variables is unknown except for Gaussian fields.

3. Shape function method
Within each element the field is replaced by an interpolation between a finite set of field values using some suitable, but otherwise arbitrary shape functions (spline functions) as interpolation functions [22]. The replacement field gets the same analytical properties as the chosen shape functions.

4. Direct linear regression method
This is an optimal version of 3 in the sense that the best shape functions (in the mean square sense) are used thus removing the arbitrariness with respect to shape function choice. The field is replaced by the linear regression of the field on a finite set of field values [12]. The replacement field has the same analytical regularity properties as the mean and covariance function of the field. The mesh of points at which the field values are taken need not be related in any particular way to the finite element mesh of the mechanical model (as is also the case for the shape function method). The joint probability distribution of the set of random variables is directly given by the definition of the field. Except at the points of the mesh, the replacement field of a non-Gaussian field is generally not mean value correct. For a Gaussian field the replacement field is a mean value correct Gaussian field.

Sometimes this procedure is called kriging (a terminology that according to its origin [3]–[6] (Section 2.1) rather should be used solely for statistical-stochastic interpolation between measured values at different spatial points. This is treated in the next section.)

5. Method of marginally backtransformed linear regression on marginally transformed field values
The field is assumed to be a marginally transformed Gaussian field (translation field [23]), for example, the lognormal field. The corresponding Gaussian field is replaced by the linear regression of the Gaussian field on a finite set of its field values. The marginal transformation is next applied to the linear regression to define the replacement field to the original field [12]. The joint probability distribution of

the set of random variables is directly given by the definition of the field. Except at the points of the mesh, the replacement field is not mean value correct. However, it reproduces the marginal median at any point.

6. Method of mean value correct backtransformed linear regression on marginally transformed field values
Same as 4 except that the linear regression of the Gaussian field is interpreted as a conditional mean and transformed back to the conditional mean of the original field on the finite set of random variables of the original field. For a lognormal field the relevant formulas are (2.82) to (2.87) and (2.93) to (2.100). More generally this method is applicable for a Nataf field [12] (Section 2.6).

7. Method of linear regression on linear functionals
The field is replaced by the linear regression of the field on a finite set of linear functionals defined on the field [18]. The linear functionals are chosen among linear functionals that are relevant for the solution of the actual problem. Thus the method is an extension of the method known from the literature as the weighted integral method [24, 25]. The joint probability distribution of the finite set of linear functionals is not known except for Gaussian fields. The replacement field is more regular with respect to analytical properties than the original field. This property makes the method applicable on fields for which none of the previously mentioned methods are applicable (Section 2.9).

8. Truncated series expansion methods
Starting from an exact expansion

$$X(t) = E[X(t)] + \sum_{n=0}^{\infty} V_n h_n(t) \quad \text{(mean square sense)} \tag{2.146}$$

of the field with respect to a complete orthonormal set $h_1(t), ..., h_n(t), ...$ of deterministic functions over $t \in \Omega$ with random variable coefficients

$$V_n = \int_{\Omega} X(t) h_n(t) dt, \quad n = 1, 2, ... \tag{2.147}$$

the replacement field is obtained by a suitable approximation to a suitable truncation of the expansion [8, 9, 26]. In case the random coefficients are required to be uncorrelated, the unique Karhunen-Loeve expansion [7, 27] is obtained as the exact expansion on which different approximation methods [8, 9] are applied as explained in Section 2.1. For non-Gaussian fields it is a difficult problem to determine the joint distribution of the random coefficients (2.147) of the expansion. For Gaussian fields the coefficients become Gaussian and mutually independent.

For any other complete orthonormal function system than that used in the Karhunen-Loeve expansion, the random coefficients become correlated [26, 27]. However, by constructing a one-to-one linear transformation of any finite subset of the random coefficients into an uncorrelated set of random variables, it can be shown that the corresponding subsum of finitely many terms in the expansion can be a good approximation to a corresponding subsum of the Karhunen-Loeve expansion containing the same number of terms [26].

The finite expansion obtained by truncation of the Karhunen-Loeve expansion after the nth term is identical to the linear regression of the field on the n random coefficients of the finite expansion (see text following (2.28)). This statement is generally not true for other expansions due to the correlation between the truncated sum and the remainder. Thus the accuracy can be improved by replacing the truncation sum by the linear regression of the field on the random coefficients of the finite expansion. In fact, since the random coefficients (2.147) are linear functionals on the field, the truncated series expansion method is a special case of the previous method 7 of linear regression on linear functionals.

Judging the accuracy on the output level of the investigated mechanical problem, the linear functionals of the Karhunen-Loeve expansion (for which the weight functions are the eigenfunctions related to the covariance function of the field) or any other orthogonal expansion may be inferior to linear functionals that are of direct relevance to the mechanical problem.

Any of the listed random field discretization methods may with varying accuracy be applied for the purpose of rendering a structural reliability analysis practicable. The replacement field to the original field X is in all the methods defined by a finite-dimensional random vector $\mathbf{F}\{X\}$ where $\mathbf{F}$ is a linear vector functional defined on the field. In all the methods the possibility of further reduction of the dimension of randomness can be studied by applying a one-to-one inhomogeneous linear transformation of $\mathbf{F}\{X\}$ into a set of uncorrelated and normalized random variables using the properties explained in the text after (2.28) [12, 26, 28].

For reliability analysis applications the mechanical finite element program is typically formulated such that it can compute approximate values of a critical vector functional defined on the set of all random variables of the problem. The adverse event is that this functional gets an outcome outside a given safe domain of functional values. This safe domain is specified on the basis of the given superior structural performance criteria. Thus the finite element program for the given safe domain defines a limit state surface in the finite-dimensional space of the vector of all random variables of the problem.

For simplicity and also sufficient for the generalization of the following explanation, let the random vector $\mathbf{F}\{X\}$ contain all the randomness of the problem. To apply the generally available computer programs for computing approximations to the probability of the adverse event (e.g., programs that are based on first or second order reliability analysis methods (FORM or SORM) as well as on simulation methods) all that is needed is to specify input information about the distribution properties of $\mathbf{F}\{X\}$. If X is a Gaussian field then $\mathbf{F}\{X\}$ is a Gaussian vector and, besides specifying this information, it is therefore only required that the mean vector and the covariance matrix of $\mathbf{F}\{X\}$ be specified. These first and second order moments can be calculated by simple numerical integration of weighted single and double integrals of the mean function and the covariance function of the random field X, respectively. However, if X is non-Gaussian, then more than the second moment characterizations of $\mathbf{F}\{X\}$ is needed. Generally it is then mandatory for obtaining calculation practicability to introduce suitable distribution approximations.

If the one-dimensional marginal distributions and the covariance matrix of $\mathbf{F}\{X\}$ are known, a possibility may be to use the Nataf distribution to approximate the distribution of $\mathbf{F}\{X\}$. If it exists, this distribution is defined as in Section 2.5 by an increasing marginal transformation of a Gaussian distribution of the same dimension as $\mathbf{F}\{X\}$. The transformation is constructed such that the covariance matrix and the one-dimensional distributions of $\mathbf{F}\{X\}$ are correctly reproduced by the transformation. However, it may in some problems not be practicable to determine the marginal distributions of $\mathbf{F}\{X\}$, except possibly as approximations obtained by statistical fitting of standard distribution types to extensive samples of simulated outcomes of $\mathbf{F}\{X\}$. Such a simulation procedure can be difficult in itself because an accurate generation of outcomes of $\mathbf{F}\{X\}$ requires an accurate simulation of finely discretized realizations of the non-Gaussian field X.

Since the assumption of having a Nataf distribution of $\mathbf{F}\{X\}$ in general is an approximation, it may be justified to make the further approximation of solely concentrating on obtaining the skewness and the kurtosis of the marginal distributions of $\mathbf{F}\{X\}$. Knowing the first four moments of each of the one-dimensional marginal distributions, one may choose some standard type distributions that reproduce these four moments, and then adopt the Nataf distribution with these marginal standard type distributions. A simpler possibility is, perhaps, to use the moment information to replace the components of the random vector $\mathbf{F}\{X\}$ by third degree polynomials of correlated Gaussian variables (Winterstein approximations [29]) such that they reproduce the first four marginal moments and all the correlation coefficients correctly [30]–[32]. If no other methods are readily applicable for the given problem, the skewness and the kurtosis may always be estimated statistically from a sample of simulated outcomes of $\mathbf{F}\{X\}$.

2.10 Classical versus Statistical-Stochastic Interpolation Formulated on the Basis of the Principle of Maximum Likelihood

In modern reliability analysis, in particular in geotechnical engineering or earthquake engineering, it can be necessary to make interpolation between measured values obtained at different spatial points and/or at different points in time [33, 34]. Following a paper of the author [35] this section first considers the classical problem of interpolation and the difficulty of error estimation in the classical theory except under conditions that only rarely are satisfied in practice. The formulation of a conditional random field model for interpolation is next introduced as a pragmatic alternative. However, it is attempted to present the statistical-stochastic interpolation method with a linking to classical interpolation because it may ease the appreciation of the method as a pragmatic interpolation method whether or not the data are generated by some physical process that behaves according to a random mechanism. Thus the method of statistical-stochastic interpolation is simply viewed as a rational model of the uncertain engineering guess on interpolated values. The basis is the idea that continuity and differentiability principles make the variation among the measured values representative for what should be expected with respect to the uncertainty of the interpolation values.

Finally it is shown that pragmatic principles of computational practicability of random field interpolation in large regularly organized tables lead to a very narrow class of applicable correlation functions of the interpolation fields.

The solution to the problem of interpolation in a table (x_0, y_0), (x_1, y_1), ...,(x_{n-1}, y_{n-1}) of values of the n times differentiable function $y = f(x)$ is in classical mathematical analysis given by Lagrange in the form

$$f(x) = y_0 Q_0(x) + y_1 Q_1(x) + ... + y_{n-1} Q_{n-1}(x) + R_n(x) \tag{2.148}$$

in which $Q_i(x)$, $i = 0, 1, ..., n-1$, is the uniquely defined polynomial of $(n-1)$th degree that takes the value 1 for $x = x_i$ and the value 0 for $x = x_0, ..., x_{i-1}, x_{i+1}, ...,$ x_{n-1}. The Lagrangian remainder $R_n(x)$ is

$$R_n(x) = (x - x_0) \cdot ... \cdot (x - x_{n-1}) \frac{f^{(n)}(\zeta)}{n!} \tag{2.149}$$

in which ζ is some number contained within the smallest interval I that contains $x, x_0, ..., x_{n-1}$. Interpolation to order $n-1$ then consists of using (2.148) with the remainder neglected. The error committed by the interpolation is bounded in absolute value by

$$| R_n(x) | \leq \frac{| (x - x_0) \cdot ... \cdot (x - x_{n-1}) |}{n!} \max_{\zeta \in I} | f^{(n)}(\zeta) | \tag{2.150}$$

Thus the error evaluation in classical interpolation is not offering a model of the error in terms of a random variable but only an upper bound that can be highly conservative. Moreover, the evaluation of the bound (2.150) requires that the nth derivative can be obtained. This, however, is in contrast to most situations in practice where interpolation is needed. Often only the table (x_0, y_0), (x_1, y_1), ..., (x_{n-1}, y_{n-1}) is given and the values of the function $f(x)$ are unknown between these points. This is the typical situation when the table is obtained by point by point measurement of some physical quantity that varies with x in an unknown way. The table may also be the result of a lengthy computation. In principle $f(x)$ may be computed for any needed value of x but it may be more economical to accept even a considerable error in the evaluation of $f(x)$ as obtained by interpolation between already computed values. The same holds in the case of values obtained by measurements. In some situations it may even be impossible to measure intermediate values (as in the obvious case of x being the time, but sometimes also in cases where x is a spatial coordinate). With only the table of points given, the interpolation procedure is usually guided by a principle of simplicity that may be supported by physical and geometrical arguments. To appreciate the philosophy of the principle of simplicity behind practical interpolation, it is only needed to note that there is an infinite set of interpolation functions passing through the given points. Besides satisfying differentiability requirements to some specified order this set is obviously characterized by the property that the difference between any two functions of the set is a function with zero points at $x_0, x_1, ..., x_n$. In practice the choice of the interpolation function from the set is usually made in view of choosing an interpolation function that in itself and in all the derivatives up to the specified order has as small a variability as possible. Even though this principle of simplicity of the interpolation may be given a precise mathematical definition and it thereafter may be demonstrated that it leads to a unique choice of the interpolation function, it is still just an arbitrary principle that does not give any indication of the interpolation error.

A solution to this error evaluation problem can be obtained by introducing an alternative but somewhat similar principle of simplicity. It is based on the principles of statistical reasoning.

Let $J(y_0, ..., y_{n-1})$ be the set of interpolation functions and introduce a probability measure over the union

$$\mathcal{F} = \bigcup_{\mathbf{z} \in \Re^n} J(\mathbf{z}) \tag{2.151}$$

in which $\mathbf{z} = (z_0, ..., z_{n-1})$. If it is solely required that the applied interpolation function is n times continuously differentiable, then $\mathcal{F}$ is the class of n times continuously differentiable functions. Consider a probability measure over $\mathcal{F}$ and assume that it depends on a number of parameters $\theta_1, ..., \theta_q$. For each value set of $\theta_1, ..., \theta_q$ the probability measure defines a random field over the range of values of x. Let the probability measure have such regularity properties that there is a probability density at the set $J(\mathbf{z})$ for each $\mathbf{z} \in \Re^n$ in the sense of being the unique limit

$$\lim_{d \to 0} \left\{ \frac{P\left[\bigcup_{\zeta \in \mathcal{N}(\mathbf{z})} J(\zeta) \mid \theta_1, ..., \theta_q \right]}{\text{Vol}[\mathcal{N}(\mathbf{z})]} \right\} \tag{2.152}$$

where $\mathcal{N}(\mathbf{z})$ is an arbitrary neighborhood of $\mathbf{z}$ with volume Vol$[\mathcal{N}(\mathbf{z})]$, d is the diameter of $\mathcal{N}(\mathbf{z})$ according to some suitable diameter definition, and $P[\cdot \mid \theta_1, ..., \theta_q]$ is a suitably defined probability measure.

The principle of simplicity now concerns a principle about how to choose the values of the parameters $\theta_1, ..., \theta_q$ that fix the probability measure on $\mathcal{F}$. Instead of the deterministic principle of least variability it is reasonable to choose $\theta_1, ..., \theta_q$ so that the probability density has its maximum for $\mathbf{z} = \mathbf{y} = (y_0, y_1, ..., y_{n-1})$, that is, at the actual set of interpolation functions. This is the well-known principle of maximum likelihood estimation in the theory of mathematical statistics. When the parameter values are chosen, for example, according to the principle of maximum likelihood, the probability measure on $\mathcal{F}$ induces a probability measure in the relevant set of interpolation functions $J(\mathbf{y})$ simply as the conditional probability measure given $J(\mathbf{y})$, that is, given $\mathbf{z} = \mathbf{y}$. Thus a conditional random field is defined that possesses the desired properties of expressing the interpolated value at x of the unknown function as a random variable. At the points of the given table $(x_0, y_0), ..., (x_{n-1}, y_{n-1})$ the random variable degenerates to be atomic, but at a point x different from the points of the table it gets a mean and a standard deviation that depend on x. The detailed nature of this dependency is laid down in the mathematical structure of the probability measure introduced in $\mathcal{F}$.

From an operational point of view the most attractive probability measure to choose is the Gaussian measure. Let the Gaussian measure on $\mathcal{F}$ have mean value function $\mu(x)$, $x \in \Re$, and covariance function $c(\xi, \eta)$, ξ, $\eta \in \Re$. Then the conditional measure on $J(\mathbf{y})$ is Gaussian with mean value function given by the linear regression (2.101), that is,

$$E[Y(x) \mid \mathbf{z} = \mathbf{y}] =$$

$$\mu(x) + [c(x, x_0)...c(x, x_{n-1})]\{c(x_i, x_j)\}^{-1} \begin{bmatrix} y_0 - \mu(x_0) \\ \vdots \\ y_{n-1} - \mu(x_{n-1}) \end{bmatrix} \tag{2.153}$$

and covariance function given by the residual covariance function corresponding to the linear regression (2.21), that is,

$$\mathrm{Cov}[Y(\zeta), Y(\eta) \mid \mathbf{z} = \mathbf{y}] =$$

$$c(\xi, \eta) - [c(\xi, x_0)...c(\xi, x_{n-1})]\{c(x_i, x_j)\}^{-1} \begin{bmatrix} c(\eta, x_0) \\ \vdots \\ c(\eta, x_{n-1}) \end{bmatrix} \tag{2.154}$$

The conditional mean (2.153), in particular, is written out explicitly in order to display an interpretation of the mean value function $\mu(x)$ and the function $c(\xi, \eta)$ that relates them to usual deterministic interpolation practice. In fact, if (2.148) and (2.153) are compared, it is seen that (2.153) is obtained from (2.148) if the polynomial $Q_i(x), i = 0, ..., n-1$, is replaced by the linear combination

$$a_{0i}c(x, x_0) + a_{1i}c(x, x_1) + ... + a_{(n-1)i}c(x, x_{n-1}) \qquad (2.155)$$

where the coefficients $a_{0i}, ..., a_{(n-1)i}$ are the elements of the ith column in the inverse to the covariance matrix $\{c(x_i, x_j)\}$. It is seen that this function like $Q_i(x)$ is 1 for $x = x_i$ and 0 for $x = x_0, ..., x_{i-1}, x_{i+1}, ..., x_{n-1}$. Indeed, if (2.155) is identified with $Q_i(x)$ for $i = 0, ..., n-1$, the functions $c(x, x_0), ..., c(x, x_{n-1})$ are uniquely determined by their values at $x_0, x_1, ..., x_{n-1}$ as

$$c(x, x_i) = c(x_0, x_i)Q_0(x) + ... + c(x_{n-1}, x_i)Q_{n-1}(x) \qquad (2.156)$$

As a function of x this is a valid covariance, that is, there exists a non-negative definite function $c(\xi, \eta)$ such that (2.156) is obtained for $(\xi, \eta) = (x, x_i)$. This is seen directly by computing the covariance function of the random field

$$Z_0Q_0(x) + ... + Z_{n-1}Q_{n-1}(x) \qquad (2.157)$$

in which $(Z_0, ..., Z_{n-1})$ is a random vector with covariance matrix $\{c(x_i, x_j)\}$. The covariance function is

$$c(\xi, \eta) = [Q_0(\xi)...Q_{n-1}(\xi)]\{c(x_i, x_j)\}\begin{bmatrix} Q_0(\eta) \\ \vdots \\ Q_{n-1}(\eta) \end{bmatrix} \qquad (2.158)$$

which gives (2.156) for $\xi = x$ and $\eta = x_i$.

It follows from this that (2.148) except for $R_n(x)$ is a special case of (2.153). The remainder becomes replaced by the term

$$\mu(x) - [c(x, x_0)...c(x, x_{n-1})]\{c(x_i, x_j)\}^{-1}\begin{bmatrix} \mu(x_0) \\ \vdots \\ \mu(x_{n-1}) \end{bmatrix} \qquad (2.159)$$

plus a Gaussian zero mean random field with covariance function given by (2.154). This interpretation of the mean value function $\mu(x)$ and the covariance function $c(\xi, \eta)$ of the random field as essentially being interpolation functions makes it easier to appreciate the consequence of a specific mathematical form of the covariance function [36]. For example, if the general appearance of the values in the table suggests the choice of a homogeneous Gaussian field, the choice of a correlation function like $\exp[-\alpha \mid \xi - \eta \mid]$ implies that the mean interpolation function given by (2.153) will be non-differentiable at $x = x_0, ..., x_{n-1}$ while a correlation function like $\exp[-\beta(\xi - \eta)^2]$ will give differentiable interpolation function also at $x = x_0, ..., x_{n-1}$. In fact, under Gaussianity the sample functions corresponding to the first correlation function will with probability 1 be continuous but not differentiable at any point while the second correlation function corresponds to sample functions that are differentiable of any order.

As mentioned in Section 2.1, the procedure of doing statistical-stochastic interpolation is often called kriging.

Example Consider the case where the points $x_0, ..., x_{n-1}$ are equidistant with $x_{i+1} - x_i = h$. Let the random interpolation field on the line be a homogeneous Gaussian field with mean μ and covariance function

$$c(\xi, \eta) = \begin{cases} \left(1 - \dfrac{\mid \xi - \eta \mid}{h}\right)\sigma^2 & \text{for } \mid \xi - \eta \mid \leq h \\ 0 & \text{otherwise} \end{cases} \tag{2.160}$$

Then (2.153) gives

$$E[Y(x) \mid \mathbf{z} = \mathbf{y}] = \frac{1}{h}[(x_1 - x)y_0 + (x - x_0)y_1] \tag{2.161}$$

for $x_0 \leq x \leq x_1$. This is equivalent to linear interpolation in the mean. The covariance function (2.154) becomes

$$\text{Cov}[Y(\xi), Y(\eta) \mid \mathbf{z} = \mathbf{y}] =$$

$$\begin{cases} \left(\dfrac{\sigma}{h}\right)^2 [h^2 - h \mid \xi - \eta \mid -(x_1 - \xi)(x_1 - \eta) - (x_0 - \xi)(x_0 - \eta)] \\ \qquad\qquad\qquad\qquad \text{for } \mid \xi - \eta \mid \leq h \\ 0 \quad \text{otherwise} \end{cases} \tag{2.162}$$

from which the standard deviation

$$D[Y(x) \mid \mathbf{z} = \mathbf{y}] = \frac{\sigma}{h}\sqrt{2(x - x_0)(x_1 - x)} \tag{2.163}$$

is obtained. It is interesting to compare this measure of the interpolation error with the Lagrangian remainder obtained from (2.149) for $n = 2$. It is seen that (2.163) varies like the square root of $R_2(x)$, that is, it predicts larger errors close to x_0 or x_1 than $R_2(x)$ does.

While the interpolation functions in the classical procedure are broken at the points of the table implying that no way is indicated of reasonable extrapolation outside the range from x_0 to x_{n-1}, the homogeneous random field procedure represents a solution to both the interpolation and the extrapolation problem in terms of a set of sample functions defined everywhere. The points of the table only play the role that all the sample functions of the set pass through the points of the table. The homogeneity of the field implies that the variability of the points within the range from x_0 to x_{n-1} is reflected in the field description outside this range. Specifically we have for $x \leq x_0$

$$E[Y(x) \mid \mathbf{z} = \mathbf{y}] = \begin{cases} \mu & \text{for } x \leq x_0 - h \\ \frac{1}{h}[(x_0 - x)\mu + (x - x_0 + h)y_0] & \\ \qquad\qquad \text{for } x_0 - h < x \leq x_0 \end{cases} \tag{2.164}$$

and similarly for $x > x_{n-1}$. The standard deviation is

$$D[Y(x) \mid \mathbf{z} = \mathbf{y}] = \begin{cases} \sigma & \text{for } x \leq x_0 - h \\ \frac{\sigma}{h}\sqrt{(x_0 - x)(x - x_0 + 2h)} & \text{for } x_0 - h < x \leq x_0 \end{cases} \tag{2.165}$$

and similarly for $x > x_{n-1}$. Outside the interval from $x_0 - h$ to $x_{n-1} + h$ the field is simply given as the homogeneous field. The variation among $y_0, \ldots, y_{n-1}$ determines the values of μ and σ^2 as the well-known maximum likelihood estimates

$$\bar{y} = \frac{1}{n}\sum_{i=0}^{n-1} \quad \text{and} \quad s^2 = \frac{1}{n}\sum_{i=0}^{n-1}(y_i - \bar{y})^2 \tag{2.166}$$

corresponding to a Gaussian sample of independent outcomes $y_0, \ldots, y_{n-1}$.

Example The maximum likelihood principle leads to a specific choice of the distribution parameters $\theta_1, \ldots, \theta_q$, that is, it leaves no room for doubt about what values to choose. Such doubt can be included in the stochastic interpolation procedure by assigning not just the distribution family to the interpolation problem but also assigning a joint probability distribution to $\boldsymbol{\theta} = (\theta_1, \ldots, \theta_q)$. The mathematical technique then becomes exactly the same as in the Bayesian statistical method.

For example, the parameter vector (μ, σ) in the previous example may be considered as an outcome of the random vector (M, Σ). Assuming a non-informative prior of $(M, \log \Sigma) \in \Re^2$ (degenerate uniform density over $\Re^2$), the special random field model of the previous example leads to the well-known Bayesian standard results in Gaussian statistics. In the interpolation case the predictive distribution of

$$\frac{Y(x) - \frac{1}{h}[(x_1 - x)y_0 + (x - x_0)y_1]}{\frac{s}{h}\sqrt{2(x - x_0)(x_1 - x)}}\sqrt{1 - \frac{1}{n}} \tag{2.167}$$

for $x \in]x_0, x_1[$ is the t-distribution with $n - 1$ degrees of freedom.

2.11 Computational Practicability of the Statistical-Stochastic Interpolation Method

Both for the maximum likelihood principle and the Bayesian principle of stochastic interpolation on the basis of a Gaussian field assumption, the governing mathematical function is the joint Gaussian density

$$f_{\mathbf{Y}}(y_0, ..., y_{n-1}, \mu, \sigma, \mathbf{P};) \propto$$

$$\frac{1}{\sigma^n\sqrt{\det(\mathbf{P})}}\exp\left[-\frac{1}{2\sigma^2}(\mathbf{y} - \mu\mathbf{e})'\mathbf{P}^{-1}(\mathbf{y} - \mu\mathbf{e})\right] \tag{2.168}$$

written here for the case of a homogeneous field with mean μ, standard deviation σ, and a correlation function that defines the correlation matrix $\mathbf{P}$ of the random vector $(Y(x_0), ..., Y(x_{n-1}))$. The elements of the correlation matrix $\mathbf{P}$ contain the unknown parameters of the correlation function. As a function of μ, σ and the correlation parameters, the right side of (2.168) defines the likelihood function. Generally neither $\mathbf{P}^{-1}$ nor the determinant $\det(\mathbf{P})$ can be expressed explicitly in terms of the correlation parameters. To maximize the likelihood with respect to the parameters by an iteration procedure both $\mathbf{P}^{-1}$ and $\det(\mathbf{P})$ must therefore be evaluated several times.

The Bayesian principle of stochastic interpolation has particular relevance for structural reliability evaluations of structures with carrying capacities that depend on the strength variation throughout a material body. For example, a direct foundation on saturated clay may fail due to undrained failure extended over a part of the clay body. The evaluation of the failure probability corresponding to a specific rupture figure requires a specifically weighted integration of the undrained shear strength across the rupture figure. If the undrained shear strength is measured only at a finite number of points in the clay body, it is required to make interpolation and or extrapolation in order to obtain the value of the integrand in any relevant point of the body. Irrespective of whether the maximum likelihood principle is applied or whether the Bayesian principle is applied in order to properly take statistical uncertainty into account, it is required that $\mathbf{P}^{-1}$ be computed iteratively for different values of the correlation parameters. For example, for the Bayesian principle of choice this is the case when the reliability is evaluated by a first or second order reliability method (FORM or SORM) in a space of basic variables that among its dimensions includes the correlation parameters. In the search for the most central limit state point by some gradient method, say, the inverse $\mathbf{P}^{-1}$ has to be computed iteratively several times for different parameter values.

From this discussion it follows that computational practicability puts limits to the order of $\mathbf{P}$ or it requires that the mathematical structure of $\mathbf{P}$ is such that $\mathbf{P}$ can be inverted analytically or such that $\mathbf{P}^{-1}$ can be obtained in terms of matrices of considerably lower order than the order of $\mathbf{P}$. A possibility of breaking down to lower order is present in case of what here is called a factorized correlation structure:

Let $\mathbf{P}$ and $\mathbf{Q} = \{q_{ij}\}_{i,j=1,...,m}$ be correlation matrices. Then the matrix defined as

$$[\mathbf{Q}] \otimes \mathbf{P} \equiv \begin{bmatrix} q_{11}\mathbf{P} & q_{12}\mathbf{P} & \cdots \\ q_{21}\mathbf{P} & q_{22}\mathbf{P} & \cdots \\ \cdots & & q_{mm}\mathbf{P} \end{bmatrix} \tag{2.169}$$

is a correlation matrix. The proof is as follows: Let $\mathbf{x}' = [\mathbf{x}_1'...\mathbf{x}_m']$ be an arbitrary vector of dimension $n \times m$, and let $\mathbf{A}$ be an orthogonal matrix such that $\mathbf{A}'\mathbf{P}\mathbf{A} = \Lambda = \lceil \lambda_1...\lambda_n \rfloor$ where $\lambda_1, ..., \lambda_n$ are the nonnegative eigenvalues of $\mathbf{P}$. Then

$$\begin{aligned} \mathbf{x}'([\mathbf{Q}] \otimes \mathbf{P})\mathbf{x} = \sum_{i=1}^{m}\sum_{j=1}^{m} q_{ij}\mathbf{x}_i'\mathbf{P}\mathbf{x}_j = \sum_{i=1}^{m}\sum_{j=1}^{m} q_{ij}\mathbf{y}_i'\Lambda\mathbf{y}_j = \\ \lambda_1\sum_{i=1}^{m}\sum_{j=1}^{m} q_{ij}y_{i1}y_{j1} + ... + \lambda_n\sum_{i=1}^{m}\sum_{j=1}^{m} q_{ij}y_{in}y_{jn} \end{aligned} \tag{2.170}$$

in which $\mathbf{y}_i = \mathbf{A}\mathbf{x}_i$, $i = 1, ..., m$. Since $\mathbf{Q}$ is a correlation matrix, the right side of (2.170) is nonnegative. Thus it follows that $[\mathbf{Q}] \otimes \mathbf{P}$ is a correlation matrix.

It follows directly by the multiplication test that if $\mathbf{P}$ and $\mathbf{Q}$ are both regular, then

$$([\mathbf{Q}] \otimes \mathbf{P})^{-1} = [\mathbf{Q}^{-1}] \otimes \mathbf{P}^{-1} \tag{2.171}$$

By application of simple row operations first diagonalizing $\mathbf{P}$ to $\mathbf{D}$ in all places in (2.169) to obtain

$$\begin{bmatrix} q_{11}\mathbf{D} & q_{12}\mathbf{D} & \cdots \\ q_{21}\mathbf{D} & q_{22}\mathbf{D} & \\ \cdots & & \end{bmatrix} \tag{2.172}$$

and next diagonalizing (2.172) by exactly those row operations that diagonalize

$\mathbf{Q}$, it follows that

$$\det([\mathbf{Q}] \otimes \mathbf{P}) = [\det(\mathbf{Q})]^n [\det(\mathbf{P})]^m \tag{2.173}$$

(m = order of $\mathbf{Q}$, n = order of $\mathbf{P}$).

Assume that the table of points corresponds to points in the plane arranged in a square mesh of equidistant points in both directions with mesh width L. Let there be k points in the first direction and l points in the second direction in total giving kl points with given values of the otherwise unknown function. The field random variables $Y_{11}, Y_{21}, ..., Y_{k1}, Y_{12}, Y_{22}, ..., Y_{k2}, ..., Y_{1l}, Y_{2l}, ..., Y_{kl}$ corresponding to the points (i, j) of the mesh are collected in a vector of dimension kl and in the order as indicated. Then it is easily shown that the correlation matrix of order kl for an arbitrary choice of L has the factorized structure as in (2.169) if and only if the random field is correlation homogeneous and the correlation function of the field can be written as the product of a correlation function solely of the coordinate along the first mesh direction and a correlation function solely of the coordinate along the second direction. In the class of such product correlation functions, the correlation functions that correspond to isotropic fields are uniquely given as

$$\exp[-\beta r^2] \tag{2.174}$$

where r is the distance between the two points at which the random variables are considered and β is a free positive parameter. Thus the assumption of isotropy of the interpolation field and the requirement of computational practicability for large kl make it almost a "must" to adopt the correlation function defined by (2.174). Writing (2.174) for $r = L$ as the correlation coefficient κ, the correlation matrix of the vector of Y-values becomes $[\mathbf{S}_1] \otimes \mathbf{S}_2$ where

$$\mathbf{S}_i = \begin{bmatrix} 1 & \kappa & \kappa^4 & \cdots & \kappa^{(\nu-1)^2} \\ \kappa & 1 & \kappa & \cdots & \kappa^{(\nu-2)^2} \\ \cdots & & & & \end{bmatrix} \tag{2.175}$$

$i = 1(\nu = k)$, $i = 2(\nu = l)$. These considerations generalize directly to 3 dimensions for a spatial equidistant mesh of klm points. Then the correlation matrix becomes $[[\mathbf{S}_1] \otimes \mathbf{S}_2] \otimes \mathbf{S}_3$ with each of the matrices $\mathbf{S}_1, \mathbf{S}_2, \mathbf{S}_3$ of form as in (2.175) in case of spatial isotropy.

For a soil body it may be reasonable to have isotropy in any horizontal plane but not necessarily in the 3-dimensional space. Then $\mathbf{S}_3$ may be any other correlation matrix obtained from the correlation function in the vertical direction. When soil strength measurements are made, for example, by the so-called CPT method (Cone Penetration Test), the distance h between measurement points is much smaller in the vertical direction than the mesh width L in the horizontal direction. Then $\mathbf{S}_3$ can be of computationally impracticable high order. However, there is at least one case of such a matrix $\mathbf{S}$ corresponding to a homogeneous field for which $\mathbf{S}^{-1}$ is known explicitly. This case is

$$\mathbf{S} = \begin{bmatrix} 1 & \rho & \rho^2 \cdots & \rho^m \\ \rho & 1 & \rho \cdots & \rho^{m-1} \\ \cdots & & & \end{bmatrix};$$

$$\mathbf{S}^{-1} = \frac{1}{1-\rho^2} \begin{bmatrix} 1 & -\rho & 0 & \cdots 0 \\ -\rho & 1+\rho^2 & -\rho & \cdots 0 \\ \cdots & & & \end{bmatrix} \tag{2.176}$$

which corresponds to a homogeneous Markov field on the line with $\rho = \exp[-\alpha h]$ where α is a free positive parameter.

This 3-dimensional homogeneous Gaussian field interpolation model with (2.174) defining the horizontal correlation properties and with Markov properties in the vertical direction has been used in a reliability study of the anchor blocks for the suspension bridge across the eastern channel of Storebælt in Denmark [37]. The size of the interpolation problem in that investigation was $k = 6, l = 10, m = 150$, giving a table of 9000 points. Without the unique structuring of the random field correlation given here, the likelihood analysis would have been totally impracticable.

2.12 Field Modeling on the Basis of Measured Noisy Data

In engineering practice measured values of fields are often obtained by use of less than perfect measuring techniques either because no better technique is available or for economical reasons. Not rarely the most simple probabilistic measuring error model is applicable for the purpose of including the influence of this error type in the interpolation problem. Referring to Section 2.11, the table of measured values $(x_0, y_0), \ldots, (x_{n-1}, y_{n-1})$ should rather be $(x_0, y_0 + z_0), \ldots, (x_{n-1}, y_{n-1} + z_{n-1})$ where $z_0, \ldots, z_{n-1}$ are measuring errors. On the basis of an analysis of the measuring method, it is often reasonable to assume that the error values $z_0, \ldots, z_{n-1}$ independently of $y_0, \ldots, y_{n-1}$ can be considered as n mutually independent realizations of a random variable Z of zero mean, or , more precisely, that the vector $(z_0, \ldots, z_{n-1})$ can be considered as a realization of a random vector $(Z_0, \ldots, Z_{n-1})$, where $Z_0, \ldots, Z_{n-1}$ are n mutually independent and identically distributed random variables of zero mean and variance ϵ^2. Clearly the deterministic interpretation method is not suited as a tool for solving this interpolation problem. However, the field method is directly applicable by assuming that $y_0, \ldots, y_{n-1}$ are the values at $x_0, \ldots, x_{n-1}$ of the field $Y(x)$ with the pragmatically chosen covariance function

$\text{Cov}[Y(\xi), Y(\eta)] = c(\xi, \eta)$. For simplicity assuming that both the field $Y(x)$ and the error vector $(Z_0, ..., Z_{n-1})$ are Gaussian and, as indicated above, that the field and the error vector are mutually independent, it follows that the maximum likelihood principle operates on the Gaussian density (likelihood function) with the covariance matrix

$$\mathbf{C} + \epsilon^2\mathbf{I} \tag{2.177}$$

where $\mathbf{C} = \{c(x_i, x_j)\}$ and $\mathbf{I}$ is the unit matrix.

Unfortunately the necessary computational efforts may increase considerably hereby, in particular for large tables. If the covariance matrix $\mathbf{C}$ can be inverted explicitly as exemplified in Section 2.12 it may be convenient to apply the quotient series expansion

$$\begin{aligned}(\mathbf{C} + \epsilon^2\mathbf{I})^{-1} &= \mathbf{C}^{-1}[\mathbf{I} - (-\epsilon^2\mathbf{C}^{-1})]^{-1} = \\ &\mathbf{C}^{-1}[\mathbf{I} - \mathbf{C}^{-1}\epsilon^2 + \mathbf{C}^{-2}\epsilon^4 - \mathbf{C}^{-3}\epsilon^6 + ...]\end{aligned} \tag{2.178}$$

cut off at some power of ϵ^2 dependent on the size of ϵ^2 as compared to the largest diagonal element of $\mathbf{C}$. It can be shown that the series (2.178) is convergent if $\epsilon^2 < 1/ \parallel \mathbf{C}^{-1} \parallel$ where the matrix norm is defined as the maximal value of the row sums in the corresponding matrix of absolute values.

In case a non-zero value of ϵ^2 is obtained, the conditional mean (2.153) used as interpolation expression is replaced by the conditional mean of $Y(x)$ given $Y(x_i) + Z_i$ for $i = 0, ..., n-1$:

$$\begin{aligned}&E[Y(x) \mid Y(x_i) + Z_i = y_i + z_i, \quad i = 0, ..., n-1] = \\ &\mu(x) + [c(x, x_0)...c(x, x_{n-1})](\mathbf{C} + \epsilon^2\mathbf{I})^{-1}\begin{bmatrix} y_0 + z_0 - \mu(x_0) \\ \vdots \\ y_{n-1} + z_{n-1} - \mu(x_{n-1}) \end{bmatrix}\end{aligned} \tag{2.179}$$

with the corresponding residual covariance function given by (2.154) after replacement of $\{c(x_i, x_j)\}$ by $\mathbf{C} + \epsilon^2\mathbf{I}$. It follows by substituting the expansion (2.178) into (2.154) that

$$\begin{aligned}&\{\text{Cov}[Y(x_i), Y(x_j) \mid Y(x_i) + Z_i = y_i + z_i, \\ &i = 0, ..., n-1]\}_{i,j=0,...,n-1} = \epsilon^2\mathbf{I} - \epsilon^4\mathbf{C}^{-1} + \epsilon^6\mathbf{C}^{-2} - ...\end{aligned} \tag{2.180}$$

Thus there is not only a residual variance at the measuring points caused by the random measuring error but also a correlation between the residuals at the different measuring points. The correlation coefficients are of order of size as in the matrix $\epsilon^2\mathbf{C}^{-1}$.

The conceptually simple measuring error model explained above has recently been successfully used for the interpretation of and interpolation between measured pressures at a series of points on the wall of a circular concrete silo for grain storing [36]. However, this model is not applicable in all situations. In particular, if the field $Y(x)$ has such properties that the random variables $Y(x_0)$, ..., $Y(x_{n-1})$ are only weakly dependent or mutually independent, then the maximum likelihood principle will not be able to separate the variability of the field $Y(x)$ from the variability induced by the measuring uncertainty. In fact, the condition for the success of the simple model is that a considerable dependency is present between the field values $Y(x_0)$, ..., $Y(x_{n-1})$. Also the method is restricted by the assumption of independent measuring errors.

Some measuring situations are such that the only way to deal rationally with the measuring uncertainty is by use of a pairing method. The simplest form of such a method is explained in the following example [38] that solely concerns random variables. An adapted application on a field measuring problem is treated after the example.

Example A sample is drawn from some unknown population Ω (object population) and each element of the sample is characterized by a measured value. The measurement procedure is assumed to be less than perfect. On each measured value it introduces an error drawn from some unknown population M_1. Without knowing anything about population M_1, it is clearly not possible on the basis of the obtained sample of values to infer anything about the properties of population Ω. However, the situation is different if each element of the sample from Ω also is characterized by a measured value obtained by use of another independent measuring method with error population M_2. It is shown in the following that if both measuring methods besides being independent are such that the two mean errors for a given object are independent of the error-free value that should be assigned to the object, then it is possible to estimate the variances of each of the value populations corresponding to Ω, M_1, M_2 on the basis of the sample of pairs of measured values. In order to obtain estimates of the mean values it is necessary to assume that at least one of the measuring methods gives unbiased measurements in the sense that the mean error is zero.

Let a random variable Z be defined on the object population Ω and let it be measured by the two independent measuring methods giving the values X_1 and X_2, respectively. Then the measuring errors are Y_1 and Y_2 and

$$Z = X_1 + Y_1 = X_2 + Y_2 \tag{2.181}$$

Obviously $\text{Cov}[Y_1, Y_2 \mid Z] = 0$, implying that $\text{Cov}[Y_1, Y_2 \mid Z] = \text{Cov}[Z - Y_1, Z - Y_2 \mid Z] = 0$.

Under the assumption that the conditional means $E[Y_1 \mid Z]$ and $E[Y_2 \mid Z]$ do not depend on Z, it then follows from the standard formula (total representation theorem [39])

$$\begin{aligned}&\mathrm{Cov}[X_1, X_2] = \\ &E[\mathrm{Cov}[X_1, X_2 \mid Z]] + \mathrm{Cov}[E[X_1 \mid Z], E[X_2 \mid Z]] = E[0] + \mathrm{Cov}[Z, Z])\end{aligned} \tag{2.182}$$

that

$$\mathrm{Cov}[X_1, X_2] = \mathrm{Var}[Z] \tag{2.183}$$

Thus the variance of the "true" quantity Z can be estimated by estimating the covariance between the results of the two measuring methods. Since

$$\begin{aligned}&\mathrm{Cov}[X_1, Y_1] = E[\mathrm{Cov}[X_1, Y_1 \mid Z]] = E[\mathrm{Cov}[X_1, Z - X_1 \mid Z]] = \\ &E[-\mathrm{Var}[X_1 \mid Z]] = -\mathrm{Var}[X_1] + \mathrm{Var}[E[X_1 \mid Z]] = \\ &-\mathrm{Var}[X_1] + \mathrm{Var}[E[Z - Y_1 \mid Z]] = -\mathrm{Var}[X_1] + \mathrm{Var}[Z]\end{aligned} \tag{2.184}$$

it follows from $\mathrm{Var}[Z] = \mathrm{Var}[X_1] + \mathrm{Var}[Y_1] + 2\mathrm{Cov}[X_1, Y_1]$ that

$$\mathrm{Var}[Z] = \mathrm{Var}[X_1] - \mathrm{Var}[Y_1] \tag{2.185}$$

Thus the variance of the measuring error Y_1 by use of (2.183) is obtained as

$$\mathrm{Var}[Y_1] = \mathrm{Var}[X_1] - \mathrm{Cov}[X_1, X_2] \tag{2.186}$$

By symmetry, $\mathrm{Var}[Y_2]$ is obtained by interchanging X_1 and X_2.

Under certain conditions the principles of the pairing method explained in the example can be adapted to destructive testing measurements. Assume that some material property varies in space as a random field. By the process of measuring the property at a point, the material is changed irreversibly or even destroyed within a certain neighborhood of the point. Therefore, the property cannot be measured by some physical measuring device applied at the same point. However, if the first measurement method is applied for one set of points and the second measurement method is applied for another set of points, these two sets of measurements can in a certain sense be paired by use of a stochastic interpolation procedure based on the random field model of the property variation. At each point of measurement by the first method, a measured value is paired with a probability distribution derived by stochastic interpolation between the measured values obtained by the second method. By suitable assumptions about the joint probabilistic structure of the measurement error fields and the "true" property field, it is then possible to pull out the "true" field from the measured field. These assumptions are generalizations of the simple independence type assumptions made in the elementary pairing method described in the example.

In the following the objects are values of a realization of a field of undrained shear strengths for a saturated clay deposit [38]. The two measuring methods are the CPT-method (Cone Penetration Test) and the vane test method, respectively, applied at two different sets of points of which the one, S_1, is more dense than the other, S_2. The pairing is made by stochastic interpolation to the points of S_2 between the measured values at the points of S_1. Instead of being a number, at least one of the elements in the pair is a probability distribution obtained by the interpolation. Also there will be stochastic dependence between the different pairs.

Let

$$\mathbf{Z} = \mathbf{X} + \mathbf{Y} \tag{2.187}$$

be a vector of logarithms of cone tip resistances obtained by ideally perfect CPT-measurements. The vector $\mathbf{X}$ corresponds to the imperfectly measured CPT values while $\mathbf{Y}$ contains the measuring error corrections. The CPT values refer to the points at which the vane tests are made. The CPT-values are therefore not obtained directly but are results of suitable interpolations between physically measured CPT-values at other points. Thus the interpolation results are considered as imperfect measurements that can be paired with the physical measurements obtained by the vane test. The logarithms of the vane test measurements are contained in the vector $\mathbf{X}_v$ and the corresponding measuring error corrections in the vector $\mathbf{Y}_v$.

The logarithmic shear strength field is assumed to be homogeneous. It is further necessary to make some reasonable assumptions about the measuring error corrections $\mathbf{Y}$ and $\mathbf{Y}_v$. Striving at simplicity of the analysis the following assumptions are made:

1. The conditional expectations $E[\mathbf{Y} \mid \mathbf{Z}]$ and $E[\mathbf{Y}_v \mid \mathbf{Z}]$ are independent of $\mathbf{Z}$.
2. Let $Z_1, ..., Z_n$ and $X_1, ..., X_n$ be the elements of $\mathbf{Z}$ and $\mathbf{X}$, respectively. The linear regression of Z_i on $\mathbf{X}$ depends solely on X_i with the same regression coefficient for all $i = 1, ..., n$.
3. The triple $(\mathbf{Z}, \mathbf{Y}, \mathbf{Y}_v)$ is jointly Gaussian with no correlation between $\mathbf{Y}$ and $\mathbf{Y}_v$ and with the elements of $\mathbf{Y}_v$ being uncorrelated with common standard deviation γ.
4. $$\mathbf{Z} = \mathbf{X}_v + \mathbf{Y}_v + c\mathbf{e} \tag{2.188}$$

   where c is some constant and $\mathbf{e}' = [1...1]$.

The last assumption is based on the fact that elasto-plastic mechanics predict that there is almost proportionality between the undrained shear strength as measured by the vane test and the cone tip resistance in the CPT-test when applied to an ideal saturated clay.

Since $\text{Cov}[\mathbf{Z}, \mathbf{Y}' \mid \mathbf{Z}] = \mathbf{0}$, it follows by use of assumption 1 in the total representation theorem (2.182) that

$$\text{Cov}[\mathbf{Z}, \mathbf{Y}'] = \text{Cov}[E[\mathbf{Z} \mid \mathbf{Z}], E[\mathbf{Y} \mid \mathbf{Z}]'] = \mathbf{0} \tag{2.189}$$

implying that

$$\text{Cov}[\mathbf{X}, \mathbf{Y}'] = -\text{Cov}[\mathbf{Y}, \mathbf{Y}'] \tag{2.190}$$

Substituting this into

$$\begin{aligned}\text{Cov}[\mathbf{Z}, \mathbf{Z}'] &= \text{Cov}[\mathbf{X}+\mathbf{Y}, (\mathbf{X}+\mathbf{Y})'] = \\ &\text{Cov}[\mathbf{X}, \mathbf{X}'] + \text{Cov}[\mathbf{Y}, \mathbf{X}'] + \text{Cov}[\mathbf{X}, \mathbf{Y}'] + \text{Cov}[\mathbf{Y}, \mathbf{Y}']\end{aligned} \tag{2.191}$$

gives

$$\text{Cov}[\mathbf{Z}, \mathbf{Z}'] = \text{Cov}[\mathbf{X}, \mathbf{X}'] - \text{Cov}[\mathbf{Y}, \mathbf{Y}'] \tag{2.192}$$

By using assumption 2 in the general expression for the linear regression

$$\hat{E}[\mathbf{Z} \mid \mathbf{X}] = E[\mathbf{Z}] + \text{Cov}[\mathbf{Z}, \mathbf{X}']\text{Cov}[\mathbf{X}, \mathbf{X}']^{-1}(\mathbf{X} - E[\mathbf{X}]) \tag{2.193}$$

it follows that

$$\text{Cov}[\mathbf{Z}, \mathbf{X}']\text{Cov}[\mathbf{X}, \mathbf{X}']^{-1} = a\mathbf{I} \tag{2.194}$$

for some constant a. Substituting $\mathbf{Z} = \mathbf{X} + \mathbf{Y}$ and using (2.190) then gives the equation

$$\mathbf{I} - \text{Cov}[\mathbf{Y}, \mathbf{Y}']\text{Cov}[\mathbf{X}, \mathbf{X}']^{-1} = a\mathbf{I} \tag{2.195}$$

from which

$$\text{Cov}[\mathbf{Y}, \mathbf{Y}'] = (1 - a)\text{Cov}[\mathbf{X}, \mathbf{X}'] \tag{2.196}$$

Since the diagonals on both sides must be non-negative, it follows that $a \leq 1$ and from (2.192) that $a \geq 0$. Setting $a = (\delta/\sigma)^2$, where σ^2 is the common variance of the elements in $\mathbf{X}$, we see from (2.192) and (2.196) that

$$\text{Cov}[\mathbf{Z}, \mathbf{Z}'] = \delta^2 \mathbf{P_X}, \; \delta^2 \leq \sigma^2 \tag{2.197}$$

where $\mathbf{P_X}$ is the correlation matrix of $\mathbf{X}$.

Assumption 3 applied to (2.188) shows that $\mathbf{X}_v$ is Gaussian with mean and covariance matrix

$$E[\mathbf{X}_v] = E[\mathbf{Z}] - E[\mathbf{Y}_v] - c\,\mathbf{e} \tag{2.198}$$

$$\mathrm{Cov}[\mathbf{X}_v, \mathbf{X}_v'] = \mathrm{Cov}[\mathbf{Z}, \mathbf{Z}'] + \mathrm{Cov}[\mathbf{Y}_v, \mathbf{Y}_v'] \tag{2.199}$$

Now let $\boldsymbol{\zeta}$ be a Gaussian vector which together with $\mathbf{X}, \mathbf{Y}, \mathbf{X}_v, \mathbf{Y}_v$ is jointly Gaussian. This vector $\boldsymbol{\zeta}$ is a vector of measurements that contains information about $\mathbf{X}$ given through the conditional means and covariances $E[\mathbf{X} \mid \boldsymbol{\zeta}]$ and $\mathrm{Cov}[\mathbf{X}, \mathbf{X}' \mid \boldsymbol{\zeta}]$. Since the two measuring methods are independent, the information contained in $\boldsymbol{\zeta}$ and obtained solely by the CPT-method carries no information about the outcome of the error correction vector $\mathbf{Y}_v$ related to the vane test. Therefore all covariances between elements of $\mathbf{Y}_v$ and elements of $\boldsymbol{\zeta}$ are zero. From the joint Gaussianity and $\mathrm{Cov}[\mathbf{Z}, \mathbf{Y}_v'] = 0$, (2.189), it then follows by use of the linear regression method that

$$\mathrm{Cov}[\mathbf{Z}, \mathbf{Y}_v' \mid \boldsymbol{\zeta}] = \mathbf{0} \tag{2.200}$$

Moreover, it follows that

$$\mathrm{Cov}[\mathbf{Y}_v, \mathbf{Y}_v' \mid \boldsymbol{\zeta}] = \mathrm{Cov}[\mathbf{Y}_v, \mathbf{Y}_v'] \tag{2.201}$$

Using the information contained in $\boldsymbol{\zeta}$, the mean vector $E[\mathbf{Z}]$ in (2.198) and the covariance matrix $\mathrm{Cov}[\mathbf{Z}, \mathbf{Z}']$ in (2.199) should therefore simply be replaced by the corresponding conditional quantities $E[\mathbf{Z} \mid \boldsymbol{\zeta}]$ and $\mathrm{Cov}[\mathbf{Z}, \mathbf{Z}' \mid \boldsymbol{\zeta}]$, respectively. To determine these conditional means and covariances we first determine $E[\mathbf{Z} \mid \mathbf{X}]$ and $\mathrm{Cov}[\mathbf{Z}, \mathbf{Z}' \mid \mathbf{X}]$ from the linear regression of $\mathbf{Z}$ on $\mathbf{X}$. According to assumption 2 we have that

$$E[\mathbf{Z} \mid \mathbf{X}] = \lambda^2 \mathbf{X} + \mu(1 - \lambda^2)\mathbf{e}, \ \lambda = \delta/\sigma \tag{2.202}$$

and the residual covariance matrix

$$\mathrm{Cov}[\mathbf{Z}, \mathbf{Z}' \mid \mathbf{X}] = \mathrm{Cov}[\mathbf{Z}, \mathbf{Z}'] - \mathrm{Cov}[\mathbf{Z}, \mathbf{X}']\mathrm{Cov}[\mathbf{X}, \mathbf{X}']^{-1}\mathrm{Cov}[\mathbf{Z}, \mathbf{X}'] =$$
$$\mathrm{Cov}[\mathbf{Z}, \mathbf{Z}'] - \lambda^2 \mathrm{Cov}[\mathbf{Z}, \mathbf{Z}'] = \delta^2(1 - \lambda^2)\mathbf{P}_{\mathbf{X}} \tag{2.203}$$

where μ is the common mean of the elements of $\mathbf{X}$ (setting $E[\mathbf{Y}] = \mathbf{0}$). The total representation theorem next gives the results

$$E[\mathbf{Z} \mid \boldsymbol{\zeta}] = \lambda^2 E[\mathbf{X} \mid \boldsymbol{\zeta}] + \mu(1 - \lambda^2)\mathbf{e} \tag{2.204}$$

$$\mathrm{Cov}[\mathbf{Z}, \mathbf{Z}' \mid \boldsymbol{\zeta}] = \lambda^2(1 - \lambda^2)\mathrm{Cov}[\mathbf{X}, \mathbf{X}'] + \lambda^4 \mathrm{Cov}[\mathbf{X}, \mathbf{X}' \mid \boldsymbol{\zeta}] \tag{2.205}$$

Thus (2.198) and (2.199) are replaced by

$$E[\mathbf{X}_v] = E[\mathbf{Z} \mid \boldsymbol{\zeta}] - c\,\mathbf{e} \tag{2.206}$$

$$\mathrm{Cov}[\mathbf{X}_v, \mathbf{X}_v'] = \mathrm{Cov}[\mathbf{Z}, \mathbf{Z}' \mid \zeta] + \gamma^2 \mathbf{I} \tag{2.207}$$

setting $E[\mathbf{Y}_v] = \mathbf{0}$. (If $E[\mathbf{Y}]$ and $E[\mathbf{Y}_v]$ are not zero, their contributions may be included in the constant c in (2.188).)

All the unknown parameters are obtained as follows. First the field that represents the measurements $\mathbf{X}$ is modeled as a homogeneous Gaussian field using a specific pragmatically chosen covariance function family. The parameters μ, σ, as well as the correlation parameters that determine the correlation matrix $\mathbf{P}_\mathbf{X}$ are obtained as explained in Section 2.11 by the maximum likelihood principle or the Bayesian method, if necessary. The likelihood function for the remaining parameters c, γ, and λ can then be formulated by use of (2.204)–(2.207) and the Gaussianity of $\mathbf{X}_v$.

Let $\mathbf{x}_v$ be the observation of $\mathbf{X}_v$ and write

$$\boldsymbol{\xi}(c, \lambda) = \mathbf{x}_v - E[\mathbf{X}_v] \tag{2.208}$$

$$\mathbf{R}(\gamma, \lambda) = \mathrm{Cov}[\mathbf{X}_v, \mathbf{X}_v']^{-1} \tag{2.209}$$

Then the Gaussian density of $\mathbf{X}_v$ computed at $\mathbf{x}_v$ is

$$f_{\mathbf{X}_v}(\mathbf{x}_v) \propto \sqrt{\det[\mathbf{R}(\gamma, \lambda)]} \exp\left[-\frac{1}{2}\boldsymbol{\xi}(c, \lambda)'\mathbf{R}(\gamma, \lambda)\boldsymbol{\xi}(c, \lambda)\right], \\ c \in \Re, \ \gamma \in \Re_+, \ \lambda \in [0, 1] \tag{2.210}$$

($\propto$ means "proportional to"). The right side of (2.210) defines the likelihood function $L(c, \gamma, \lambda; \mathbf{x}_v)$ of c, γ, and λ. Let (C, Γ, Λ) be the set of Bayesian random variables corresponding to (c, γ, λ), and adopt the non-informative prior for which $(C, \log\Gamma, \log(\Gamma^2 + \sigma^2\Lambda^2))$ has a diffuse prior over $\Re \times \{\log\gamma, \log(\gamma^2 + \sigma^2\lambda^2) \mid \gamma \in \Re_+, 0 \le \lambda \le 1\}$. Then the prior density of (C, Γ, Λ) is proportional to $\lambda/[\gamma(\gamma^2 + \sigma^2\lambda^2)]$ and we get the posterior density $f_{C,\Gamma,\Lambda}(c, \gamma, \lambda \mid \mathbf{x}_v)$ by multiplying this prior with the likelihood function.

Before the likelihood function (2.210) can be used to infer about the parameters c, γ, and λ there is a further step to be taken because the set of vane test undrained shear strength observations in practice often will be imperfect in the sense that some of the test results are reported not by their values but by the information that the values are larger than the measuring capacity of the applied vane (censored data). This is expressed by saying that each of the random variables representing the vane test measurements is *clipped* (or censored) at a given value x_{0i}, $i = 1, \ldots, n$. Thus the sample is given as $x_1 = x_{01}, x_2 = x_{02}, \ldots, x_r = x_{0r}, x_{r+1} < x_{0r+1}, \ldots, x_n < x_{0n}$, where the sample has been ordered such that the first r vane test shear strengths are larger than the respective measuring capacities while the remaining $n - r$ tests are "well-behaved". For this clipped sampling case the likelihood function is obtained by integrating the joint density of $\mathbf{X}_v$ in (2.210) with respect to x_i from x_{0i} to ∞ for $i = 1, ..., r$. Thus the likelihood function becomes

$$L(c, \gamma\lambda; \mathbf{x}_v) \propto \sqrt{\det[\mathbf{R}(\gamma, \lambda)]}$$
$$\int_{x_{01}}^{\infty} \cdots \int_{x_{0r}}^{\infty} \exp\left[-\frac{1}{2}\boldsymbol{\xi}(c, \lambda; \mathbf{x})'\mathbf{R}(\gamma, \lambda)\boldsymbol{\xi}(c, \lambda; x)\right] dx_1 \cdot ... \cdot dx_r \tag{2.211}$$

The numerical studies of the posterior density of (C, Γ, Λ) obtained from (2.211) after multiplication by the prior density must in practice be based on Monte Carlo integration except if $r = 1$ or 2.

For given values of all the parameters the formulas (2.204) and (2.205) define the interpolated "true" field $Z(\mathbf{t})$ "cleaned" for measuring uncertainty as a Gaussian field with conditional mean value function

$$E[Z(\mathbf{t}) \mid \zeta] = \lambda^2 E[X(\mathbf{t}) \mid \zeta] + \mu(1 - \lambda^2) \tag{2.212}$$

and conditional covariance function

$$\begin{aligned}&\text{Cov}[Z(\mathbf{t}_1), Z(\mathbf{t}_2) \mid \zeta] = \\ &\lambda^2(1 - \lambda^2)\text{Cov}[X(\mathbf{t}_1), X(\mathbf{t}_2)] + \lambda^4\text{Cov}[X(\mathbf{t}_1), X(\mathbf{t}_2) \mid \zeta]\end{aligned} \tag{2.213}$$

For the purpose of making reliability analysis with respect to plastic collapse, this interpolation technique has actually been applied to obtain a random field model for the undrained shear strengths of the clay under the Anchor Block West of the Great Belt Bridge in Denmark conditioned on given CPT and vane test results measured in situ [37]. The field model for $X(\mathbf{t})$ was chosen as mentioned at the end of the previous section.

2.13 Discretization Defined by Linear Regression on Derivatives at a Single Point

In certain problems, in particular when t is time so that $X(t)$ is a random process, it is useful to make a discretization in terms of $X(t)$ and its first n derivatives $dX(t)/dt, ..., d^n X(t)/dt^n$ at a given point $t = t_0$. It is sufficient to let $t_0 = 0$. More generally we will consider the linear regression of a random vector process $\mathbf{X}(t) = (X_1(t), ..., X_m(t))$ on $X_1(0), X_1^{(1)}(0), ..., X_1^{(n)}(0)$ where $X_1^{(j)}(0) = d^j X_1(0)/dt^j$.

For continuity reasons we have

$$\hat{E}[\mathbf{X}(t)|X_1(0),...,X_1^{(n)}(0)] = \lim_{h\to 0}\hat{E}[\mathbf{X}(t)|X_1(0),X_1(h),...,X_1(nh)] \quad (2.214)$$

because there is a one-to-one linear mapping between the vector of difference quotients $\mathbf{Q}$ up to order n corresponding to the time step h and the vector $\mathbf{Z} = (X_1(0), X_1(h), ..., X_1(nh))$, that is, there is a regular difference quotient operator matrix Δ such that $\mathbf{Q} = \Delta\mathbf{Z} \to [\mathbf{D}X_1(s)]_{s=0}$ as $h \to 0$, $\mathbf{D} = (1, \frac{\partial}{\partial s}, ..., \frac{\partial^n}{\partial s^n})$. It follows from (2.101) that

$$\begin{aligned}\hat{E}[X_k(t)|\mathbf{Z}] &= E[X_k(t)] + \text{Cov}[X_k(t),\mathbf{Z}']\text{Cov}[\mathbf{Z},\mathbf{Z}']^{-1}(\mathbf{Z}-E[\mathbf{Z}]) = \\ &E[X_k(t)] + \text{Cov}[X_k(t),\mathbf{Z}']\Delta'(\Delta\text{Cov}[\mathbf{Z},\mathbf{Z}']\Delta')^{-1}(\Delta\mathbf{Z}-E[\Delta\mathbf{Z}])\end{aligned} \quad (2.215)$$

so that as $h \to 0$:

$$\begin{aligned}&\hat{E}[X_k(t)|X_1(0),...,X_1^{(n)}(0)] = E[X_k(t)]+ \\ &\left\{\left[\frac{\partial^i}{\partial s^i}\text{Cov}[X_k(t),X_1(s)]\right]_{s=0}\right\}_{i=0,...,n} \\ &\left\{\left[\frac{\partial^{i+j}}{\partial u^i\partial v^j}\text{Cov}[X_1(u),X_1(v)]\right]_{u=v=0}\right\}^{-1}_{i,j=0,...,n} \\ &\left\{X_1^{(i)}(0) - E[X_1^{(i)}(0)]\right\}_{i=0,...,n}\end{aligned} \quad (2.216)$$

It is seen that the linear regression of $\mathbf{X}(t)$ on $X_1(0), ..., X_1^{(n)}(0)$ is determined by the mixed partial derivatives of the covariance function of $X_1(t)$ at $(s,t) = (0,0)$ and the partial derivatives of $\text{Cov}[\mathbf{X}(t), X_1(s)]$ at $s = 0$.

The vector process $\mathbf{X}(t)$ is identical to the process

$$\hat{E}[\mathbf{X}(t) \mid X_1(0),...,X_1^{(n)}(0)] + \mathbf{R}(t) \quad (2.217)$$

where the linear regression term for $X_1(t)$ dominates in the vicinity of $t = 0$ while the residual vector process $\mathbf{R}(t)$ approaches $\mathbf{X}(t) - E[\mathbf{X}(t)]$ as the distance from $t = 0$ increases, given that the covariance functions of $\mathbf{X}(t)$ all approach zero as $\mid t \mid \to \infty$.

It follows from (2.217) that $R_1(0)$ and the first n derivatives $R_1^{(1)}(0), ..., R_1^{(n)}(0)$ of the first component of the residual vector process $\mathbf{R}(t)$ are all zero. Therefore, the linear regression of $X_k(t)$ on $X_1(0), ..., X_1^{(n)}(0)$ gives a good approximation for $k = 1$. Whether approximations that are sufficiently good for the considered applications are obtained also for $k \neq 1$ depends on the cross-correlations. Clearly, if the cross-correlations are small or zero, the information contained in given values of $X_1(0), ..., X_1^{(n)}(0)$ has small or no predictive effect on $X_k(t)$ for $k \neq 1$.

The case where $\mathbf{X}(t)$ is a stationary vector process has important applications treated in later sections. For a stationary $\mathbf{X}(t)$ the covariance function matrix depends solely on the time difference $s - t$:

$$\text{Cov}[\mathbf{X}(t), \mathbf{X}(s)'] = \{c_{kl}(s-t)\}_{k,l=1,\ldots,m} \tag{2.218}$$

The linear regression (2.216) consequently specializes to

$$\hat{E}[X_k(t)|X_1(0), ..., X_1^{(n)}(0)] = \\ \left\{(-1)^i c_{1k}^{(i)}(t)\right\}_{i=0,\ldots,n} \left\{(-1)^j c_{11}^{(i+j)}(0)\right\}^{-1}_{i,j=0,\ldots,n} \left\{X_1^{(i)}(0)\right\}_{i=0,\ldots,n} \tag{2.219}$$

in which without loss of generality $\mathbf{X}(t)$ is assumed to have the zero vector as expectation. Since $c_{k1}(-t) = c_{1k}(t)$, we have $c_{1k}^{(i)}(t) = (-1)^i c_{k1}^{(i)}(-t)$. This has been used to obtain the first row matrix in (2.219). Moreover, since $c_{11}(t) = c_{11}(-t)$ we have $c_{11}^{(i+j)}(0) = 0$ for $i + j$ odd.

For $n = 0, 1, 2$ it is seen that (2.219) reads (using the standard dot notation for the derivatives of $X_1(t)$):

$$\hat{E}[X_k(t)|X_1(0)] = c_{1k}(t)\lambda_0^{-1} X_1(0) \tag{2.220}$$

$$\hat{E}[X_k(t)|X_1(0), \dot{X}_1(0)] = c_{1k}(t)\lambda_0^{-1} X_1(0) - c_{1k}'(t)\lambda_2^{-1}\dot{X}_1(0) \tag{2.221}$$

$$\hat{E}[X_k(t)|X_1(0), \dot{X}_1(0), \ddot{X}_1(0)] = \\ [c_{1k}(t) - c_{1k}'(t)\; c_{1k}''(t)] \begin{bmatrix} \lambda_0 & 0 & -\lambda_2 \\ 0 & \lambda_2 & 0 \\ -\lambda_2 & 0 & \lambda_4 \end{bmatrix}^{-1} \begin{bmatrix} X_1(0) \\ \dot{X}_1(0) \\ \ddot{X}_1(0) \end{bmatrix} = \\ = (\lambda_0\lambda_4 - \lambda_2^2)^{-1} \Big\{ c_{1k}(t)[\lambda_4 X_1(0) + \lambda_2 \ddot{X}_1(0)] + \\ c_{1k}''(t)[\lambda_2 X_1(0) + \lambda_0 \ddot{X}_1(0)] \Big\} - \lambda_2^{-1} c_{1k}'(t)\dot{X}_1(0) \tag{2.222}$$

where $\lambda_0 = c_{11}(0)$, $\lambda_2 = -c''_{11}(0)$, $\lambda_4 = c_{11}^{(4)}(0)$ are the spectral moments

$$\lambda_n = \int_0^\infty \omega^n dG(\omega) \tag{2.223}$$

for $n = 0, 2, 4$, corresponding to the spectral representation

$$c_{11}(t) = \int_0^\infty \cos\omega t \, dG(\omega) \tag{2.224}$$

of the covariance function of the process $X_1(t)$. It is implicitly assumed above that the considered derivatives exist. The spectral moment λ_{2n} is the variance of the nth derivative process $X^{(n)}(t)$ (which follows in the same way as (2.216) follows from (2.215), noting that $\Delta\text{Cov}[\mathbf{Z}, \mathbf{Z}']\Delta' = \text{Cov}[\Delta\mathbf{Z}, (\Delta\mathbf{Z})'] \to \text{Cov}[\mathbf{D}X_1(s), \mathbf{D}'X_1(t)]_{(s,t)=(0,0)} =$ $\left\{\text{Cov}\left[\frac{d^i X_1(s)}{ds^i}, \frac{d^j X_1(t)}{dt^j}\right]\right\}_{(s,t)=(0,0)}$ as $h \to 0$).

Even though $\lambda_4 = \infty$ implies that the second derivative process $\ddot{X}_1(t)$ does not exist in an elementary sense (may exist, though, e.g., in the sense of white noise), it is seen that

$$\hat{E}[X_k(t)|X_1(0), \dot{X}_1(0), \ddot{X}_1(0)] \to \hat{E}[X_k(t)|X_1(0), \dot{X}_1(0)] \tag{2.225}$$

as $\lambda_4 \to \infty$ (λ_0, λ_2 bounded). Similarly

$$\hat{E}[X_k(t)|X_1(0), \dot{X}_1(0)] \to \hat{E}[X_k(t)|X_1(0)] \tag{2.226}$$

as $\lambda_2 \to \infty$ (λ_0 bounded).

The (k, l)-element of the residual covariance function matrix is obtained as the covariance $\text{Cov}[X_k(t), X_l(s)]$ minus the covariance between the linear regressions of $X_k(t)$ and $X_l(s)$ on $X_1(0), \ldots, X_1^{(n)}(0)$ that in turn is obtained by replacing $\{X_1^{(i)}(0)\}$ in (2.219) by the column vector $\{(-1)^j c_{1l}^{(j)}(s)\}_{j=0,\ldots,n}$.

For $n = 0, 1, 2$ we have the residual covariance expressions:

$$n = 0 : c_{kl}(s-t) - c_{1k}(t)c_{1l}(s)\lambda_0^{-1} \tag{2.227}$$

$$n = 1 : c_{kl}(s-t) - c_{1k}(t)c_{1l}(s)\lambda_0^{-1} - c'_{1k}(t)c'_{1l}(s)\lambda_2^{-1} \tag{2.228}$$

$$\begin{aligned} n = 2 : c_{kl}(s-t) &- (\lambda_0\lambda_4 - \lambda_2^2)^{-1}\{\lambda_2[c_{1k}(t)c''_{1l}(s) + c''_{1k}(t)c_{1l}(s)] + \\ &\lambda_4 c_{1k}(t)c_{1l}(s) + \lambda_0 c''_{1k}(t)c''_{1l}(s)\} - \lambda_2^{-1}c'_{1k}(t)c'_{1l}(s) \end{aligned} \tag{2.229}$$

It is seen that the residual covariance for $n = 1$ is the limit of the residual covariance for $n = 2$ as $\lambda_4 \to \infty$ (λ_0, λ_2 bounded), and similarly for $n = 0$ and $n = 1$ as $\lambda_2 \to \infty$ (λ_0 bounded).

2.14 Conditioning on Crossing Events

Let $\mathbf{X}(t)$ be an m-dimensional Gaussian vector process and let $\mathbf{Y}(t)$ be an s-dimensional ($s \leq m$) subvector of $\mathbf{X}(t)$. Then the conditional vector process $\mathbf{X}(t)$ given $\mathbf{Y}(0)$ is Gaussian with the linear regression $\hat{E}[\mathbf{X}(t) \mid \mathbf{Y}(0)]$ as vectorial mean value function and the corresponding residual covariance function as covariance function. Of course, by unconditioning with respect to the Gaussian distribution of $\mathbf{Y}(0)$, the set of finite dimensional distributions of the Gaussian vector process $\mathbf{X}(t)$ is regained according to the total probability rule:

$$f_{\mathbf{X}(t_1),\ldots,\mathbf{X}(t_q)}(\mathbf{x}_1, \ldots, \mathbf{x}_q) = \int_{\Re^s} f_{\mathbf{X}(t_1),\ldots,\mathbf{X}(t_q)}(\mathbf{x}_1, \ldots, \mathbf{x}_q \mid \mathbf{Y}(0) = \mathbf{y}) f_{\mathbf{Y}(0)}(\mathbf{y}) d\mathbf{y} \tag{2.230}$$

for any $q > 0$ and $t_1, \ldots, t_q \in \Re$. However, other vector processes than $\mathbf{X}(t)$ can be defined on the basis of (2.230) by assigning some other distribution to $\mathbf{Y}(0)$ than the original Gaussian distribution. Clearly this type of modeling can be considered for any Gaussian field with conditioning on any set of random variables but, as it will be shown in the following, such modeling has particular relevance for Gaussian vector processes.

Equivalently to using (2.230) as the basis for defining a vector process $\mathbf{V}(t)$ different from $\mathbf{X}(t)$ we may define $\mathbf{V}(t)$ directly as

$$\mathbf{V}(t) = \hat{E}[\mathbf{X}(t) \mid \mathbf{Y}(0)] + \mathbf{R}(t) \tag{2.231}$$

where $\mathbf{R}(t)$ is the Gaussian residual vector process. Thus $\mathbf{V}(t)$ becomes modeled as $\mathbf{R}(t)$ added under the assumption of independence to the given inhomogeneous linear function $\hat{E}[\mathbf{X}(t) \mid \mathbf{Y}(0)]$ of the random vector $\mathbf{Y}(0)$ that may have any applicationally relevant probability distribution assigned to it.

Several interesting problems can be analysed by use of (2.231) in connection with phenomena that depend on the event that the trajectory of $\mathbf{X}(t)$ crosses out through the hyperplane orthogonal to the x_1-axis in the distance u from the origin. With sufficient generality, consider crossings that happen at time $t = 0$, and assume that the process $X_1(t)$ possesses differentiable sample functions. Then the crossing event C can be defined as $C = \{X_1(0) = u, \dot{X}_1(0) > 0\}$. Obviously C has probability zero so that conditioning on C requires special care in order to ensure uniqueness of the obtained conditional probabilities. In particular it is necessary to have an applicationally relevant definition of the conditional distributions of the vector process $\mathbf{X}(t)$ given the event C.

The uniqueness problem can best be illustrated by a study of two alternative (among several) definitions of the conditional probability density of $\dot{X}_1(0)$ given C. For shortage of notation omit the index and write X for X_1. Instead of considering the sample curve upcrossings of level u with slope z in the time interval $[0, dt]$ it is sufficient to consider the upcrossings with slope z of the tangents to the trajectories at $t = 0$ as $dt \to 0$ [2]. It is seen from Figure 2.4 (top) that these tangents are those that cross through the x-axis within the interval $[u - z dt, u]$. Thus the probability (asymptotically as $dt \to 0$) of having a sample curve of any positive slope at $t = 0$ crossing through level u within the time interval $[0, dt]$ is equal to the probability content of the angular domain shown to the right in Figure 2.4 (top) calculated by integration of the joint density of $X = X(0)$ and $Z = \dot{X}(0)$ over the domain. Since the area element is $z\,dz\,dt$ this probability becomes

$$\kappa_h(0)dt = \left[\int_0^\infty f_{X,Z}(u, z) z\,dz\right] dt \tag{2.232}$$

and the conditional density of Z given the crossing event $C = C_h$ obviously becomes

$$f_Z(z \mid C_h) = \frac{1}{\kappa_h(0)} z f_{X,Z}(u, z) \tag{2.233}$$

Alternatively to counting the number of crossings of level u in the time interval $[0, dt]$ we may count the number of crossings with slope z through the interval $[u - dx, u]$ on the x-axis. Surely, as $dx \to 0$ these sample curves cross level u in the time interval $[0, dx/z]$ that approaches $[0, 0]$. However, the probability (asymptotically as $dx \to 0$) of having a sample curve of any positive slope at $t = 0$ crossing through the x-axis within the interval $[u - dx, u]$ becomes, Figure 2.4 (bottom),

$$\kappa_v(0)dx = \left[\int_0^\infty f_{X,Z}(u, z) dz\right] dx \tag{2.234}$$

and the conditional density of Z given the crossing, even $C = C_v$ becomes

$$f_Z(z \mid C_v) = \frac{1}{\kappa_v(0)} f_{X,Z}(u, z) \tag{2.235}$$

Thus different results are obtained depending on how the crossing event C of zero probability is interpreted as the limit of non-zero probability events. The indices h and v on C refer to "horizontal window" and "vertical window", respectively. For applications that refer to the number of crossings per time unit, the following considerations show that it is the horizontal window crossings that are of interest.

Even though the event of having an upcrossing at a specified time point has probability zero, upcrossings of level u occur with non-zero probability at some time points within any time interval of non-zero length. We may count these

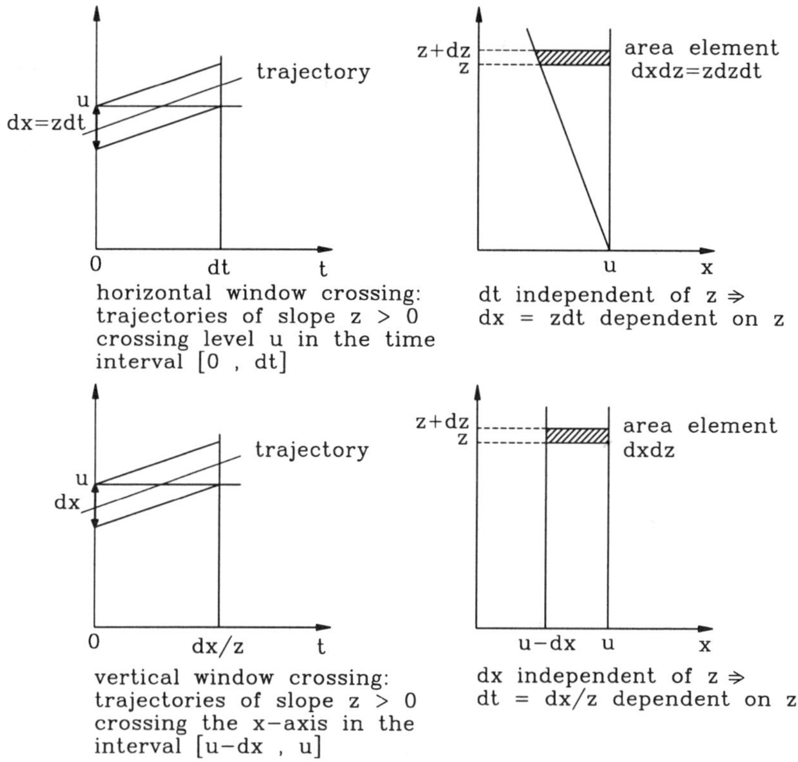

FIGURE 2.4
Definition of upcrossings of level u through horizontal window (top) and vertical window (bottom).

upcrossings of level u in the time interval $[0, T]$ by dividing the interval in n disjoint intervals of length T/n and assign the indicator random variable

$$\mathbf{1}_i = \begin{cases} 1 \text{ if upcrossing in } &]\frac{i-1}{n}T, \frac{i}{n}T] \\ 0 \text{ otherwise} & \end{cases} \tag{2.236}$$

Then

$$\left\{\sum_{i=1}^{n} \mathbf{1}_i\right\}_{n\in\{1,2,\ldots\}} \uparrow N(T) \tag{2.237}$$

is an increasing sequence of random variables with limit equal to the random number of upcrossings $N(T)$ in the interval $[0, T]$. Since according to (2.232) (shifting 0 to t) we obviously have

$$E[\sum_{i=1}^{n}\mathbf{1}_i] = \sum_{i=1}^{n}E[\mathbf{1}_i] \to \int_0^T \kappa_h(t)dt \tag{2.238}$$

as $n \to \infty$, it follows that the mean number of upcrossings of level u is

$$E[N(T)] = \int_0^T \left[\int_0^\infty f_{X(t),\dot{X}(t)}(u,z)z\,dz\right]dt \tag{2.239}$$

The integrand in (2.239)

$$\nu_+(t,u) = \int_0^\infty f_{X(t),\dot{X}(t)}(u,z)z\,dz \tag{2.240}$$

can thus be interpreted as the mean rate of upcrossings of level u at time t. Formula (2.240) is the celebrated Rice formula [40] derived here by simple heuristic arguments.

Under the assumption that $X(t)$ is Gaussian, the joint distribution of $X(t)$ and $\dot{X}(t)$ is Gaussian and both of zero mean if the process $X(t)$ has zero mean. Let $\lambda_0, \lambda_1, \lambda_2$ denote, see (2.216),

$$\lambda_0 = \text{Var}[X(t)] \tag{2.241}$$

$$\lambda_1 = \text{Cov}[X(t), \dot{X}(t)] = \frac{\partial}{\partial v}\text{Cov}[X(t), X(v)]_{v=t} \tag{2.242}$$

$$\lambda_2 = \text{Var}[\dot{X}(t)] = \frac{\partial^2}{\partial u \partial v}\text{Cov}[X(u), X(v)]_{u=v=t} \tag{2.243}$$

Then

$$E[\dot{X}(t) \mid X(t)] = \frac{\lambda_1}{\lambda_0}X(t) \tag{2.244}$$

$$\text{Var}[\dot{X}(t) \mid X(t)] = \lambda_2\left(1 - \frac{\lambda_1^2}{\lambda_0\lambda_2}\right) \tag{2.245}$$

calculated as the linear regression of $\dot{X}(0)$ on $X(0)$ and the corresponding residual variance, respectively. Thus the joint density of $X = X(t)$ and $Z = \dot{X}(t)$ becomes

$$f_{X,Z}(x,z) = \frac{1}{\sqrt{\lambda_0\lambda_2 - \lambda_1^2}}\varphi\left(\frac{x}{\sqrt{\lambda_0}}\right)\varphi\left(\frac{z - x\lambda_1/\lambda_0}{\sqrt{\lambda_2 - \lambda_1^2/\lambda_0}}\right) \tag{2.246}$$

so that the mean upcrossing rate of level u at time t according to (2.240) becomes

$$\nu_+(t,u) = \frac{1}{\lambda_0}\sqrt{\lambda_0\lambda_2 - \lambda_1^2}\,\varphi\left(\frac{u}{\sqrt{\lambda_0}}\right)\psi\left(\frac{\lambda_1}{\sqrt{\lambda_0\lambda_2-\lambda_1^2}}\frac{u}{\sqrt{\lambda_0}}\right) \tag{2.247}$$

in which $\psi(\cdot)$ is the function

$$\psi(x) = \varphi(x) + x\Phi(x) \tag{2.248}$$

When $X(t)$ is stationary, we have $\lambda_1 = 0$, and (2.247) reduces to

$$\nu_+(t,u) = \sqrt{\frac{\lambda_2}{2\pi\lambda_0}}\,\varphi\left(\frac{u}{\sqrt{\lambda_0}}\right) \tag{2.249}$$

independent of t.

The density (2.233) follows from (2.246) and (2.247) as

$$f_Z(z \mid C_h) = \frac{\lambda_0}{\lambda_0\lambda_2 - \lambda_1^2}\,\frac{z\varphi\left(\dfrac{z-u\lambda_1/\lambda_0}{\sqrt{\lambda_2-\lambda_1^2/\lambda_0}}\right)}{\psi\left(\dfrac{\lambda_1}{\sqrt{\lambda_0\lambda_2-\lambda_1^2}}\dfrac{u}{\sqrt{\lambda_0}}\right)}, \quad z > 0 \tag{2.250}$$

For $\lambda_1 = 0$ this is the so-called Rayleigh density

$$f_Z(z \mid C_h) = \frac{z}{\lambda_2}\exp\left[-\frac{1}{2}\left(\frac{z}{\sqrt{\lambda_2}}\right)^2\right], \quad z > 0 \tag{2.251}$$

The formulas (2.232)–(2.240) are valid also for non-Gaussian processes with such smoothly varying sample curves that exclude the occurrence of everywhere dense crossing points in any time interval and also exclude events of having sample function tangents in common with the level u as being events of probability zero [2]. Under these assumptions, it follows from the step from (2.238) to (2.239) that the asymptotic probability $\kappa_h(t)dt$ of having an upcrossing of level u in the time interval $[t, t+dt]$ equals the mean number of upcrossings $\nu_+(t,u)dt$ in $[t, t+dt]$ as $dt \to 0$.

If not all upcrossings in $[0, T]$ are counted but only those for which the slopes of the sample curve at the crossing points are in the interval $[z, z+dz]$, the right side of (2.238) becomes replaced by

$$\int_0^T [f_{X(t),\dot{X}(t)}(u, z) z \, dz] dt \tag{2.252}$$

and is thus the mean number of crossings of this kind. The ratio of (2.252) and $E[N(T)]$ given by (2.239) is seen to approach $f_Z(z \mid C_h)dz$ as $T \to 0$. Thus the conditional probability that $Z = \dot{X}(0) \in [z, z+dz]$ given that there is a horizontal window upcrossing of level u at time $t = 0$ is identical to the ratio between the mean rate at $t = 0$ of upcrossings under slope in the interval $[z, z + dz]$ and the mean rate at $t = 0$ of all upcrossings (asymptotically as $dz \to 0$).

For stationary and ergodic processes it is possible to give a useful interpretation of the conditional probability of any event given the event C_h of a horizontal window upcrossing of level u. In fact, for $X(t)$ stationary the ratio of (2.252) and (2.239) becomes $f_Z(z \mid C_h)dz$ independent of T. Under the assumption of ergodicity (that is, ensemble averaging can be replaced by time averaging) the ratio of the number of crossings with given properties (like the sample function at the upcrossing having slope in the interval $[z, z+dz]$) to the total number of crossings within $[0, T]$ is an estimator of the conditional probability of having the property given C_h. By use of the law of large numbers it can be shown that this estimator converges with probability 1 to the conditional probability as $T \to \infty$ [41].

Returning now to (2.231) an example of an application is to let $X_2(t)$ be the derivative process $\dot{X}_1(t)$ of the Gaussian process $X_1(t)$ and define the vector process

$$\mathbf{V}(t) = E[\mathbf{X}(t) \mid X_1(0) = u, \; \dot{X}_1(0) = Z] + \mathbf{R}(t) \tag{2.253}$$

where Z is a random variable with the density (2.250), and the linear regression is given by (2.216) for $n = 1$ or, for $\mathbf{X}(t)$ stationary, by (2.221). This vector process $\mathbf{V}(t)$ gives detailed information about the properties of the sample trajectories of $\mathbf{X}(t)$ in the vicinity of an upcrossing (hereafter omitting reference to horizontal window) of level u at $t = 0$. This is particularly useful when $\mathbf{X}(t)$ is stationary and ergodic because then the description is valid without prespecifying the time point of the upcrossing. Moreover, the description can be empirically verified by sampling from the same trajectory of $\mathbf{X}(t)$ at each upcrossing point.

The topics explained on a heuristic level in this section were first studied by Kac and Slepian [42, 43]. Lindgren has made extensive use of Kac's and Slepian's results in studies of all sorts of both purely mathematical problems and applications related to crossings of levels and to local extremes in Gaussian processes. Lindgren introduced the name "Slepian model (vector) process" for a process defined like $\mathbf{V}(t)$ in (2.253) [44]. In this and the following sections only the case of conditioning on level crossings of a scalar process is considered. However, the Slepian model process concept may be generalized to conditioning on outcrossings of a vector process through the boundary of some suitably regular domain [45] and even to make sense for random fields [46, 47].

2.15 Slepian Model Vector Processes

Let $\mathbf{X}(t)$ be a stationary Gaussian vector process of zero mean vector, and let $\mathbf{Y}(t)$ be a subvector of $\mathbf{X}(t)$. Moreover, let two of the elements in $\mathbf{X}(t)$ be a process $X(t)$ and its derivative process $\dot{X}(t)$, and let none of these two processes be elements of $\mathbf{Y}(t)$. Slightly more general than in (2.253) we may then define the Slepian model vector process

$$\begin{aligned} \mathbf{V}(t) &= E[\mathbf{X}(t) \mid X(0) = u,\ \dot{X}(0) = Z,\ \mathbf{Y}(0) = \mathbf{Y}] + \mathbf{R}(t) = \\ &\mathbf{a}(t)u + \mathbf{b}(t)Z + \mathbf{C}(t)\mathbf{Y} + \mathbf{R}(t) \end{aligned} \tag{2.254}$$

associated to upcrossings of level u by the scalar process $X(t)$. The random variable Z represents the slope of the sample curve of $X(t)$ at an arbitrary upcrossing of level u, and the probability density of Z (interpreted in the sense of long run sampling as explained in the previous section) is the Rayleigh density (2.251).

By marking the upcrossings not only by the observed value of Z but also by the value of $\mathbf{Y}$ at the upcrossings, it is seen exactly as in the previous section that the long run probability density of $(Z, \mathbf{Y})$ is given by

$$f_{Z,\mathbf{Y}}(z, \mathbf{y} \mid C_h) = f_Z(z \mid C_h) f_{\mathbf{Y}}(\mathbf{y} \mid X(0) = u,\ \dot{X}(0) = z,\ C_h) \tag{2.255}$$

in which the first factor is the Rayleigh density (2.251) and the second factor is independent of the crossing event and is the normal density of mean

$$E[\mathbf{Y} \mid X(0) = u,\ \dot{X}(0) = z] = \frac{\text{Cov}[\mathbf{Y}, X(0)]}{\text{Var}[X(0)]} u + \frac{\text{Cov}[\mathbf{Y}, \dot{X}(0)]}{\text{Var}[\dot{X}(0)]} z \tag{2.256}$$

and covariance matrix

$$\begin{aligned} &\text{Cov}[\mathbf{Y}, \mathbf{Y}' \mid X(0),\ \dot{X}(0)] = \\ &\text{Cov}[\mathbf{Y}, \mathbf{Y}'] - \text{Cov}[\mathbf{Y}, X(0)]\text{Cov}[X(0), \mathbf{Y}']/\text{Var}[X(0)] \\ &-\text{Cov}[\mathbf{Y}, \dot{X}(0)]\text{Cov}[\dot{X}(0), \mathbf{Y}^{\,prime}]/\text{Var}[\dot{X}(0)] \end{aligned} \tag{2.257}$$

using that $\text{Cov}[X(0), \dot{X}(0)] = 0$ because $X(t)$ is stationary.

The process $\mathbf{R}(t)$ is non-stationary Gaussian and independent of $(Z, \mathbf{Y})$, and it has the same finite dimensional distributions as the residual process corresponding to the linear regression of $\mathbf{X}(t)$ on $X(0)$, $\dot{X}(0)$, $\mathbf{Y}(0)$.

The coefficients $\mathbf{a}(t), \mathbf{b}(t), \mathbf{C}(t)$ are submatrices of the regression coefficient matrix

$$\mathbf{B}(t) = [\mathbf{a}(t)\ \mathbf{b}(t)\ \mathbf{C}(t)] = \mathrm{Cov}[\mathbf{X}(t),\ [X(0)\ \dot{X}(0)\ \mathbf{Y}(0)]\cdot$$

$$\begin{bmatrix} \mathrm{Var}[X(0)] & 0 & \mathrm{Cov}[X(0), \mathbf{Y}(0)'] \\ 0 & \mathrm{Var}[\dot{X}(0)] & \mathrm{Cov}[\dot{X}(0), \mathbf{Y}(0)'] \\ \mathrm{Cov}[\mathbf{Y}(0), X(0)] & \mathrm{Cov}[\mathbf{Y}(0), \dot{X}(0)] & \mathrm{Cov}[\mathbf{Y}(0), \mathbf{Y}(0)'] \end{bmatrix}^{-1} \tag{2.258}$$

and the covariance function matrix of $\mathbf{R}(t)$ is

$$\mathrm{Cov}[\mathbf{R}(s), \mathbf{R}(t)'] = \mathrm{Cov}[\mathbf{X}(s), \mathbf{X}(t)'] -$$

$$\mathbf{B}(s)\mathrm{Cov}[\begin{bmatrix} X(0) \\ \dot{X}(0) \\ \mathbf{Y}(0) \end{bmatrix}, \mathbf{X}(t)'] \tag{2.259}$$

Example The local minimum values of a sample function of a scalar process $Y(t)$ are the values of $Y(t)$ at the upcrossings of the level $u = 0$ by the derivative process $X(t) = \dot{Y}(t)$. Assume that $Y(t)$ is Gaussian, $E[Y(t)] = 0$ and that $Y(t)$ has the four times differentiable covariance function $\mathrm{Cov}[Y(u), Y(v)] = r(v-u)$. Then the value at $t = 0$ of the Slepian model process for $Y(t)$ corresponding to level $u = 0$ becomes $Y_0(0) = a(0)0 + b(0)Z + R(0)$. With $\lambda_0 = \mathrm{Var}[Y(t)] = r(0)$, $\lambda_2 = \mathrm{Var}[\dot{Y}(t)] = -r''(0)$, $\lambda_4 = \mathrm{Var}[\ddot{Y}(t)] = r^{(4)}(0)$ we find according to (2.258) that

$$b(0) = \frac{1}{\lambda_4}\mathrm{Cov}[Y(0), \ddot{Y}(0)] = -\frac{\lambda_2}{\lambda_4} \tag{2.260}$$

and

$$\mathrm{Var}[R(0)] = \lambda_0 - \frac{\lambda_2^2}{\lambda_4} \tag{2.261}$$

Thus the local minimum value of the normalized process $Y(t)/\sqrt{\lambda_0}$ can be written as

$$\frac{Y_0(0)}{\sqrt{\lambda_0}} = -\frac{\lambda_2}{\sqrt{\lambda_0\lambda_4}}W + \sqrt{1 - \frac{\lambda_2^2}{\lambda_0\lambda_4}}\,U \tag{2.262}$$

where W is a standard Rayleigh variable, i.e., $f_W(x) = x\exp[-\frac{1}{2}x^2]$, $x > 0$, and U is standard Gaussian variable, i.e., $f_U(x) = \varphi(x)$, and where W and U are mutually independent. By changing $-$ to $+$, the analogous expression for the local maximum values is obtained. The well-known formula of Rice [40] for the long run density of the local maximum values follows by a convolution calculation to obtain the density of $\beta W + \sqrt{1-\beta^2}U$, $\beta = \lambda_2/\sqrt{\lambda_0\lambda_4}$. This probabilistic proof was proposed by J. de Maré [48].

The factor

$$\beta = \frac{\lambda_2}{\sqrt{\lambda_0\lambda_4}} \tag{2.263}$$

is called the regularity factor of $Y(t)$, and $\sqrt{1-\beta^2}$ is called the spectral width parameter of $Y(t)$. For $\beta \to 1$ the process $Y(t)$ is called a "narrow band" process, and the local maximal values of $Y(t)/\sqrt{\lambda_0}$ have asymptotically standard Rayleigh distribution. For $\beta \to 0$ the process is called a "wide band" process, and the local maximal values of $Y(t)/\sqrt{\lambda_0}$ have asymptotically standard normal distribution, that is, the same distribution as $Y(t)$. It is seen that if λ_2 has a finite limit as $\lambda_4 \to \infty$, then $\beta \to 0$. As we shall see later this gives rise to a paradox in the theory of random vibration.

Example In the previous example consider the linear regression of $Y(t)$ on $Y(0)$, $\dot{Y}(0)$, and $\ddot{Y}(0)$ followed by conditioning on an upcrossing of level $u = 0$ at $t = 0$ by the derivative process $\dot{Y}(t)$. Then the probabilistic behavior of the sample functions in the vicinities of the local minima is described by the Slepian model

$$Y(0) = a(t)0 + b(t)Z + c(t)Y + R(t) \tag{2.264}$$

where the coefficient functions are obtained from (2.255) as

$$\begin{aligned}&[a(t)\ b(t)\ c(t)] = \\ &[-r'(t) \quad r''(t) \quad r(t)]\begin{bmatrix}\lambda_2 & 0 & 0\\ 0 & \lambda_4 & -\lambda_2\\ 0 & -\lambda_2 & \lambda_0\end{bmatrix}^{-1}\end{aligned} \tag{2.265}$$

giving

$$b(t) = \frac{1}{\lambda_0\lambda_4 - \lambda_2^2}[\lambda_2 r(t) + \lambda_0 r''(t)] \tag{2.266}$$

$$c(t) = \frac{1}{\lambda_0\lambda_4 - \lambda_2^2}[\lambda_4 r(t) + \lambda_2 r''(t)] \tag{2.267}$$

These functions also follow from (2.222) after suitable interpretation of the symbols.

Let T be the time to the first maximum of $b(t)Z + c(t)Y$ after $t = 0$ and let

$$H = b(T)Z + [c(T) - 1]Y \tag{2.268}$$

Neglecting the contribution from the residual process $R(t)$, the random variable T is thus the time distance between an arbitrary local minimum and the following local maximum, and the random variable H is the increment from the minimum value to the maximum value. It can be shown that if there is no positive t for which $r'(t) = r'''(t) = 0$, then there is a one-to-one mapping between (Z, Y) and (T, H) that allows (Z, Y) to be explicitly expressed by (T, H) in terms of the functions $b(t)$ and $c(t)$ [49]. This makes it possible to calculate the probability density of (T, H) from the density of (Z, Y) as given by (2.255). The result seems to be the only known empirically verified closed form approximation to the joint density of wave length and amplitude that uses the entire covariance function $r(t)$ of the Gaussian process $Y(t)$ [49, 50].

Example If Y is put to a given value y in (2.264) the probability density of Z becomes conditional on $Y = y$. Thus we have from (2.255) that

$$f_Z(z \mid Y = y, C_h) \propto f_{Z,Y}(z, y \mid C_h) \propto$$

$$z\varphi\left(\frac{z}{\sqrt{\lambda_4}}\right)\varphi\left(\frac{y + z\lambda_2/\lambda_4}{\sqrt{\lambda_0 - \lambda_2^2/\lambda_4}}\right) \propto z\varphi\left(\frac{z + y\lambda_2/\lambda_0}{\sqrt{\lambda_4 - \lambda_2^2/\lambda_0}}\right), \quad z > 0 \tag{2.269}$$

The obtained Slepian model

$$Y_0(t) = b(t)Z + c(t)y + R(t) \tag{2.270}$$

is useful for simulating a sequence of consecutive local extremes of $Y(t)$. Starting by simulating a local minimum y from (2.262) an outcome z of Z is generated from the population defined by (2.269). The time τ to the first maximum of $b(t)y+c(t)z$ is calculated and an outcome r of the Gaussian residual $R(t)$ for $t = \tau$ is generated. Then $b(\tau)y + c(\tau)z + r$ may be taken as an approximation to the next maximum after the minimum y. Due to the symmetry of the Gaussian process $Y(t)$ the same procedure applies after obvious modification to simulate the next minimum after the obtained maximum, etc. An explicit expression for the density of the random time T from the minimum of given value y to the next maximum can be obtained [48]. Thus the outcome $T = \tau$ may be directly generated. The described simulation procedure is of interest in fatigue life studies assuming that the essential effect of random process stressing is determined by the sequence of local extremes rather than the detailed stress path between these extremes.

Example Warnings on a near future upcrossing of a critical level u by the process $X(t)$ may be given on the basis of the behavior of the sample trajectory of $\mathbf{Y}(t)$ observed up to the present time. An alarm policy can be determined on the basis of the Slepian model vector process for the behavior of $\mathbf{Y}(t)$ given that X has an upcrossing of the critical level u at a given time. On this basis there is a theory of optimal alarm policy [45, 51].

It is seen from these examples and the examples of the next chapter that Slepian vector model processes are well suited for insight giving reasoning concerning sample function behavior at level upcrossings. The reasoning is often approximate in particular with respect to the way the residual process is treated, of simply because the residual process in some applications is completely neglected. Therefore, it is important to be able to verify the results by other methods or by simulation. Simple procedures for simulating sample functions of Slepian model processes for both stationary and non-stationary Gaussian processes are described in the literature [52, 53]. Simulations corresponding to the non-stationary case may serve to test the sensitivity of the results to departure from the stationarity assumption underlying most reported applications of the Slepian model process concept.

2.16 Applications of Slepian Model Processes in Stochastic Mechanics

In stochastic mechanics there are several interesting problems that can be successfully analysed by Slepian model processes. In this chapter some examples concerning random vibrations of elastic and elasto-plastic oscillators will be treated in detail. Before this, some other useful examples of applications will be summarized.

Example Of relevance for the design conditions for vehicles aimed for travelling on rough roads there have been made probabilistic studies of the jumps and bumps of an idealized vehicle consisting of a single wheel of zero mass connected through a linear spring and a viscous damper to a concentrated mass [54]. Since the wheel can only be in compressive contact with the random road surface, it is a critical crossing event when the wheel loses contact and the mass starts a free flight until the wheel bumps into the road again. Modeling the road surface as a Gaussian field over which the one-wheel vehicle travels with constant speed, the Slepian model vector process is the obvious tool for such a study.

Example Close approximations to the distribution of the duration of a wide band Gaussian process visit to an interval can be obtained by use of Slepian process models [55]. This problem is equivalent to the problem of calculating the distribution of the time to first passage out of the interval. Its relevance in structural reliability is therefore obvious. A specific application concerns fatigue accumulation in elastic bars that may vibrate strongly within disjoint random time intervals due to the vortex shedding effect of the natural wind. Between these intervals only weak vibrations occur. This is an effect of the turbulent nature of the natural wind and the so-called "lock-in" phenomenon originating from the coupling between the movements of the bar and the surrounding velocity field. As a simplified engineering model it may be assumed that damped harmonic vibrations start up when the velocity in front of the bar crosses into some specified interval determined by a natural frequency of the bar and the mechanical damping. Further it may be assumed that the vibrations damp out when the velocity crosses out of the interval again. For this model it is of interest to be able to determine the distribution of the duration of an arbitrary visit of the wind velocity process to the critical interval. By use of the distribution and the mean number of incrossings to the critical interval, it is possible to calculate the expected accumulated damage during a given time under the adoption of a linear accumulation law such as the Palmgren-Miner rule [56, 57].

Example The level crossing behavior of envelope processes associated to narrow band random vibration response processes is of interest for the study of the long run distribution of the distances between the clumps of response process exceedances outside critical response levels.

Envelope excursions above a level u may take place without there being any upcrossings of the response process during the time of these excursions. Such envelope excursions are said to be empty. Otherwise they are said to be qualified [58]. For an ergodic narrow-band Gaussian process the distances between the clumps of response process exceedances are approximately the same as the distances between the qualified envelope excursions for any reasonably useful envelope definition. Thus it is relevant to estimate the long run fraction of qualified envelope excursions among all the envelope excursions [58].

Let $X(t)$ be an ergodic narrow-band Gaussian process and let $\tilde{X}(t)$ be its Hilbert transform simply defined by making a phase shift of $\pi/2$ in the spectral representation of $X(t)$. Then a convenient envelope process is defined as the norm $\sqrt{X(t)^2 + \tilde{X}(t)^2}$ of the vector process $(X(t), \tilde{X}(t))$ [2]. The excursions of the envelope above level u obviously become coincident with the excursions of the vector process outside the circle of radius u and centre at the origin (the level circle). Among these excursions the qualified excursions are those for which $X(t) > u$. Since the probabilistic structure of the vector process is invariant to a rotation of the co-ordinate system, it follows that the points of crossing out of the level circle are uniformly distributed on the circle. Therefore, the long run fraction of empty excursions can be approximately estimated by calculating the probability that a randomly chosen tangent to the level circle does not have points in common with the trajectory of the linear regression part of the Slepian model vector process of $(X(t), \tilde{X}(t))$ corresponding to an outcrossing of the level circle taking place at the point $(u, 0)$ to time $t = 0$.

For a given outcrossing velocity vector, this linear regression trajectory is a fixed curve that in polar coordinates can be conveniently approximated by the second degree Taylor expansion with respect to t of the radius vector $r(t)$ and of the polar angle $\theta(t)$. By use of this approximation it can be shown that the excursion trajectory asymptotically as $\lambda_1^2/(\lambda_0\lambda_2) \to 1$ ($\lambda_n = n$th spectral moment of $X(t)$, (2.223)) can be approximated by a simple translation of the level circle to fit tangentially to the outcrossing velocity vector. In this asymptotic limit it is thus a simple geometric consideration to obtain the aforementioned probability conditional on the outcrossing velocity vector. By conditioning considerations analogous to those in Section 2.14 it can be seen that the velocity vector $(\dot{X}(0), \dot{\hat{X}}(0))$ conditional on the outcrossing taking place through the time window $[0, dt]$ can be expressed by

$$\dot{X}(0) = W\sqrt{\lambda_2 - \lambda_1^2} \tag{2.271}$$

$$\dot{\hat{X}}(0) = U\sqrt{\lambda_2 - \lambda_1^2} + \lambda_1 u \tag{2.272}$$

where W is a standard Rayleigh variable and U is a standard normal variable. Finally, unconditioning leads to the long run fraction of qualified envelope excursion above level u ($\alpha = 1/2$), or outside the interval $[-u, u]$ ($\alpha = 1$), to be approximately [59, 60]

$$r = 1 - 2\int_0^u \varphi(\eta)\left[1 - \alpha\sqrt{2\pi}\,\frac{2\Phi\left(\gamma\pi\dfrac{u^2-\eta^2}{u}\right) - 1}{\gamma\pi\dfrac{u^2-\eta^2}{u}}\right] d\eta \tag{2.273}$$

in which $\gamma = \sqrt{(\lambda_0\lambda_2/\lambda_1)[(\lambda_0\lambda_2/\lambda_1) - 1]}$.

Of relevance for random vibrations of strain hardening elasto-plastic oscillators, these Slepian model process applications can be extended to excursions outside unsymmetric intervals $[-u_l, u_u]$, considering the possibility of having empty excursions simultaneously at both ends of the interval or only at the one or the other end of the interval [60].

Example The equation of motion of a linear single degree of freedom oscillator can be written on dimensionless form as

$$\ddot{X}(\tau) + 2\alpha\dot{X}(\tau) + (1 + \alpha^2)X(\tau) = N(\tau) \tag{2.274}$$

such that the stationary dimensionless response process $X(\tau)$ has zero mean, unit standard deviation, and correlation function

$$r(\tau) = e^{-\alpha|\tau|}(\cos\tau + \alpha\sin|\tau|) \tag{2.275}$$

under stationary white noise excitation $N(\tau)$ of intensity $4\alpha(1+\alpha^2)$, that is, $\mathrm{Cov}[N(\tau_1), N(\tau_2)] = 4\alpha(1+\alpha^2)\delta(\tau_1-\tau_2)$, where $\delta(\cdot)$ is Dirac's delta function. (Standard formulation: $\ddot{x} + 2\zeta\omega_0\dot{x} + \omega_0^2 x = f(t)/m$, x = displacement, t = time, m = mass, ζ = damping ratio ($0 \le \zeta \le 1$), ω_0 = natural frequency for $\zeta = 0$, $f(t)$ = stationary white noise of intensity πS. Dimensionless variables: $\tau = \omega_0\sqrt{1-\zeta^2}t$, $X(\tau) = x(t)/\sigma$, $\sigma^2 = \mathrm{Var}[X(t)] = \pi S/(4\zeta\omega_0^3 m^2)$ for $x(t)$ stationary, $\alpha = \zeta/\sqrt{1-\zeta^2}$, $N(\tau) = f(t)/[(1-\zeta^2)m\omega_0^2\sigma]$).

The spectral increment of the response process $X(\tau)$ is obtained directly by operating on the equation of motion (2.268) or by Fourier transformation of $r(\tau)$ given by (2.269) as

$$dG(\omega) = \frac{dF(\omega)}{[\omega^2 - (1-\alpha^2)]^2 + 4\alpha^2} \tag{2.276}$$

where $dF(\omega)$ is the spectral increment of the excitation process $H(\tau)$, that is,

$$dF(\omega) = \frac{4}{\pi}\alpha(1+\alpha^2)d\omega \tag{2.277}$$

It follows from (2.275) (or from (2.276) using (2.223)) that $\lambda_0 = r(0) = 1$, $\lambda_2 = -r''(0) = 1+\alpha^2$, and $\lambda_n = \infty$ for $n \ge 3$.

According to (2.254) and (2.258) (or (2.221)) the Slepian model process $X_u(\tau)$ for the response $X(\tau)$ in the vicinity of an upcrossing of the level u at $\tau = 0$ becomes

$$\begin{aligned} X_u(\tau) &= r(\tau)u - r'(\tau)Z/\lambda_2 + R_1(\tau) = \\ &[(\cos\tau + \alpha\sin|\tau|)u + Z\sin\tau]e^{-\alpha|\tau|} + R_1(\tau) \end{aligned} \tag{2.278}$$

Substitution of (2.275) in (2.228) leads to

$$\begin{aligned} &\mathrm{Cov}[R_1(\tau_1), R_1(\tau_2)] = e^{-\alpha|\tau_1-\tau_2|}[\cos(\tau_1-\tau_2) + \alpha\sin(|\tau_1-\tau_2|)] \\ &-e^{-\alpha(|\tau_1|+|\tau_2|)}[(1+\alpha^2)\cos(\tau_1-\tau_2) + \alpha\sin(|\tau_1|+|\tau_2|) \\ &-\alpha^2\cos(|\tau_1|+|\tau_2|)] \end{aligned} \tag{2.279}$$

By letting $\lambda_4 \to \infty$ the Slepian model process (2.264) can be used herein to obtain the Slepian model process $X_{min}(\tau)$ for the response $X(\tau)$ in the vicinity of a local minimum at $\tau = 0$. Since $b(\tau) \to 0$, (2.266), and $c(\tau) \to r(\tau)$, (2.267), for $\lambda_4 \to \infty$, we get the limit Slepian model process

$$\left.\begin{matrix} X_{min}(\tau) \\ X_{max}(\tau) \end{matrix}\right\} = X(0)r(\tau) + R_2(\tau) \tag{2.280}$$

that since Z disappears from the Slepian model for $\lambda_4 \to \infty$ also is the Slepian model $X_{max}(\tau)$ for the response $X(\tau)$ in the vicinity of a local maximum at $\tau = 0$. The absence of Z in the limit $\lambda_4 \to \infty$ is related to the fact that the derivative process $\dot{X}(\tau)$ is not differentiable. The covariance function of $R_2(\tau)$ is identical to that of $R_1(\tau)$ given by (2.279).

From the first example in the previous paragraph, we have the paradoxical result that the local maximal values and the local minimal values have normal distribution identical to the one-dimensional distribution of $X(\tau)$ itself. This result is obtained in spite of the fact that the sample functions of $X(\tau)$ for small damping ($\alpha << 1$) appear to have narrow band nature such as described by the Slepian models (2.278) or (2.280) (and as it is also seen from the spectrum (2.276)). Nevertheless, since $\beta \to 0$ for $\lambda_4 \to \infty$, the process is categorized to be a wide band process.

This paradox can be cleared up by the following consideration. According to the normality of X_{max}, negative values of X_{max} are just as probable as positive values. Assume that $-y < 0$ is given as such a local maximum value. Then the Slepian model (2.280) with $X(0) = -y$ shows that the expectation of $X_{max}(\tau)$ has a local minimum for $\tau = 0$, that is, it curves upwards while the realization of the process itself curves downwards in the vicinity of this point. Since the residual standard deviation as obtained from (2.279) for $|\tau| < \pi$

$$D[R_2(\tau)] = \sqrt{1 - e^{-2\alpha|\tau|}} \tag{2.281}$$

is small for $\alpha << 1$, it follows that there is a large probability that there are minima next to the maximum at $-y$ at almost the same value level. Thus the local maxima and minima correspond to small buckles on the sample function. If these buckles are imagined to be smoothed out, the sample function by and large behaves as the trajectory of an oscillatory motion between crests and troughs with period 2π and slowly varying amplitude and phase.

The following argument shows that this amplitude of the smoothed realization has a distribution that asymptotically for $\alpha \to 0$ is a Rayleigh distribution. Consider the Slepian model process $X_u(\tau)$ defined by (2.278) for any level $u > 0$. For simplicity of the argument, assume that α is sufficiently small to be neglected at least for τ varying from the upcrossing of level u at $\tau = 0$ until the mean of $X_u(\tau)$ reaches its maximum. Moreover, assume that the residual process $R_1(\tau)$ can be neglected in this time interval. Then (2.278) becomes

$$X_u(\tau) = u \cos \tau + Z \sin \tau, \quad \tau \geq 0 \tag{2.282}$$

for which the first local maximum is at the smallest value of τ for which $\tan \tau = Z/u$ and the corresponding maximum value is

$$M = u\sqrt{1 + \left(\frac{Z}{u}\right)^2} \tag{2.283}$$

This value is denoted as a crest value. Since Z has a standard Rayleigh distribution asymptotically for $\alpha \to 0$ we have

$$P(M > x \mid M > u) = P(Z > \sqrt{x^2 - u^2}) = e^{-(1/2)x^2}/e^{-(1/2)u^2} \qquad (2.284)$$

which shows that the conditional distribution of M given $M > u$ is a standard Rayleigh distribution truncated at level u.

The fact that the mean function of the Slepian model process $X_{max}(\tau)$ is symmetric in τ and that the residual standard deviation is small in the vicinity of $\tau = 0$ shows that the small buckles must predominantly be located at the troughs and the crests of the sample curve.

Consistency between the normal distribution of the local maxima and the Rayleigh distribution of the crest values is then only achieved if the random number of local maxima or local minima (buckles) at a crest or a trough are relatively distributed in the proportion $1/x$ where x is the amplitude.

Example Assume that the linear oscillator considered in the previous example is made nonlinear by the introduction of symmetric elasticity limits at dimensionless elastic displacement levels $u > 0$ and $-u$, and that the yielding beyond these limits is ideal plastic. (In terms of the variables of the standard formulation we have $u = 2x^* m\omega_0\sqrt{\zeta\omega_0}/\sqrt{\pi S}$, where x^* is the elastic displacement limit of the physical oscillator.) At any time τ measuring the response relative to the current value of the accumulated plastic displacement, we then have the dimensionless equation of motion for the symmetric elasto-plastic oscillator (EPO) as the following modification of (2.274):

$$\begin{array}{ll} \ddot{X}(\tau) + (1+\alpha^2)u = N(\tau) & \text{for } X(\tau) > u \\ \ddot{X}(\tau) + 2\alpha\dot{X}(\tau) + (1+\alpha^2)X(\tau) = N(\tau) & \text{for } -u \le X(\tau) \le u \\ \ddot{X}(\tau) - (1+\alpha^2)u = N(\tau) & \text{for } X(\tau) < u \end{array} \qquad (2.285)$$

The viscous damping is not considered to be important in the plastic domain, and for simplicity it is neglected in (2.285). The oscillator with (2.274) as equation of motion is denoted as the associated linear oscillator (ALO) to the EPO.

If the elasticity limit u is large as compared to 1, the stationary response of the EPO will most of the time behave quite similar to the stationary response of the ALO, that is, almost as a stationary Gaussian process of zero mean, and covariance function (2.275). However, once in a while the EPO response will cross the elasticity limit u. This results in a transient behavior of bouncing between the upper and the lower plastic domain a random number of times before the response returns for a shorter or longer time to the elastic domain.

Since this shorter or longer time of elastic vibrations must start with zero velocity either at the displacement $-u$ or at the displacement u relative to the current plastic displacement, it behaves like a non-stationary response to the ALO as long as it stays within the elasticity limits. More precisely this can be expressed as follows: Given that the EPO response is within the elasticity limits in a time interval of length at least T after the start of the interval, then the probabilistic structure of the EPO response is in the interval up to T identical to the conditional probabilistic structure of the ALO response with the same initial conditions, given that the ALO response is within the elasticity limits for a longer time than T.

The result of the bouncing of the EPO response between the upper and the lower plastic domain is a net plastic displacement increment

$$\pm(\Delta_1 - \Delta_2 + \Delta_3 - ... + (-1)^N \Delta_{N+1}) \tag{2.286}$$

where Δ_i is the absolute value of the displacement increment from the ith visit in the same clump to the plastic domain. The sign $+/-$ is used if the first visit is to the upper/lower plastic domain. The first displacement increment Δ_1 in the clump has as a random variable another distribution than the following mutually independent and identically distributed random variables Δ_2, Δ_3.... This follows from the fact that $\Delta_2, \Delta_3, ...$ all come from an initial start of zero velocity at the elasticity limit, which is not the case for Δ_1. The number of terms in the sum (2.286) is determined by a random variable N that has the geometric distribution

$$P(N = n) = (1 - p)p^n, \quad n = 0, 1, ... \tag{2.287}$$

in which p is the probability that the EPO response moves directly from zero velocity at the elasticity limit u to the opposite plastic domain beyond $-u$. Thus N is the random "waiting time" until a plastic displacement clump is completed.

In the following it will be shown that Slepian model process considerations applied to the ALO response can be used to obtain closed form approximate expressions for the distributions of Δ_i for $i = 1$ and $i > 1$ and also to obtain the probability p in equation (2.287). Closed form approximations to the distribution of the accumulated net plastic displacement increment (2.286) and the accumulated absolute displacement increment

$$\Delta_1 + \Delta_2 + ... + \Delta_{N+1} \tag{2.288}$$

from a clump can be obtained on the basis of these distributions [61].

The Slepian process model $X_{max}(\tau)$ defined in (2.280) gives the conditional probability

$$P[X(-\pi) > -u \mid X(0) = z] = P[R_2(-\pi) - \mu z > -u] = \Phi\left(\frac{u - \mu z}{\sigma}\right) \tag{2.289}$$

in which

$$\mu = e^{-\alpha\pi}, \quad \sigma = \sqrt{1-\mu^2} \tag{2.290}$$

according to (2.275) and (2.281).

Identifying $X(0)$ with the standard Rayleigh variable M and applying (2.289) in Bayes' formula then gives the conditional density

$$f_M(z \mid X(-\pi) > -u) \propto P[X(-\pi) > -u \mid X(0) = z] f_M(z) =$$
$$\Phi\left(\frac{u - \mu z}{\sigma}\right) z e^{-1/2z^2}, \quad z > 0 \tag{2.291}$$

This density function is directly interpreted as the density function of the crest values sampled along a realization of the stationary ALO response but censored to include only those crest values that half a period π before the crest are accompanied by a value of the realization above level $-u$. Taking this censoring rule supplemented with lower truncation at level u as a characterization of the beginning of an outcrossing clump that starts out at level u, (2.291) gives for $z > u$ the shape of the density function of the first crest value in a clump given that the clump starts at the upper level u.

The density function (2.291) is for $z > u$ also approximately the density function of the first crest value in the first clump of crest values above level u of the transient response that corresponds to the initial condition of zero velocity and displacements equal to $-u$ or u with probability 1/2 on each possibility. This is a consequence of having slowly varying amplitude over the period 2π because then the last crest/trough value in a clump of crest/trough values above/below level $u/-u$ is not much different from the initial value $u/-u$. Thus the density function (2.291) can be applied even for small values of the yield limit to determine a good approximation to the distribution of the excess energy $(M^2 - u^2)/2$ of the ALO available for the EPO to be dissipated into plastic work after the first yield limit crossing in a clump of crossings. Assuming that the plastic work $u\Delta_1$ is equal to the excess energy, the plastic displacement becomes

$$\Delta_1 = \frac{1}{2u}(M^2 - u^2) \tag{2.292}$$

For white noise excitation the energy balance assumption (2.292) is a close approximation to what is obtained by solving the part of the differential equation (2.285) that corresponds to movement in the plastic domain and with the proper initial conditions (matching conditions between ALO and EPO) at yield limit impact [62]. Then from (2.291) and (2.292) we get the approximating density function

$$\begin{aligned} f_{\Delta_1}(x) &= f_M(\sqrt{u^2+2ux})\frac{u}{\sqrt{u^2+2ux}} \\ &\propto \Phi\left[\frac{1}{\sigma}(u-\mu\sqrt{u^2+2ux})\right]e^{-ux}, \quad x>0 \end{aligned} \tag{2.293}$$

The normalizing constant to be multiplied to the right side of (2.293) is obtained by integration by parts to be $u/[1-(1+\mu)\Phi(-\gamma u)]$ in which

$$\gamma = \frac{1-\mu}{\sigma} = \sqrt{\frac{1-\mu}{1+\mu}}, \quad \mu = \frac{1-\gamma^2}{1+\gamma^2} \tag{2.294}$$

The distribution of the plastic displacement that results from start of the EPO with zero velocity at the level $-u$, given that it passes through the yield level at u in the first cycle, is obtained by changing the conditioning in (2.291) from $X(-\pi) > -u$ to $X(-\pi) = -u$. This simply implies that (2.291) is replaced by

$$\begin{aligned} f_M(z \mid X(-\pi) = -u) &\propto \varphi\left(\frac{u-\mu z}{\sigma}\right) z e^{-(1/2)z^2} \\ &\propto z\varphi\left(\frac{z-\mu u}{\sigma}\right), \quad z>0 \end{aligned} \tag{2.295}$$

which in the same way as in (2.293) gives the common density function of $\Delta_2, \Delta_3, \ldots$ as

$$f_{\Delta_2}(x) \propto \varphi\left[\frac{1}{\sigma}(\sqrt{u^2+2ux}-\mu u)\right], \quad x>0 \tag{2.296}$$

The normalizing constant to be multiplied to the right hand side of equation (2.296) is $u/[\sigma^2\varphi(\gamma u) + \sigma\mu u\Phi(-\gamma u)]$.

The probability $p(u)$ that exceedance of level u actually occurs within the first period is directly obtained from the complementary distribution function of M given $X(-\pi) = -u$. According to (2.295) we have

$$1 - F_M[x \mid X(-\pi) = u] = \frac{\displaystyle\int_x^\infty z\varphi\left(\frac{z-\mu u}{\sigma}\right) dz}{\displaystyle\int_0^\infty z\varphi\left(\frac{z-\mu u}{\sigma}\right) dz} = \frac{\sigma\varphi\left(\frac{x-\mu u}{\sigma}\right) + \mu u \Phi\left(-\frac{x-\mu u}{\sigma}\right)}{\sigma\varphi\left(\frac{\mu u}{\sigma}\right) + \mu u \Phi\left(\frac{\mu u}{\sigma}\right)} \tag{2.297}$$

which for $x = u$ gives

$$p(u) = \frac{2\gamma\varphi(\gamma u) + (1-\gamma^2)u\Phi(-\gamma u)}{2\gamma\varphi\left(\frac{1-\gamma^2}{2\gamma}u\right) + (1-\gamma^2)u\Phi\left(\frac{1-\gamma^2}{2\gamma}u\right)} \tag{2.298}$$

It is seen that $p(u) \to \Phi(-\gamma u)$ asymptotically as $u \to \infty$.

These Slepian model based results have been successfully verified to be very good approximations to the empirical distributions obtained by simulation of the response of the EPO [61]. Moreover, there is a very convincing approximate numerical coincidence of the density (2.296) of Δ_2 with a density of a completely different analytical appearance obtained by solving a diffusion equation (Fokker-Planck equation) for the slowly varying amplitude formulated by the method of stochastic averaging [63].

2.17 Conclusions and References

This chapter mainly deals with dimension reductions and field discretizations based directly on the concept of linear regression. Several examples of interesting applications in stochastic mechanics are also given.

References

[1] Kolmogorov, A.N., Grundbegriffe der Wahrscheinlichkeitsrechnung, *Erg. Mat.* 2(3), 1933.

[2] Cramer, H. and Leadbetter, M.R., *Stationary and Related Stochastic Processes*, Wiley, New York, 1967.

[3] Journel, A.B. and Huijbregts, C.J., *Mining Geostatistics*, Academic Press, New York, 1978.

[4] Krige, D.G., Two-dimensional Weighted Moving Averaging Trend Surfaces for Ore Evaluation. *Proc. Symposium on Mathematical Statistics and Computer Applications for Ore Evaluation*, Johannesburg, S.A., 1966, pp. 13–38.

[5] Matheron, G., Principles of Geostatistics, *Economic Geology*, Vol. 58, 1962, pp. 1246–1266.

[6] Matheron, G., *Estimating and Choosing* (transl. from French by A.M. Hasofer), Springer Verlag, Berlin-Heidelberg, 1989.

[7] Loeve, M., *Probability Theory*, Van Nostrand, Princeton, N.J., 3rd ed., 1963.

[8] Lawrence, M., Basis Random Variables in Finite Element Analysis. *Int. J. of Numerical Methods in Engineering*, 24(10), 1987, pp. 1849–1863.

[9] Spanos, P.D. and Ghanem, R.G., Stochastic Finite Element Expansion for Random Media. *Journal of Engineering Mechanics*, ASCE, 115(5), 1989, pp. 1035–1053.

[10] Ghanem, R.G. and Spanos, P.D., Spectral Stochastic Finite-element Formulation for Reliability Analysis. *Journal of Engineering Mechanics*, ASCE, 117(10), 1991, pp. 2351–2372.

[11] Ghanem, R.G. and Spanos, P.D., *Stochastic Finite Elements: a Spectral Approach*, Springer Verlag, New York, N.Y., 1991.

[12] Li, C.-C. and Der Kiureghian, A., Optimal Discretization of Random Fields. *Journal of Engineering Mechanics*, ASCE, 119(b), 1993, pp. 1136–1154.

[13] Ditlevsen, O. and Munch-Andersen, J., Empirical Stochastic Silo Load Model. Part 1: Correlation Theory. *Journal of Engineering Mechanics*, ASCE, 121(9), 1995, pp. 973–980.

[14] Roberts, J.B. and Spanos, P.D., *Random Vibration and Statistical Linearization*, Wiley, Chichester, U.K., 1990.

[15] Sondhi, M.M., Random Processes with Specified Spectral Density and First-order Probability Density. *The Bell System Technical Journal*, 62(3), 1983, pp. 679–701.

[16] Der Kiureghian, A. and Liu, P.-L., Structural Reliability under Incomplete Probability Information, *Journal of Engineering Mechanics*. ASCE, 112, 1986, pp. 857–865.

[17] Ditlevsen, O., Christensen, C., and Randrup-Thomsen, S., Reliability of Silo Ring under Lognormal Stochastic Pressure using Stochastic Interpolation. *Proc. of IUTAM Symposium: Probabilistic Structural Mechanics: Advances in Structural Reliability Methods*, San Antonio, Texas, June 1993. (Eds.: P.D. Spanos and Y.-T. Wu), Springer Verlag, 1994, pp. 134–162.

[18] Ditlevsen, O., Discretization of Random Fields in Beam Theory. *Proc. of ICASP7: Applications of Statistics and Probability. Civil Engineering Reliability and Risk Analysis* (Eds.: Lemaire, M., Favre, J.-L., and Mebarki, A.). Balkema, Rotterdam, 1995, pp. 983–990.

[19] Ditlevsen, O., Christensen, C., and Randrup-Thomsen, S., Empirical Stochastic Silo Load Model. Part 3: Reliability Applications. *Journal of Engineering Mechanics*, ASCE, 121(9), 1995, pp. 987–993.

[20] Der Kiureghian, A. and Ke, J-B., The Stochastic Finite Element Method in Structural Reliability. *Probabilistic Engineering Mechanics*, 3(2), 1988, pp. 83–91.

[21] Vanmarcke, E.H. and Grigoriu, M., Stochastic Finite Element Analysis of Simple Beams. *Journal of Engineering Mechanics*, ASCE, 109(5), 1983, pp. 1203–1214.

[22] Liu, W.K., Mani, A., and Belytschko, T., Random Field Finite Elements. *Int. J. Numerical Methods in Engineering*, 23(10), 1986, pp. 1831–1845.

[23] Grigoriu, M., Crossing of Non-Gaussian Translation Processes, *Journal of Engineering Mechanics*, ASCE, 110(4), 1984, pp. 610–620.

[24] Takada, T., Weighted Integral Method in Multi-dimensional Stochastic Finite Element Analysis. *Probabilistic Engineering Mechanics*, 5(4), 1990, pp. 158–166.

[25] Deodatis, G. and Shinozuka, M., Weighted Integral Method I: Stochastic Stiffness matrix, *Journal of Engineering Mechanics*. ASCE, 117, 1991, pp. 1851–1864.

[26] Zhang, J. and Ellingwood, B., Orthogonal Series Expansion of Random Fields in First-order Reliability Analysis. *Journal of Engineering Mechanics*, ASCE, 120(12), 1994, pp. 2660–2677.

[27] Christensen, C.F. and Ditlevsen, O., Lock-in Interval Measurements and Fabigne Calculations of Tubular Structural Elements Subject to Wind, *Journal of Wind Engineering and Industrial Aerodynamics,* submitted, 1995.

[28] Bucher, C.G. and Brenner, C.E., Stochastic Response of Uncertain Systems. *Archive of Appl. Mech.*, 62, 1992, pp. 507–516.

[29] Winterstein, S.R., Nonlinear Vibration Models for Extremes and Fatigue, *Journal of Engineering Mechanics*, ASCE, 114, 1988, pp. 1772–1790.

[30] Ditlevsen, O. and Madsen, H.O., *Structural Reliability Methods*, Wiley, 1996.

[31] Ditlevsen, O., Mohr, G., and Hoffmeyer, P., Integration of Non-Gaussian Fields. *Probabilistic Engineering Mechanics*, 1995.

[32] Mohr, G. and Ditlevsen, O., Partial Summations of Stationary Sequences of Non-Gaussian Random Variables. *Probabilistic Engineering Mechanics*, 1995.

[33] Keaveny, J.M., Nadim, F., and Lacasse, S., Autocorrelation Functions for Offshore Geotechnical Data, in *Structural Safety & Reliability*, Vol. 1 (Ed. Ang, A.H.-S., Shinozuka, M., and Schuëller, G.I.), Proceedings of ICOSSAR'89, the 5th International Conference on Structural Safety and Reliability, San Francisco, August 7-11, 1989. ASCE, New York, 1990, pp. 263–270.

[34] Kameda, H. and Morikawa, H., An Interpolating Stochastic Process for Simulation of Conditional Random Fields. *Probabilistic Engineering Mechanics*, 7(4), 1992, pp. 243-254.

[35] Ditlevsen, O., Random Field Interpolation between Point by Point Measured Properties. *Computational Stochastic Mechanics* (P.D. Spanos, C.A. Brebbia, Eds.), Computational Mechanics Publications, Southampton, Boston, Elsevier Applied Science, London, New York, 1991, pp. 801–812.

[36] Munch-Andersen, J., Ditlevsen, O., Christensen, C., Randrup-Thomsen, S., and Hoffmeyer, P., Empirical Stochastic Silo Load Model. Part 2: Data Analysis. *Journal of Engineering Mechanics*, ASCE, 121(9), 1995, pp. 981–986.

[37] Ditlevsen, O. and Gluver, H., Parameter Estimation and Statistical Uncertainty in Random Field Representations of Soil Strength. *Proc. CERRA-ICASP6: Sixth international conference on applications of statistics and*

probability in civil engineering (L. Esteva, S.E. Ruiz, Eds.), Institute of Engineering, UNAM, Mexico City, 1991, pp. 691–704.

[38] Ditlevsen, O., Measuring Uncertainty Correction by Pairing. *Proceedings of ICOSSAR'93 - The 6th International Conference on Structural Safety & Reliability*, (Eds.: Schuëller, Shinozuka & Yao), Balkema, Rotterdam, 1994, pp. 1977-1983.

[39] Ditlevsen, O., *Uncertainty Modeling*. McGraw-Hill, New York, 1981.

[40] Rice, S.O., Mathematical Analysis of Random Noise, *The Bell System Technical Journal*, 23, 1944, pp. 282–332, and 24, 1944, pp. 46-156. Reprinted in Selected Papers in Noise and Stochastic Processes (Ed.: N. Wax), Dover, New York, 1954.

[41] Leadbetter, M.R., Lindgren, G., and Rootzén, H., *Extremes and Related Properties of Random Sequences and Processes*, Springer-Verlag, New York, 1983.

[42] Kac, M., and Slepian, D., Large Excursions of Gaussian Processes. *Annals of Mathematical Statistics*, 30, , 1959, pp. 1215–1228.

[43] Slepian, D., On the Zeros of Gaussian Noise, in *Time Series Analysis* (Ed.: M. Rosenblatt), Wiley, New York, 1962, pp. 143–149.

[44] Lindgren, G., Functional Limits of Empirical Distributions in Crossing Theory, *Stochastic Process. Appl.*, 5, 1977, pp. 143–149.

[45] Lindgren, G., Use and Structure of Slepian Model Processes for Prediction and Detection in Crossing and Extreme Value Theory. CODEN: LUNFD6/(NFMS-3079)/1-30/(1983), Univ. of Lund, Dept. of Math. Sta., Box 725, S-220 07 Lund, Sweden, 1983.

[46] Lindgren, G., Local Maxima of Gaussian Fields. *Ark. Mat.*, 10, 1972, pp. 195–218.

[47] Wilson, R.J. and Adler, R.J., The Structure of Gaussian Fields Near a Level Crossing. *Adv. Appl. Probability*, 14, 1982, pp. 543–565.

[48] Lindgren, G., On the Use of Effect Spectrum to Determine a Fatigue Life Amplitude Spectrum. ITM-Symposium on Stochastic Mechanics, Lund, 1983, CODEN: LUTFD2/(TFMS-3031)/1-169/(1983), Univ. of Lund, Dept. of Math. Stat., Box 725, S-220 07 Lund, Sweden, 1983.

[49] Lindgren, G. and Rychlik, I., Wave Characteristic Distributions for Gaussian Waves - Wave-length, Amplitude and Steepness. *Ocean Engineering* 9, 1982, pp. 411–432.

[50] Rychlik, I., Regression Approximations of Wave Length and Amplitude Distributions. *Advances in Applied Probability*, 19, 1987, pp. 396–430.

[51] Lindgren, G., Model Processes in Nonlinear Prediction with Application to Detection and Alarm. *Ann. Probab.* 8, 1980, pp. 775–792.

[52] Hasofer, A.M., Slepian Process of a Non-stationary Process. *Probabilistic Mechanics and Structural and Geotechnical Reliability* (Ed.: Y.K. Lin), ASCE, 1992, pp. 296–299.

[53] Hasofer, A.M., Representation for the Slepian Process with Applications. *Journal of Sound and Vibration*, 124(3), 1988, pp. 435–441.

[54] Lindgren, G., Jumps and Bumps on Random Roads. *Journal of Sound and Vibration*, 78, 1981, pp. 383–395.

[55] Ditlevsen, O., Duration of Visit to Critical Set by Gaussian Process. *Probabilistic Engineering Mechanics*, 1(2), 1986, pp. 82–92.

[56] Ditlevsen, O., Non-Gaussian Vortex Induced Aeroelastic Vibrations under Gaussian Wind. A Slepian Model Approach to "Lock In". *Probabilistic Mechanics and Structural and Geotechnical Reliability* (Ed.: Y.K. Lin), ASCE, 1992, pp. 292–295.

[57] Christensen, C.F. and Ditlevsen, O., Fatigue Damage from Random Vibration Pulse Process of Tubular Structural Elements subject to Wind. *Proc. Third International Conference on Stochastic Structural Dynamics*. San Juan, Puerto Rico, Jan. 1995.

[58] Vanmarcke, E.H., On the Distribution of the First-passage Time for Normal Stationary Random Processes. *American Society of Mechanical Engineers, Journal of Applied Mechanics*, 42, 1975, pp. 215–220.

[59] Ditlevsen, O. and Lindgren, G., Empty Envelope Excursions in Stationary Gaussian Processes. *Journal of Sound and Vibration*, 112, 1988, pp. 571–587.

[60] Ditlevsen, O., Qualified Envelope Excursions out of an Interval for Stationary Narrow-band Gaussian Processes. *Journal of Sound and Vibration*, 173(3), 1994, pp. 309–327.

[61] Ditlevsen, O. and Bognár, L., Plastic Displacement Distributions of the Gaussian White Noise Excited Elasto-plastic Oscillator. *Probabilistic Engineering Mechanics*, 8, 1993, pp. 209–231.

[62] Ditlevsen, O., Elasto-plastic Oscillator with Gaussian Excitation. *Journal of Engineering Mechanics*, ASCE, 112(4), 1986, pp. 386–406.

[63] Roberts, J.B., The Yielding Behaviour of a Randomly Excited Elasto-plastic Structure. *Journal of Sound and Vibration*, 72, 1980, pp. 71–85.

3

Reliability of Randomly Excited Hysteretic Systems

J.B. Roberts
University of Sussex, Brighton, U.K.

3.1 Introduction

In designing systems to withstand random environmental loading, such as that arising from earthquakes, the wind, or ocean waves, it is often vitally important to take full account of the effect of any non-linearity which is present. Many studies have shown that non-linearity can have a profound influence on the probability distribution of the response and hence on the probability of failure. A linear analysis may, at best, lead to an excessively conservative design. At worst it can lead to the design of a system which is liable to catastrophic failure.

Hysteretic behavior is a common source of non-linearity, in many structural and mechanical systems. Despite its apparently complex nature, which involves non-linearity with respect to both energy dissipation and restoring force, it is often possible to model hysteresis in terms of a single, first order differential equation [1]–[3]. This can be combined with the normal system equations to yield an extended set of equations. The problem of predicting the probability distribution of the response, and hence the system reliability, can then be tackled by using one of a number of well-established analytical methods in the field of non-linear random vibration.

The most popular technique for dealing with hysteretic systems is that of statistical linearisation [4]. This method was first applied to a mechanical hysteretic system excited by random excitation by Caughey [5] in a pioneering study.

Subsequently, with differential models of hysteresis, many researchers have studied the response of systems with various kinds of hysteresis, using statistical linearisation (e.g., see [5]–[21]). As demonstrated by numerous comparisons with digital simulation results, statistical linearisation usually yields good estimates of mean square statistics. However, it is not a useful method for dealing with the probability distribution of the response, and hence with reliability estimation. This is because the minimisation of the mean square of the equation error, which is at the core of the method, requires the assumption that the response is Gaussian. This is, of course, an approximation which becomes increasingly less accurate as the degree of non-linearity increases. Thus the standard statistical linearisation technique cannot give information on the departure from Gaussianity, due to non-linearity. As indicated above, failure to account for such departures, which can be very significant in the "tails" of the distribution, can have serious consequences.

A number of attempts have been made to improve the statistical linearisation method, with respect to reliability estimation, by using other minimisations. For example, Naess [22] has proposed a minimisation with response to higher order moments of the equation error. Comparison with respect to simulation studies, in some specific cases, shows that improved estimation of reliability estimates can be achieved. However, a difficulty with this approach is to establish that such modifications will always lead to improvements.

A natural extension of the statistical linearisation approach is the moment closure method [4]. The basic idea here is to generate a set of moment statistics for the response and then to use these moments to provide an estimate of the response probability distribution. A well-known difficulty here is that, if the system is non-linear, a closed (i.e., soluble) set of moment equations cannot be found. Thus it is necessary to invoke some closure approximation. The simplest such approximation is to assume a Gaussian response; then higher order moments (order greater than two) can be related to second order moments, using standard relationships for Gaussian processes. This will normally lead to results which are identical to those found by the statistical linearisation method [4]. It is natural to seek improvements to this by closing the moment equations at a higher level. This is often done by assuming that cumulants of the response above a certain level are zero; cumulant relationships can then be employed to achieve closure [4]. An alternative approach, which can be equivalent to cumulant closure, is to assume a parametric form for the distribution of the response. On the basis of this form all moments can be evaluated in terms of the parameters and thus a soluble set of equations can be generated. Examples of applications of the moment closure to hysteretic systems can be found in [23]–[32]. Such work has revealed the usual problems associated with this approach. Principal amongst these is the difficulty of obtaining stable estimation in the tails of the response probability density function. For example, it is quite possible to obtain negative estimates of this function, at high amplitudes. Another disadvantage of the method is its shear complexity, a feature which casts doubt on its value as a design tool.

A completely different approach is to employ the theory of continuous Markov processes. If the excitation is sufficiently broad-band it is possible to model the response as a multi-dimensional Markov process. For example, the response of an oscillator with a hysteretic restoring force, driven by broad-band random excitation, is approximately a three-dimensional Markov process, the appropriate variables being the displacement and velocity of the response, together with an auxiliary variable arising from the differential equation representation of the restoring force. On the basis of this Markov model, which is essentially a continuous diffusion process, one can formulate governing diffusion equations for the evolution in time of the probability distribution of the response. These can be either of the forward (Fokker-Planck-Kolmogorov — FPK) or backward type, the latter being an adjoint form of the former [33]–[35]. Being three-dimensional in character these equations are not easy to solve, even in the special case where the excitation is stationary and only a stationary response distribution is sought. In fact, analytical solutions are not available and recourse must be made to approximations [33] or to the use of numerical solution techniques [34, 35]. The computation effort required to obtain solutions of FPK equations (or their adjoint form) increases very sharply with increase in dimension. Thus even three-dimensional equations require considerable effort. This fact effectively limits the applicability of this approach to single degree of freedom systems.

If the response bandwidth is significantly smaller than the excitation bandwidth (this generally occurs when the rate of energy dissipation is small and the excitation is broad-band) then the standard stochastic averaging method [36] can be applied, as a means of reducing the number of dimensions required when constructing a Markov model of the response. This reduction leads to drastic simplifications and thus has a considerable attraction. For example, in the case of an oscillator with a hysteretic restoring force it is possible, by means of stochastic averaging to obtain a one-dimensional FPK equation for the amplitude of the response [37]–[40]. A solution to this equation can easily be obtained numerically and, moreover, enables very useful simple asymptotic results to be found.

Generalised versions of stochastic averaging can also be applied [41]–[47]. These are normally based on a consideration of the energy envelope of the response and may also allow, at least to some extent, for the shape of the input spectrum [45]–[47]. Again, in the case of oscillators with a hysteretic restoring force, these lead to a one-dimensional FPK equation for the response (here the energy envelope). Although the use of the energy envelope does, at first sight, take better account of the non-linearity of a hysteretic restoring force, with respect to stiffness, an in-depth analysis shows that, in many cases, it will not lead to an improvement over results obtainable by using standard stochastic averaging [37].

Finally, mention is made here of an interesting alternative approach, based on the use of Slepian model processes, by Ditlevsen and co-workers [48]–[50]. By this approach it is possible, for example, to find the distribution of the first yielding of an elasto-plastic oscillator.

In this Chapter we present, for the first time, a unified treatment of the application of the standard stochastic averaging method to hysteretic systems. Attention is focused on oscillators with both bilinear and curvilinear hysteretic restoring forces. An application of the generalised averaging method, based on the energy envelope, to an oscillator with curvilinear hysteresis is also discussed. Some comparisons with both simulated and real experimental data are given.

3.2 Models of Hysteresis

Many materials and structures, when subjected to large fluctuating loadings, will exhibit hysteretic restoring forces of a non-linear character (e.g., see [3]). A general representation of such forces is as follows

$$P(x,t) = k\,[rx + (1-r)z] = kG(x,t) \tag{3.1}$$

where P is the total restoring force, x is the displacement, z is the non-linear hysteretic component of P, k is a stiffness parameter, and r is the "rigidity ratio"; the latter provides a measure of the relative contribution of the linear component krx.

Constitutive laws governing z cannot, in general, be expressed simply in terms of x and $\dot{x}$. However, such laws can be formulated in terms of first-order differential equations. Here two particular kinds of hysteretic behavior will be considered in some detail: bilinear and curvilinear hysteresis.

3.2.1 Bilinear Hysteresis

The simplest piecewise linear model for z is normalised perfectly elasto-plastic hysteresis, as shown in figure 3.1(a). It may be expressed by the differential equation

$$\dot{z} = \dot{x}\,[1 - U(\dot{x})U(z-1) - U(-\dot{x})U(-z-1)] \equiv g_b(\dot{x}, z) \tag{3.2}$$

where $U(\cdot)$ denotes the step function — i.e., $U(x) = 1$ for $x \geq 0$ and $U(x) = 0$ for $x < 0$. On combining equations (3.1) and (3.2) one obtains the bilinear characteristic shown in figure 3.1(b). Here the primary slope has a gradient k whereas the secondary slope has a gradient rk.

Under cyclic loading conditions the elasto-plastic loop will be closed, in the steady state, as shown in figure 3.2(a). The corresponding closed bilinear characteristic is shown in figure 3.2(b).

A number of other, more complex, piecewise linear hysteretic models have been discussed by Suzuki and Minai [23].

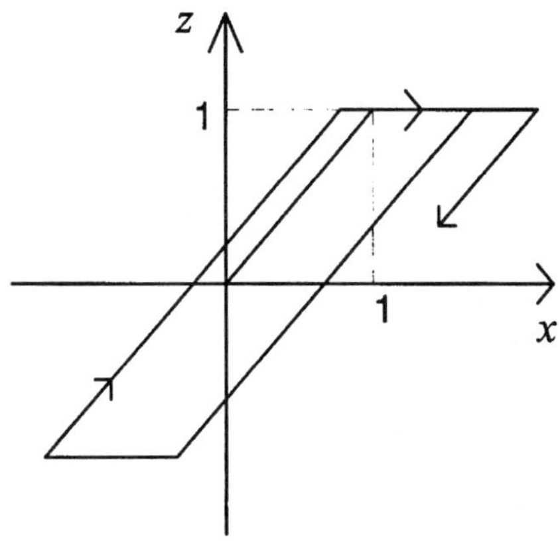

a)

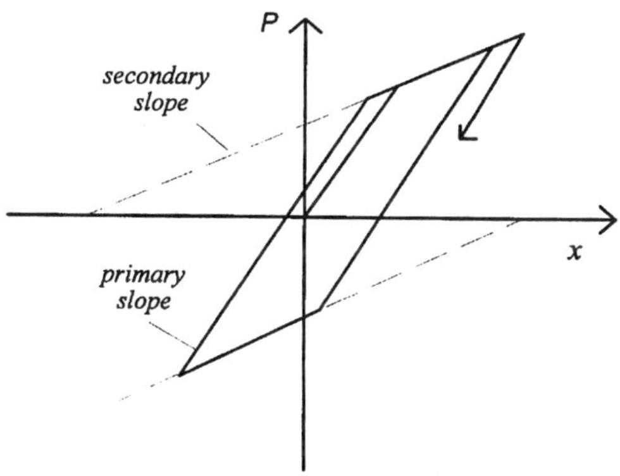

b)

FIGURE 3.1
(a) Normalised elasto-plastic hysteresis — z characteristic; (b) bilinear hysteresis — P characteristic.

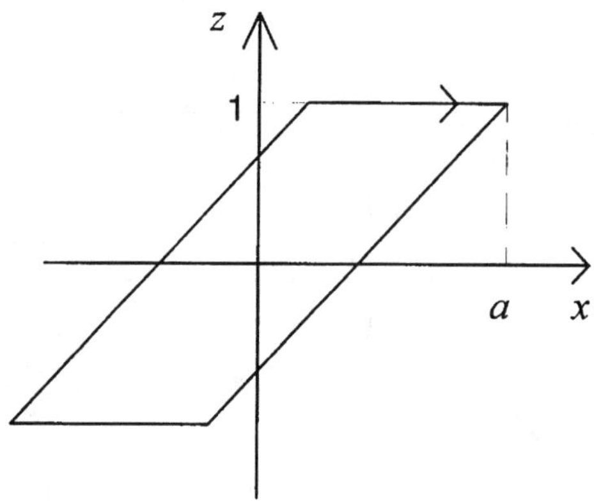

a)

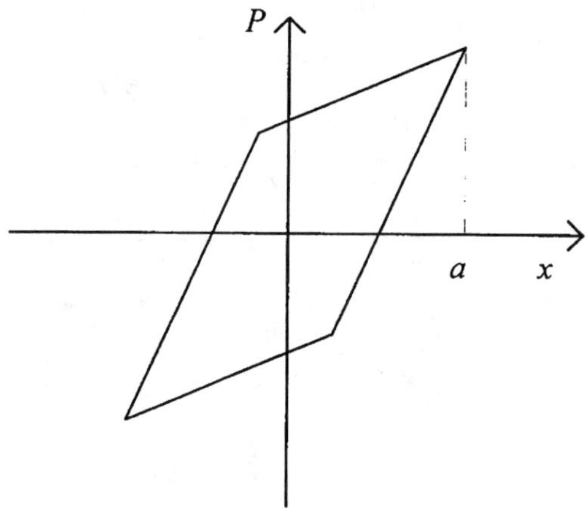

b)

FIGURE 3.2
(a) Closed elasto-plastic hysteresis loop — z characteristic; (b) closed elasto-plastic hysteresis loop — P characteristic.

3.2.2 Curvilinear Hysteresis

A wide variety of curvilinear hysteretic loops can be obtained by using the simple differential model originally proposed by Bouc [1] and subsequently generalised by Wen [2]. Here z is represented by

$$\dot{z} = -\gamma \mid \dot{x} \mid z \mid z \mid^{n-1} - \nu \dot{x} \mid z \mid^{n} + A\dot{x} \equiv g_c(\dot{x}, z) \tag{3.3}$$

Here the quantities γ, ν, A and n are "loop parameters" which control the shape and magnitude of the hysteresis loop. A typical closed loop for z, obtained for cyclic loading, is shown in figure 3.3(a).

The corresponding characteristic obtained by combining equations (3.1) and (3.3) is shown in figure 3.3(b).

3.2.3 Backbone Models

If the hysteretic loops are slim then the notion of a "backbone" is useful. Figure 3.4 shows a family of hysteresis loops, which could be obtained by using various levels of amplitude in the cyclic loading process. The backbone may be defined as the locus of extremities of the loops, and denoted $B(x)$, say. The hysteretic force, z, can then be decomposed into a sum of a non-hysteretic, conservative component and a purely dissipative term, $q(x, t)$, say. Thus

$$z(x, t) = B(x) + q(x, t) \tag{3.4}$$

$B(x)$ will, in general, be a non-linear function.

3.3 The Stochastic Averaging Method

In this section it is shown how the standard method of stochastic averaging [36] may be applied to oscillators with a hysteretic restoring force. Special attention is paid to the conditions under which this method is applicable.

3.3.1 The Equation of Motion

An oscillator with a hysteretic restoring force, excited by a random process, $F(t)$, is governed by an equation of motion of the following form:

$$m\ddot{x} + \varepsilon^2 H(x, \dot{x}) + kG(x, t) = F(t) \tag{3.5}$$

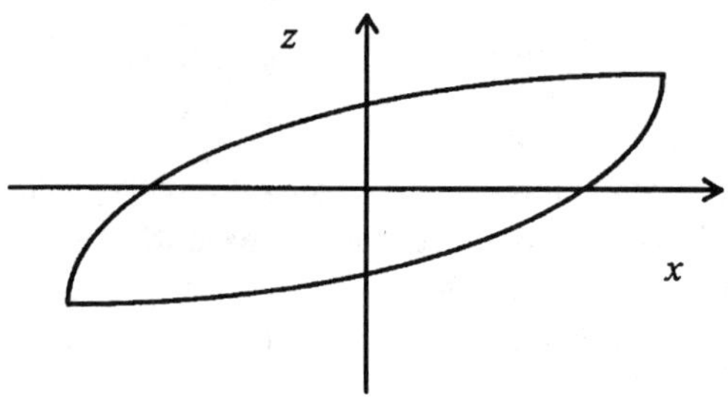

a)

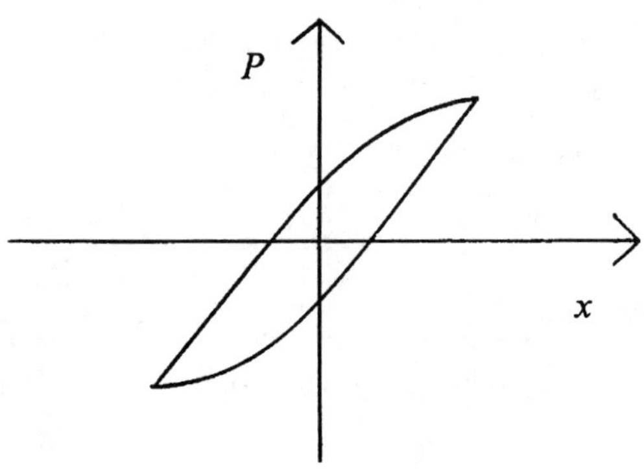

b)

FIGURE 3.3
(a) Closed curvilinear hysteresis loop — z characteristic; (b) closed curvilinear hysteresis loop — P characteristic.

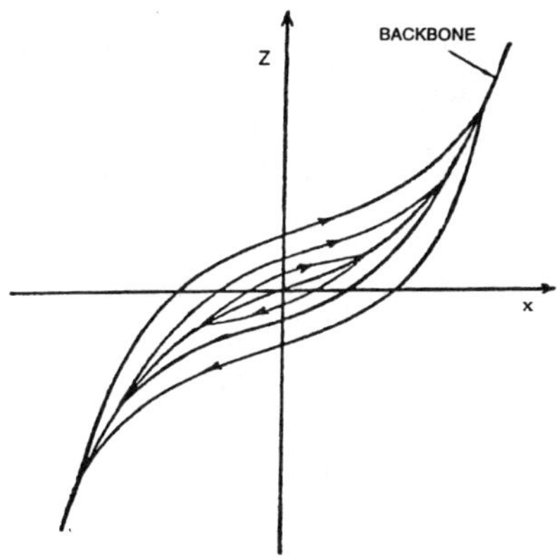

FIGURE 3.4
A family of closed hysteresis loop, and the backbone.

Here m is the mass, $H(x, \dot{x})$ is a non-linear damping term of arbitrary form, and $G(x, t)$, as before (see equation (3.1)) represents a hysteretic restoring term. ε^2 and k are damping and stiffness scaling parameters, respectively. It is useful, at the outset, to non-dimensionalise the equation of motion. If

$$y = x/x^* \ , \ \tau = \omega_0 t \tag{3.6}$$

where

$$\omega_0 = (k/m)^{1/2} \tag{3.7}$$

is the "linear natural frequency" and x^* is a characteristic length scale associated with the hysteretic force, then equation (3.5) may be rewritten as

$$\ddot{y} + \varepsilon^2 h(y, \dot{y}) + g(y, \tau) = f(\tau) \tag{3.8}$$

where differentiation is now with respect to τ,

$$h(y, \dot{y}) = \frac{H(x, \dot{x})}{m x^* \omega_0^2} \ , \quad f(\tau) = \frac{F(t)}{m x^* \omega_0^2} \tag{3.9}$$

and

$$g(y, \tau) = G(x, t)/x^* \tag{3.10}$$

As discussed earlier, in section 3.2, the non-dimensional hysteretic term $g(y, \tau)$ can be expressed as a combination of a linear, non-hysteretic term and a "purely hysteretic" component, as follows:

$$g(y, \tau) = ry + (1 - r)z(y, \tau) \tag{3.11}$$

where $z(y, \tau)$ is the constitutive law of the hysteretic component. The parameter r measures the relative contribution of the non-hysteretic term to $g(y, \tau)$. On combining equations (3.8) and (3.11), the governing equation of motion becomes

$$\ddot{y} + \varepsilon^2 h(y, \dot{y}) + ry + (1 - r)z(y, \tau) = f(\tau) \tag{3.12}$$

The excitation process, $f(\tau)$, will be assumed, initially, to be a wide-band stationary process, with zero mean and a power spectrum, $S_f(\omega)$, defined by

$$S_f(\omega) = \frac{1}{2\pi} \int_{-\infty}^{\infty} R_f(s) e^{-i\omega s} ds \tag{3.13}$$

where $R_f(s)$ is the correlation function for $f(\tau)$ defined by

$$R_f(s) = E\{f(\tau) f(\tau + s)\} \tag{3.14}$$

If the energy dissipated in a typical cycle of the response is low, compared with the total energy in that cycle, the response process, $y(\tau)$, will be narrow-band (see next section for a discussion of the precise conditions under which this applies). It can then be associated with an amplitude process, $a(\tau)$, and a phase process, $\phi(\tau)$. These are uniquely defined by

$$y = a \cos(\omega_c \tau + \phi), \quad \dot{y} = -a\omega_c \sin(\omega_c \tau + \phi) \tag{3.15}$$

where ω_c is a "characteristic frequency" of the response process, $y(\tau)$. As shown later, ω_c will, in general, depend on the strength of the excitation.

On transforming from $(y, \dot{y})$ to (a, ϕ) variables, equation (3.12) can be re-expressed as the following pair of coupled, first-order equations

$$\omega_c \dot{a} = d \sin \Phi - f(\tau) \sin \Phi, \quad \omega_c a \dot{\phi} = d \cos \Phi - f(\tau) \cos \Phi \tag{3.16}$$

where $\Phi = \omega_c \tau + \phi$ and

$$d = \varepsilon^2 h(a \cos \Phi, -a\omega_c \sin \Phi) + (r - \omega_c^2) a \cos \Phi + (1 - r)z \tag{3.17}$$

3.3.2 Averaging the Energy Dissipation Terms

Since y is assumed to be narrow-band, a and ϕ are slowly varying processes, i.e., their magnitude changes little over one cycle of oscillation, where the period of such cycles, T, is given by $\omega_c T = 2\pi$. The first terms on the right hand sides of equations (3.16) and (3.17), which relate to energy dissipation, can therefore be averaged over one cycle, on the basis that a and ϕ are constant. Hence one obtains the following partially averaged equations:

$$\omega_c \dot{a} = -\left[\varepsilon^2 \alpha(a) + \beta(a)\right] - f(\tau)\cos\Phi \tag{3.18}$$

$$a\omega_c \dot{\phi} = \frac{a}{2}\left[\omega^2(a) - \omega_c^2\right] - \varepsilon^2 \gamma(a) - f(\tau)\cos\Phi \tag{3.19}$$

where

$$\alpha(a) = -\frac{1}{2\pi}\int_0^{2\pi} h(a\cos\Phi, -a\omega_c \sin\Phi)\sin\Phi d\Phi \tag{3.20}$$

$$\beta(a) = \frac{(1-r)}{2\pi a}A(a), \quad \omega^2(a) = r + (1-r)B(a) \tag{3.21}$$

$$\gamma(a) = -\frac{1}{2\pi}\int_0^{2\pi} h(a\cos\Phi, -a\omega_c \sin\Phi)\cos\Phi d\Phi \tag{3.22}$$

and

$$A(a) = -\int_0^{2\pi} za\sin\Phi d\Phi = \oint z dy, \quad B(a) = \frac{1}{\pi a}\int_0^{2\pi} z\cos\Phi d\Phi \tag{3.23}$$

Here $\omega(a)$ is an amplitude dependent equivalent natural frequency, in the Krylov-Bogoliubov sense, and $A(a)$ is the area enclosed by a hysteretic loop, of amplitude a.

For the response to be narrow-band the energy dissipated per cycle, due to damping, must be a relatively small fraction of the energy associated with that cycle. The energy loss is given by

$$E(a) = \oint \left[\varepsilon^2 h(y, \dot{y}) + (1-r)z\right] dy = \varepsilon^2 2\pi a\omega_c \alpha(a) + (1-r)A(a) \tag{3.24}$$

Thus the energy lost per cycle, as a fraction of the energy in the cycle ($= \dot{y}_{\max}^2/2 = a^2\omega_c^2$) is given by

$$R(a) = R_1(a) + R_2(a) \tag{3.25}$$

where

$$R_1(a) = \frac{4\pi\varepsilon^2\omega_c\alpha(a)}{a} \quad R_2(a) = \frac{2(1-r)A(a)}{a^2\omega_c^2} \tag{3.26}$$

Now $R_1(a)$, the contribution from $h(y, \dot{y})$, is small if ε is a small parameter. For a narrow-band response it is necessary for $R_2(a)$ to be of order ε^2, as well. This will occur if at least one of the following conditions is met:

1. The parameter, r, (the secondary slope) is close to unity, such that

$$\varepsilon^2 = 1 - r \tag{3.27}$$

   In these circumstances the contribution of the purely hysteretic component to $g(y, \tau)$ is small (see equation (3.21) and thus the characteristic frequency, ω_c, is close to unity. Hysteretic loops will then be slim, as shown in figure 3.5(a), for a curvilinear model.

2. $A(a)$ is of order ε^2. Generally the loop area will increase monotonically with increasing a, so this condition implies a weak level of excitation, such that a is small. Thus ω_c will again be close to unity. Hysteretic loops will again be slim, as shown in figure 3.5(b). For a bilinear characteristic $A(a)$ is actually zero for $\alpha < 1$. For curvilinear loops, of the Bouc-Wen type, it is necessary that $A(a)/a^2$ is small, for small a (see equation (3.26).

3. The excitation level is relatively high, so that much of the response is associated with the near horizontal portions of the z characteristic. For large amplitude response $A(a)$ will grow linearly with amplitude (see section 3.5); hence $A(a)/a^2$ will decrease as a increases and $R_2(a)$ will become progressively smaller. Hysteretic loops will appear as shown in figure 3.5(c) in these circumstances. It is natural here to set ω_c to equal the natural frequency associated with the secondary stiffness, i.e., $\omega_c^2 = r$.

For all three of the above conditions, the term $\omega^2(a) - \omega_c^2$ appearing in the phase equation (3.19) will also be small. Thus for conditions 1 and 2, both $\omega(a)$ and ω_c will be close to unity. Condition 3 $B(a)$ will, according to equation (3.23) become small as a increases. Thus both $\omega^2(a)$ and ω_c^2 will approach r in magnitude.

It will be assumed, henceforth, that at least one of the above conditions is met. One can then make the smallness of the dissipation terms explicit by writing equations (3.18) and (3.19) as

$$\dot{a} = -\varepsilon^2\frac{\eta(a)}{\omega_c} - \frac{f(\tau)\sin\Phi}{\omega_c}, \quad \dot{\phi} = -\varepsilon^2\frac{\rho(a)}{a\omega_c} - \frac{f(\tau)\cos\Phi}{a\omega_c} \tag{3.28}$$

where

$$\eta(a) = \alpha(a) + \beta'(a), \quad \rho(a) = \gamma(a) + \mu(a) \tag{3.29}$$

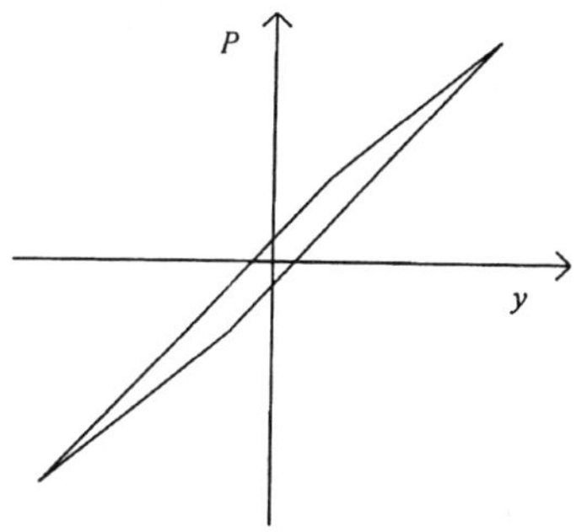

a)

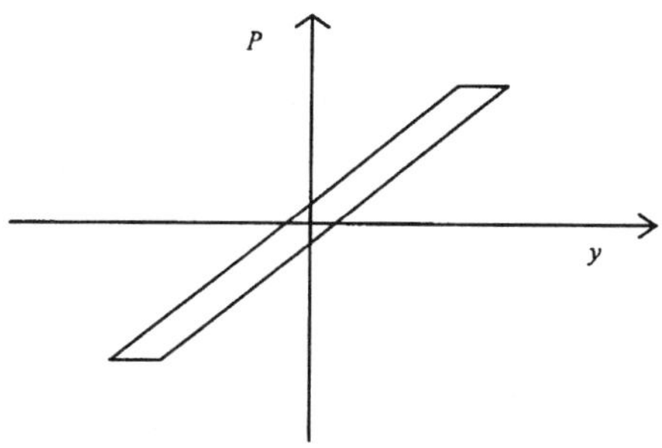

b)

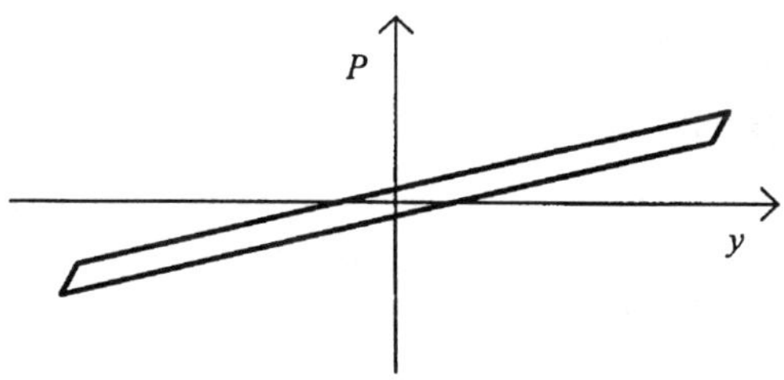

c)

FIGURE 3.5

(a) Bilinear hysteretic loops — *r* close to unity (b) Bilinear hysteretic loops — weak excitation (c) Bilinear hysteretic loops — strong excitation

$$\beta'(a) = \beta(a)/\varepsilon^2, \quad \mu(a) = -\frac{a}{2\varepsilon^2}\left[\omega^2(a) - \omega_c^2\right] \tag{3.30}$$

Here both $\beta'(a)$ and $\mu(a)$ are of order ε^0.

From equation (3.28) it can be inferred that

$$\sigma_y = O\left(\varepsilon^{-1}\right) \tag{3.31}$$

Clearly $\sigma_y \to \infty$ as $\varepsilon \to 0$. For analysis purposes this behavior is inconvenient and it is desirable to scale the excitation so that the level of the response, as measured by σ_y, is $O\left(\varepsilon^0\right)$. This implies that the excitation spectrum should be of order ε^2, i.e.,

$$\sup_{\text{over } \omega} \left\{S_f(\omega)\right\} = O\left(\varepsilon^2\right) \tag{3.32}$$

To make this scaling explicit it is convenient to introduce the process $z(\tau)$ where

$$f(\tau) = \varepsilon z(\tau) \tag{3.33}$$

and the spectrum of $z(\tau)$, $S_z(\omega)$, is of order ε^0. The equations for a and ϕ now become

$$\dot{a} = -\varepsilon^2 \frac{\gamma(a)}{\omega_c} - \varepsilon \frac{z(\tau)\sin\Phi}{\omega_c}, \quad \dot{\phi} = -\varepsilon^2 \frac{\rho(a)}{a\omega_c} - \varepsilon \frac{z(\tau)\cos\Phi}{a\omega_c} \tag{3.34}$$

It is emphasised that this step does not usually imply any restrictions regarding the "strength" of the excitation process. When ε is small the level of the excitation will normally be low compared with the maximum level of the response. Equations (3.34) simply make this fact appear explicitly. It is also now quite clear that both a and ϕ are slowly varying quantities since all terms on the right-hand side of equations (3.34) are explicitly small.

3.3.3 Averaging the Excitation Term

So far the averaging method corresponds exactly to the deterministic method of Bogoliubov and Mitropolsky. The next step, however, which involves averaging the last terms in equation (3.34) is rather more difficult. The correct procedure for averaging those excitation terms is due to Stratonovich [53]. Here we give a simplified approach to the averaging method which, whilst not rigorous, does give a physical insight into the nature of the approximations involved.

The basic difficulty in averaging the terms

$$y_1(\tau) = \frac{z(\tau)\sin\Phi}{\omega_c}, \quad y_2(\tau) = \frac{z(\tau)\cos\Phi}{a\omega_c} \tag{3.35}$$

which appear in equation (3.34) lies in the fact that the phase process, $\phi(\tau)$, is correlated with $z(\tau)$. Thus the means of $y_1(\tau)$ and $y_2(\tau)$ are generally non-zero, even though the process $z(\tau)$ has zero mean.

Suppose that the excitation process, $z(\tau)$, has a correlation time scale, τ_{cor}, which is so small that a time interval, $\Delta\tau$, can be defined such that the following two conditions are satisfied simultaneously:

(i) $\Delta\tau >> \tau_{cor}$,
(ii) $a(\tau)$ and $\phi(\tau)$ do not change appreciably from τ to $\tau + \Delta\tau$.

It is noted that these conditions are always met if ε is sufficiently small, for then the correlation time scale associated with the response will always become large, compared with that of the excitation. Put another way, the bandwidth of the response will become small, compared with that of the excitation. From a practical viewpoint, however, it is desirable that the excitation should have a large bandwidth (so that τ_{cor} is, absolutely, small) otherwise ε may have to be extremely small (and may be unrealistically so) in order that the above requirements are met.

If the conditions hold then, with an error of order $(\Delta\phi)^2$ one can write

$$\sin(\omega_c\tau + \phi) = \sin(\omega_c\tau + \phi_1) + \cos(\omega_c\tau + \phi_1)\Delta\phi \tag{3.36}$$

where

$$\phi_1 = \phi(\tau - \Delta\tau)\ ;\ \phi = \phi(\tau)\ ;\ \Delta\phi = \phi - \phi_1 \tag{3.37}$$

If this expression is substituted into the first of equations (3.35) one obtains

$$y_1(\tau) = \frac{1}{\omega_c} z(\tau)\,[\sin(\omega_c\tau + \phi_1) + \cos(\omega_c\tau + \phi_1)\Delta\phi] \tag{3.38}$$

Now, since ϕ_1 is uncorrelated with $z(\tau)$ ($\Delta\tau >> \tau_{cor}$) the average value of $z(\tau)\sin(\omega_c\tau + \phi_1)$ is zero. Hence the mean of $y_1(\tau)$ is given by

$$E\{y_1(\tau)\} = \frac{1}{\omega_c}\cos(\omega_c\tau + \phi_1)E\{z(\tau)\Delta\phi\} \tag{3.39}$$

An expression for $\Delta\phi$ can be found by integrating the phase equation, as given in equations (3.34). Thus

$$\Delta\phi = -\frac{\varepsilon^2}{a\omega_c}\rho(a)\Delta\tau - \frac{\varepsilon}{a\omega_c}\int_{\tau-\Delta\tau}^{\tau} z(s)\cos(\omega_c s + \phi)ds \tag{3.40}$$

where the fact that $a(\tau)$ and $\phi(\tau)$ are sensibly constant over the interval $\Delta\tau$ has been used. A combination of equation (3.39) and (3.40) now gives

$$E\{y_1(\tau)\} = -\frac{\varepsilon}{a\omega_c^2}\int_{\tau-\Delta\tau}^{\tau} E\{z(\tau)z(s)\}\cos(\omega\tau_c + \phi)\cos(\omega_c s + \phi)ds \tag{3.41}$$

With a change of variable, $u = \tau - s$, equation (3.41) can be re-expressed as

$$E\{y_1(\tau)\} = -\frac{\varepsilon}{a\omega_c^2}\int_{-\infty}^{0} w_z(u)\cos(\omega_c\tau + \phi)\cos(\omega_c\tau + \omega_c u + \phi)du \tag{3.42}$$

where

$$w_1(s) = E\{z(\tau)z(\tau + s)\} \tag{3.43}$$

is the correlation function for $z(\tau)$ and the bottom limit of the integral, $\Delta\tau$, has been replaced by $-\infty$, since the correlation function is effectively zero when $u = \Delta\tau$ ($\Delta\tau >> \tau_{cor}$, again).

The product of the two cosines in equation (3.42) can be written as

$$\frac{1}{2}\cos(\omega_c u) + \text{ oscillatory terms, with frequency } \omega_c \tag{3.44}$$

Substituting this into equation (3.42) gives

$$E\{y(\tau_1)\} = -\frac{\varepsilon}{2a\omega_c^2}\int_{-\infty}^{0} w_z(u)\cos(\omega_c u)du \; + \text{ oscillatory terms} \tag{3.45}$$

The oscillatory terms will, when averaged over the period defined by $T = 2\pi/\omega_c$, be zero. Moreover, the power spectrum, $S_z(\omega)$, of $z(\tau)$ is defined by

$$S_z(\omega) = \frac{1}{2\pi}\int_{-\infty}^{\infty} w_z(s)e^{i\omega s}ds \tag{3.46}$$

It follows that

$$\overline{y}_1 = E\{y_1(\tau)\} = -\frac{\varepsilon\pi S_z(\omega_c)}{2a\omega_c^2} \tag{3.47}$$

Next the correlation function of the zero mean process

$$y_1^* = y_1 - \overline{y}_1 \tag{3.48}$$

will be found. Since the second term on the right-hand side of equation (3.38) influences only the mean value, $\overline{y}_1$, one need only consider the contribution from the first term in evaluating the correlation function of y_1. Thus

$$y_1^* = z(\tau)\sin(\omega_c\tau + \phi) \tag{3.49}$$

and

$$w_{y_1}(s) = E\left\{y_1^*(\tau)y_1^*(\tau+s)\right\}\sin(\omega_c\tau+\phi)\sin(\omega_c\tau+\omega_c s+\phi) \tag{3.50}$$

where the fact that ϕ is sensibly constant, over an interval where the correlation function of $z(\tau)$ is significant, has been used.

Now since τ_{cor} is much smaller than the correlation time scale of the response, the process y_1^* behaves approximately like an uncorrelated white noise process. Thus one can approximate $w_{y_1}(\tau)$ as

$$w_{y_1}(s) = I\delta(s) \tag{3.51}$$

where the "strength" of the white noise process is given by

$$I = \int_{-\infty}^{\infty} w_{y_1}(s)ds \tag{3.52}$$

Combining equations (3.50) and (3.52) one has

$$\begin{aligned} I &= \int_{-\infty}^{\infty} w_z(s)\sin(\omega_c\tau+\phi)\sin(\omega_c\tau+\omega_c s+\phi)ds \\ &= \frac{1}{2}\int_{-\infty}^{\infty} w_z(s)\cos(\omega_c s)ds \end{aligned} \tag{3.53}$$

Once again, the oscillatory terms will disappear if averaging is carried out, over the period T. Also, the interval in the above expression can be related to the power spectrum, $S_z(\omega)$, of $z(\tau)$. Thus, from equations (3.46) and (3.53)

$$I = \pi S_z(\omega_c) \tag{3.54}$$

The above analysis has shown that the process $y_1(\tau)$ can be approximated as a white noise process, with a strength, I, given by equation (3.54) and a non-zero mean, given by equation (3.47). Hence one can express $y_1(\tau)$ as

$$y_1(\tau) = -\frac{\varepsilon\pi S_z(\omega_c)}{2a\omega_c^2} - \left[\pi S_z(\omega_c)\right]^{1/2} \xi_1(\tau) \tag{3.55}$$

where $\xi_1(\tau)$ is a zero-mean white noise process, with unit strength. Combining equations (3.34), (3.35), and (3.55) one obtains

$$\dot{a} = -\frac{\varepsilon^2}{\omega_c}\eta(a) + \frac{\varepsilon^2\pi S_z(\omega_c)}{2a\omega_c^2} + \frac{\left[\varepsilon^2\pi S_z(\omega_c)\right]^{1/2}}{\omega_c}\xi_1(\tau) \tag{3.56}$$

This is the final "simplified" equation for $a(\tau)$.

A similar analysis can be carried out for the phase process, $\phi(\tau)$, to show that

$$\dot{\phi} = -\frac{\varepsilon^2}{a\omega_c}\rho(a) + \frac{\left[\varepsilon^2\pi S_z(\omega_c)\right]^{1/2}}{a\omega_c}\xi_2(\tau) \tag{3.57}$$

where $\xi_2(\tau)$ is another white noise process, with zero mean and unit strength, independent of $\xi_1(\tau)$. It is observed that the mean of the process $y_2(\tau)$ defined in equations (3.35) is zero. Equation (3.57) is the "simplified equation" for $\phi(\tau)$.

From equation (3.33) it follows that the spectrum of $f(\tau)$ is given by

$$S_f(\omega) = \varepsilon^2 S_z(\omega) \tag{3.58}$$

Thus, in terms of $S_f(\omega)$ one can write the simplified equations as

$$\dot{a} = -\frac{\varepsilon^2}{\omega_c}\eta(a) + \frac{\pi S_f(\omega_c)}{2a\omega_c^2} + \frac{\left[\pi S_f(\omega_c)\right]^{1/2}}{\omega_c}\xi_1(\tau) \tag{3.59}$$

$$\dot{\phi} = -\frac{\varepsilon^2}{a\omega_c}\rho(a) + \frac{\left[\pi S_f(\omega_c)\right]^{1/2}}{a\omega_c}\xi_2(\tau) \tag{3.60}$$

Written rather more formally, as Itô equations, equations (3.56) and (3.57) become

$$d\boldsymbol{X} = \varepsilon^2\boldsymbol{a}(\boldsymbol{X})d\tau + \varepsilon\boldsymbol{B}d\boldsymbol{W} \tag{3.61}$$

where

$$\boldsymbol{X} = \begin{bmatrix} X_1 \\ X_2 \end{bmatrix} = \begin{bmatrix} a \\ \phi \end{bmatrix} \tag{3.62}$$

$$\boldsymbol{a} = \begin{bmatrix} a_1(\boldsymbol{X}) \\ a_2(\boldsymbol{X}) \end{bmatrix} = \begin{bmatrix} -\dfrac{\eta(a)}{\omega_c} + \dfrac{\pi S_z(\omega_c)}{2a\omega_c^2} \\ -\dfrac{\rho(a)}{a\omega_c^2} \end{bmatrix} \tag{3.63}$$

$$\boldsymbol{B} = \begin{bmatrix} B_{11} & 0 \\ 0 & B_{22} \end{bmatrix} , \quad d\boldsymbol{W} = \begin{bmatrix} dW_1 \\ dW_2 \end{bmatrix} \tag{3.64}$$

where

$$B_{11} = -\frac{\left[\pi S_z(\omega_c)\right]^{1/2}}{\omega_c} = aB_{22} \tag{3.65}$$

W_1 and W_2 are independent Wiener processes and formally

$$\xi_1 = \frac{dW_1}{d\tau} , \quad \xi_2 = \frac{dW_2}{d\tau} \tag{3.66}$$

It can be shown [53] that the approximations involved in averaging the stochastic terms, as outlined above, are entirely consistent with the approximations inherent in averaging the dissipation terms. In fact the averaging approximation is, overall, correct to order ε^2.

Since $\boldsymbol{X}$, as defined by equation (3.62), is governed by a first order equation, with white noise excitation, it follows that the joint process $[a(\tau), \phi(\tau)]$ is (at least approximately) a joint Markov process. Strictly, $[a(\tau), \phi(\tau)]$ converges to a Markov process as ε^2 tends to zero. Thus one can regard the simplified equations for $[a(\tau), \phi(\tau)]$ as asymptotically exact, as $\varepsilon \to 0$. The mathematical basis for this result is contained in the Stratonovich-Khasminskii limit theorem.

This theorem is applicable to stochastic differential equations of the following form:

$$\dot{\boldsymbol{X}} = \varepsilon^2(\boldsymbol{f}\boldsymbol{X}, \tau) + \varepsilon \boldsymbol{g}[\boldsymbol{X}, \tau, \boldsymbol{Y}(\tau)] \tag{3.67}$$

Here $\boldsymbol{X}(\tau)$ is an n-vector stochastic process, usually relating to the response, and $\boldsymbol{Y}(\tau)$ is an m-vector stochastic process, usually relating to the excitation. The elements of $\boldsymbol{Y}(\tau)$ are broad-band processes, with zero means and the vectors $\boldsymbol{f}$ and $\boldsymbol{g}$ satisfying certain requirements [54] which are almost invariably met in practice. It can be shown that $\boldsymbol{X}(\tau)$ may be uniformly approximated over a time interval of order ε^{-1} by an n-dimensional Markov process, which satisfies the Itô equation

$$d\boldsymbol{X} = \varepsilon^2 \boldsymbol{a}(\boldsymbol{X}) + \varepsilon \boldsymbol{B}(\boldsymbol{X}) d\boldsymbol{W} \tag{3.68}$$

Here the symbol $\boldsymbol{W}$ denotes, as before, a vector of independent unit Wiener processes and $\boldsymbol{a}$ and $\boldsymbol{B}$ are, respectively, the "drift vector" and "diffusion matrix". In fact, according to the limit theorem, $\boldsymbol{X}(t)$ converges weakly to a Markov process as $\varepsilon \to 0$. Hence the approximation of $\boldsymbol{X}(t)$ as a Markov process, governed by equation (3.61), is asymptotically exact as $\varepsilon \to 0$. The quantities $\boldsymbol{a}$ and $\boldsymbol{B}$ can be evaluated according to the expressions:

$$\boldsymbol{a} = T^{av}\left[E\{\boldsymbol{f}\} + \int_{-\infty}^{0} E\left\{\frac{\partial \boldsymbol{g}}{\partial \boldsymbol{X}_\tau}\boldsymbol{g}_{\tau+s}\right\} ds\right] \tag{3.69}$$

and

$$\boldsymbol{B}\boldsymbol{B}^T = T^{av} \int_{-\infty}^{\infty} E\left\{\boldsymbol{g}_\tau \boldsymbol{g}_{\tau+s}^T\right\} ds \tag{3.70}$$

Here

$$T^{av}\{\cdot\} = \lim \frac{1}{T} \int_{\tau_0}^{\tau_0+T} \{\cdot\}\, ds \tag{3.71}$$

the integration being performed over explicit τ only. Also the subscripts τ and $\tau+s$ in the above indicate that the appropriate quantities are to be evaluated at those times. In evaluating the expectations in equations (3.69) and (3.70) the elements of $\boldsymbol{X}(\tau)$ are treated as constants. If the expected quantities in equations (3.69) and (3.70) are periodic, with period T_0, say, then T^{av} is just an averaging operation over one period and the result is independent of τ_0.

Application of the limit theorem leads to results identical to equations (3.56) and (3.57).

3.3.4 The FPK Equations

It has been shown that the joint process $[a(\tau), \phi(\tau)]$ converges to a Markov process, with the variables governed by Itô equations, in the form of equation (3.68). It follows that $[a(\tau), \phi(\tau)]$ has a transition density function $p(a, \phi \mid a_0, \phi_0; \tau)$ such that $p(a, \phi \mid a_0, \phi_0; \tau)da\, d\phi$ is the probability that $a < a(\tau) < a + da$, $\phi < \phi(\tau) < \phi + d\phi$ at time τ, given that $a = a_0, \phi = \phi_0$ at $\tau = 0$. The transition density function is governed by a partial differential equation, of diffusion type, known as a Fokker-Planck-Kolmogorov (FPK) equation: thus,

$$\begin{aligned}\frac{\partial p}{\partial \tau} = \frac{\partial}{\partial a}\left[\left\{\frac{\varepsilon^2 \eta(a)}{\omega_c} - \frac{\pi S_f(\omega_c)}{2a\omega_c^2}\right\} p\right] + \frac{\varepsilon^2 \rho(a)}{a\omega_c}\frac{\partial p}{\partial a} \\ + \frac{\pi S_f(\omega_c)}{2\omega_c^2}\left[\frac{\partial^2 p}{\partial a^2} + \frac{1}{a^2}\frac{\partial^2 p}{\partial \phi^2}\right]\end{aligned} \tag{3.72}$$

An inspection of the differential equations for $a(\tau)$ and $\phi(\tau)$ (see equations (3.59) and (3.60)) shows that the amplitude process, $a(\tau)$, is *uncoupled* from the phase process, $\phi(\tau)$, as in the deterministic case, where the excitation is absent. It follows that $a(\tau)$ is a *one-dimensional* Markov process. The transition density function for $a(\tau)$, $p(a) \mid a_0; \tau)$ is governed by the following FPK equation:

$$\frac{\partial p}{\partial \tau} = \frac{\partial}{\partial a}\left[\left\{\frac{\varepsilon^2 \eta(a)}{\omega_c} - \frac{\pi S_f(\omega_c)}{2a\omega_c^2}\right\} p\right] + \frac{\pi S_f(\omega_c)}{2\omega_c^2}\frac{\partial^2 p}{\partial a^2} \tag{3.73}$$

Equations (3.72) and (3.73) must be solved subject to the following initial conditions:

$$p(a \mid a_0; \tau) \to \delta(a - a_0) \text{ as } \tau \to 0 \tag{3.74}$$

and

$$p(a, \phi \mid a_0, \phi_0; \tau) \to \delta(a - a_0)\delta(\phi - \phi_0) \text{ as } \tau \to 0 \tag{3.75}$$

It is noted that, in the case where $\varepsilon^2 h(x, \dot{x})$ is a linear damping term, it may be written as

$$\varepsilon^2 h(x, \dot{x}) = 2\varsigma\omega_c \dot{x} \tag{3.76}$$

One then has

$$\gamma(a) = \varsigma\omega_c^2 a \quad , \quad \rho(a) = 0 \tag{3.77}$$

More generally, if $h(x, \dot{x})$ is a function of x only then, from equation (3.22), it follows that $\gamma(a) = 0$. Similarly, if $h(x, \dot{x})$ depends on $\dot{x}$ only, then $\alpha(a) = 0$.

3.3.5 Stationary Solutions

Since the excitation is assumed, at present, to be stationary, the response will approach stationarity as time elapses, i.e.,

$$\lim_{\tau \to \infty} p(a, \phi \mid a_0, \phi_0; \tau) = w(a, \phi) \tag{3.78}$$

where $w(a, \phi)$ is the stationary density function for $a(\tau)$ and $\phi(\tau)$. Similarly

$$\lim_{\tau \to \infty} p(a, \mid a_0; \tau) = w(a) \tag{3.79}$$

where $w(a)$ is the stationary density function for $a(\tau)$.

Equations for $w(a, \phi)$ and $w(a)$ may be obtained by setting $\partial p / \partial t = 0$ in equations (3.72) and (3.73). Hence, for $w(a, \phi)$,

$$0 - \frac{\partial}{\partial a}\left[\left\{\frac{\varepsilon^2 \eta(a)}{\omega_c} - \frac{\pi S_f(\omega_c)}{2a\omega_c^2}\right\} w\right] + \frac{\varepsilon^2 \rho(a)}{a\omega_c}\frac{\partial w}{\partial a} + \frac{\pi S_f(\omega_c)}{2\omega_c^2}\left[\frac{\partial^2 w}{\partial a^2} + \frac{1}{a^2}\frac{\partial^2 w}{\partial \phi^2}\right] \tag{3.80}$$

and

$$0 = \frac{\partial}{\partial a}\left[\left\{\frac{\varepsilon^2\eta(a)}{\omega_c} - \frac{\pi S_f(\omega_c)}{2a\omega_c^2}\right\} w\right] + \frac{\pi S_f(\omega_c)}{2\omega_c^2}\frac{\partial^2 w}{\partial a^2} \tag{3.81}$$

A comparison of equations (3.80) and (3.81) reveals that the stationary density function, $w(a, \phi)$, must be related to $w(a)$ as follows:

$$w(a, \phi) = \frac{1}{2\pi} w(a) \tag{3.82}$$

Here the factor $1/2\pi$ arises from the normalisation conditions

$$\int_0^\infty w(a)da = \int_0^{2\pi}\int_0^\infty w(a, \phi)da\, d\phi = 1 \tag{3.83}$$

Equation (3.82) shows that, when stationarity is reached, the phase angle ϕ is uniformly distributed between 0 and 2π.

Through a transformation from a, ϕ variables to the original y, $\dot{y}$ variables an expression for the stationary density function of y and $\dot{y}$, $w(y, \dot{y})$, can be deduced from equation (3.82). Thus

$$w(y, \dot{y}) = \frac{1}{2\pi\omega_c a} w(a) \tag{3.84}$$

where

$$a = \left(y^2 + \frac{\dot{y}^2}{\omega_c^2}\right)^{1/2} \tag{3.85}$$

and $w(a)$ is the solution of equation (3.81). This solution is readily found to be

$$w(a) = Ca \exp\left\{-\frac{2\varepsilon^2\omega_c}{\pi S_f(\omega_c)}\int_0^a \eta(\xi)d\xi\right\} \tag{3.86}$$

where C is a normalisation constant.

3.3.6 The Characteristic Frequency

An expression for the frequency ω_c can be found by replacing equation (3.12) by the partially linearised equation

$$\ddot{y} + \varepsilon^2 h(y, \dot{y}) + \omega_c^2 y = f(\tau) \tag{3.87}$$

If the difference between equations (3.12) and (3.87) is minimised, in a least square sense, one obtains the following expression for ω_c^2 (see [4])

$$\omega_c^2 = r + (1-r)\frac{E\{yz\}}{E\{y^2\}} \tag{3.88}$$

Using the fact that y is assumed here to be a narrow band process, with an amplitude, a, and a phase, ϕ, defined by equation (3.15), one can rewrite equation (3.88) as

$$\omega_c^2 = r + (1-r)\frac{E\{a\cos\Phi z\}}{E\{a^2\cos^2\Phi\}} \tag{3.89}$$

For a response within a sufficiently small bandwidth one can assume, as before, that a and ϕ are sensibly constant over one cycle of period T. One can then approximate the expectations in equation (3.89) by integrating over one cycle. This leads to

$$\omega_c^2 = r + (1-r)\frac{E\{\oint a\cos\Phi z d\Phi\}}{E\{\oint a^2\cos^2\Phi d\Phi\}} \tag{3.90}$$

Since

$$E\left\{\oint a^2\cos^2\Phi d\Phi\right\} = \pi E\left\{a^2\right\} \tag{3.91}$$

and, from equation (3.23),

$$E\left\{\oint a\cos\Phi z d\Phi\right\} = \pi a B(a) \tag{3.92}$$

If a is sensibly constant over a cycle period, equation (3.90) can be re-expressed as

$$\omega_c^2 = r + (1-r)\frac{E\{a^2B(a)\}}{E\{a^2\}} \tag{3.93}$$

or, using equation (3.21),

$$\omega_c^2 = \frac{E\{a^2\omega^2(a)\}}{E\{a^2\}} \tag{3.94}$$

The expectations in equation (3.93) (or (3.94)) can be evaluated by using the stationary solution given by equation (3.86). Thus

$$E\left\{a^2\omega^2(a)\right\} = \int_0^\infty a^2\omega^2(a)w(a)da \tag{3.95}$$

and

$$E\left\{a^2\right\} = \int_0^\infty a^2 w(a) da \tag{3.96}$$

Combining equations (3.94) to (3.96) will lead to an algebraic equation for ω_c.

3.4 Stationary Response of the Bilinear Oscillator

For purely elasto-plastic hysteresis as shown in figure 3.2(a), the area enclosed by the loop, of amplitude a, is found, by simple geometry, to be

$$\begin{aligned} A(a) &= 4(a-1)\,, \quad && a \geq 1 \\ &= 0\,, && a \leq 1 \end{aligned} \tag{3.97}$$

Thus the function $\eta(a)$ defined by equation (3.29) is here given by

$$\begin{aligned} \varepsilon^2\eta(a) &= \varepsilon^2\alpha(a) + \frac{2(1-r)(a-1)}{\pi a}\,, \quad && a \geq 1 \\ &= \varepsilon^2\alpha(a)\,, && a \leq 0 \end{aligned} \tag{3.98}$$

On substituting these expressions into equations (3.80) to (3.81) it is possible to find expressions for the probability density function of the response. Thus from equations (3.86) and (3.98) one has

$$\begin{aligned} w(a) &= ca^{1+1/\beta}\exp\left\{-\frac{1}{\beta}\left[(a-1) + \frac{\pi G(a)}{2(1-r)}\right]\right\}, \quad && a \geq 1 \\ &= ca\,\exp\left\{-\frac{\pi G(a)}{2\beta(1-r)}\right\}, && a \leq 1 \end{aligned} \tag{3.99}$$

where

$$\beta = \frac{\pi^2 S(\omega_c)}{4(1-\alpha)\omega_c}\,, \quad G(a) = \varepsilon^2 \int_0^a \alpha(\xi) d\xi \tag{3.100}$$

and c is a normalisation constant.

For the joint process $[y, \dot{y}]$ it is found, from equations (3.84) and (3.99), that

$$w(y, \dot{y}) = c' a^{1/\beta} \exp\left\{-\frac{1}{\beta}\left[(a-1) + \frac{\pi G(a)}{2(1-r)}\right]\right\} \quad a \geq 1$$

$$= c' a \exp\left\{-\frac{\pi G(a)}{2\beta(1-r)}\right\} \quad a \leq 1 \quad (3.101)$$

where a is related to y and $\dot{y}$ according to equation (3.85) and c' is another normalisation constant.

In the particular case where the energy dissipation arises solely from the hysteretic term, g (i.e., $h(y, \dot{y}) = 0$) one has $G(a) = 0$ and equations (3.98) and (3.101) reduce to

$$w(a) = c a^{1+1/\beta} \exp\left\{-\frac{(a-1)}{\beta}\right\} \quad a \geq 1$$

$$= ca \quad a \leq 1 \quad (3.102)$$

and

$$w(y, \dot{y}) = c' a^{1/\beta} \exp\left\{-\frac{(a-1)}{\beta}\right\} \quad a \geq 1$$

$$= c' \quad a \leq 1 \quad (3.103)$$

These density functions reduce to simple forms for very weak excitation ($\beta \to 0$) and very strong excitation ($\beta \to \infty$). For $\beta \to 0$

$$w(a) \to \frac{a}{2} \quad a \leq 1$$

$$\to 0 \quad a > 0 \quad (3.104)$$

and for $\beta \to \infty$

$$w(y, \dot{y}) \to \frac{1}{\pi \omega_c} \quad a \leq 1$$

$$\to 0 \quad a > 1 \quad (3.105)$$

Thus, for $w(y, \dot{y})$, one obtains, asymptotically, a uniform distribution within the region $a \leq 1$ and the probability density is zero outside this region. As $\beta \rightarrow \infty$

$$w(A) \rightarrow A \exp(-A) \tag{3.106}$$

where $A = a/\beta$, and

$$w(Y, \dot{Y}) \rightarrow \frac{1}{2\pi\omega_c} \exp(-A) \tag{3.107}$$

where $Y = y/\beta$ and $\dot{Y} = \dot{y}/\beta$.

3.4.1 Response Statistics

A number of important statistics can be deduced from the joint density function given by equation (3.101). Thus, the mean square response is given by

$$\sigma^2 = E\left\{y^2\right\} = \int_{-\infty}^{\infty}\int_{-\infty}^{\infty} w(y, \dot{y}) y^2 dy\, d\dot{y} \tag{3.108}$$

On substituting equation (3.101) into equation (3.108), and evaluating the integration, σ^2 can be computed. In the case where $G(a)$ is zero, analytical results can be obtained [37]. Thus

$$\sigma^2 = \frac{0.25 + F_3(\beta)}{1 + 2\, F_1(\beta)} \tag{3.109}$$

where

$$F_n(\beta) = \int_0^{\infty} (1 + u)^{n+1/\beta} \exp\left(-\frac{u}{\beta}\right) du$$

$$= \left(\frac{1}{\beta}\right)^{-n-1-1/\beta} \exp\left(\frac{1}{\beta}\right) \Gamma\left(n + 1 + \frac{1}{\beta}, \frac{1}{\beta}\right) \tag{3.110}$$

and $\Gamma(\alpha, x)$ is the incomplete Gamma function, defined as

$$\Gamma(\alpha, x) = \int_0^{\infty} \exp(-\xi)\xi^{\alpha-1} d\xi \tag{3.111}$$

As $\beta \to 0$ the $F_n(\beta)$ approach zero. Hence one has the asymptotic result

$$\sigma \to \frac{1}{2} \quad \text{as } \beta \to 0 \tag{3.112}$$

At the other extreme, for strong excitation,

$$\sigma \to \sqrt{3}\beta \quad \text{as } \beta \to \infty \tag{3.113}$$

The first order density function for y, $w(y)$, is given by

$$w(y) = \int_{-\infty}^{\infty} w(y, \dot{y}) d\dot{y} \tag{3.114}$$

On combining equations (3.101) and (3.114) one can obtain $w(y)$ by integration. Again, for the case where $G(a) = 0$, simple asymptotic results can be found [37]. As $\beta \to 0$

$$\begin{aligned} w(\lambda) &\to \frac{1}{\pi}\left(1 - \frac{\lambda^2}{4}\right)^{\frac{1}{2}} & |\lambda| < 2 \\ &\to 0 & |\lambda| > 2 \end{aligned} \tag{3.115}$$

where $w(\lambda)$ is the density function for $\lambda = y/\sigma$. Also, as $\beta \to \infty$,

$$w(\lambda) \to \frac{3\lambda}{\pi} K_1(\sqrt{3}\lambda) \tag{3.116}$$

where K is the modified Bessel function.

The characteristic frequency, w_c, can be evaluated from a combination of equations (3.21), (3.23), and (3.94). The function $B(a)$ is here given by [37].

$$\begin{aligned} B(a) &= \frac{1}{\pi}\left(\wedge - \frac{1}{2}\sin 2\wedge\right) & a > 1 \\ &= a & a \leq 1 \end{aligned} \tag{3.117}$$

where

$$\cos\wedge = 1 - \frac{2}{a} \tag{3.118}$$

For the case where $G(a) = 0$, some numerical results are presented by Roberts [37]. As expected, $w_c \to 1$ for low levels of excitation ($\beta \to 0$) and $w_c \to \sqrt{r}$, the natural frequency corresponding to the secondary slope, for high levels of excitation ($\beta \to \infty$).

3.4.2 Comparison with Simulation Results

Figure 3.6 shows the theoretical variation of σ with β, as computed from equation (3.109) ($G(a) = 0$).

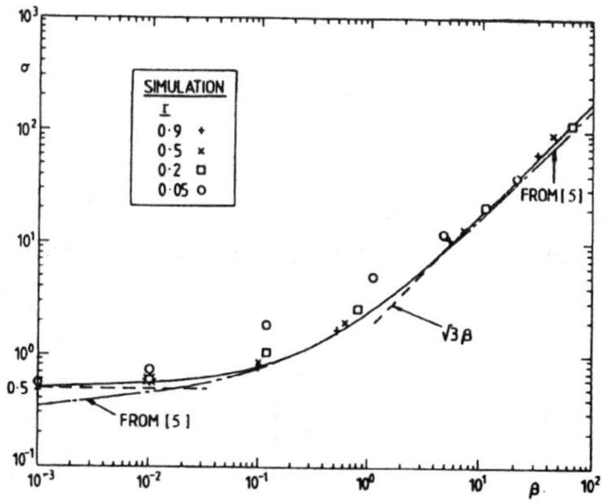

FIGURE 3.6
Variation of σ with β and a comparison with digital simulation results, for various values of α: Bilinear hysteresis. (From Roberts, J.B., *Journal of Applied Mechanics,* 1978. With permission).

The low and high asymptotes, given by equations (3.112) and (3.113), respectively, are also shown. Digital simulation estimates of σ are shown for $r = 0.9$, 0.5, 0.2, and 0.005. As the basic assumptions of the theory would lead one to expect, there is a close agreement with the simulation results when r is close to unity. However, even for values of r as low as 0.2, the theory still provides a fairly good guide to σ, over the whole range of β. The response is narrow-band when β is low and when β is high, whilst in the intermediate range of β, the response is wide-band when r is small. This accounts for the poor agreement when $r = 0.05$ and $10^{-2} < \beta < 10^1$. The theory still compares well with the simulation results for $r = 0.05$ if $\beta > 10^1$ or $\beta < 10^{-2}$. It is noted that if $r = 0$, the response is non-stationary and thus a comparison between theory and simulation is not possible, in this case.

Also shown in figure 3.6 is the theoretical variation of σ with β, as calculated by Caughey's equivalent linearisation method [5]. In the present notation, Caughey's result can be written as

$$\beta = \frac{1}{2}\sqrt{\frac{\pi}{2}}\sigma \operatorname{erfc}\left\{\frac{1}{\sqrt{2}\sigma}\right\} \tag{3.119}$$

It is seen from figure 3.6 that this result gives values of τ which are virtually identical to those computed by the present method, in the range $10^{-1} < \beta < 10^{-1}$. As $\beta \to \infty$, equation (3.119) gives the asymptote

$$\sigma \to 2\sqrt{\frac{2}{\pi}}\beta = 1.60\beta \tag{3.120}$$

which is about 8% below the asymptote found by the present method (equation (3.113). The discrepancy between equation (3.119) and the present theory at low values of β can be attributed to the fact that the equivalent linearisation method assumes a Rayleigh distribution for $a(\tau)$ whereas the present theory shows that there is a pronounced deviation as $\beta \to 0$. Figure 3.7(a) shows the first-order density function for $y(\tau)$, as calculated from equations (3.101) and (3.114), for a low level of excitation ($\beta = 0.01$). Again $G(a) = 0$. Here the amplitude limiting effect of the yielding phenomenon results in a highly non-Gaussian density function. The simulation results for $r = 0.9$ agree very closely with the theoretical density function and, for $r = 0.5$, the agreement is still reasonable. Figure 3.7(b) shows the first-order density function for $y(\tau)$ for a high level of excitation ($\beta = 22.7$). At this large value of β the asymptotic result given by equation (3.116) is a very close approximation and the theoretical line shown in figure 3.7 was obtained from this equation. The simulation results shown, for $r = 0.05$, are in excellent agreement with the asymptotic theory.

3.4.3 Yield Statistics

Consider one cycle of the response process, of amplitude a, where $a > 1$. The corresponding restoring force-displacement hysteresis loop for such a cycle is shown in figure 3.8. For a sufficiently small excitation level, as measured by β, the hysteresis loop will be symmetric about the line $g(y, \tau) = y$ since the amplitude process, $a(\tau)$ changes only slightly during one cycle. Under these circumstances, the plastic displacement, Y, shown in figure 3.8, is simply related to a by the equation

$$Y = 2(a - 1) \tag{3.121}$$

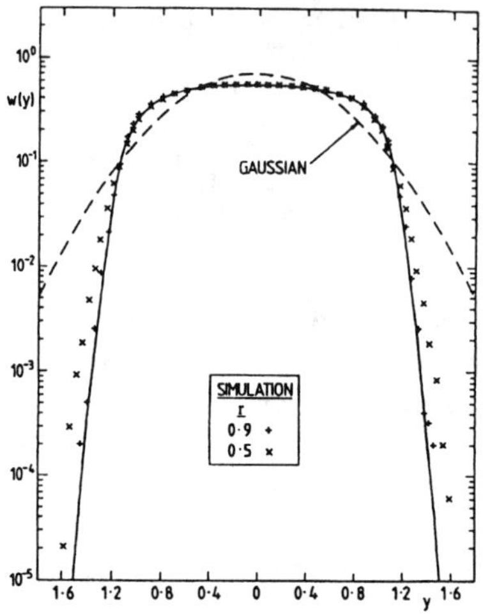

a)

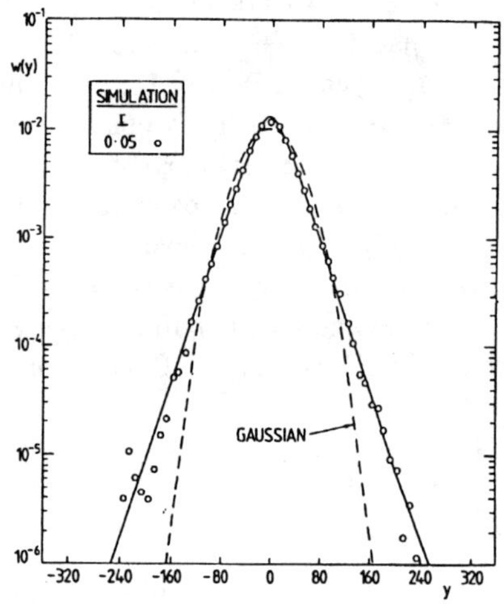

b)

FIGURE 3.7

(a) The first order density function of the response $y(\tau)$, and a comparison with digital simulation estimates: Bilinear hysteresis — $\beta = 0.01$; (b) The first order density function of the response $y(\tau)$, and a comparison with digital simulation estimates: Bilinear hysteresis — $\beta = 22.7$. (From Roberts, J.B., *Journal of Applied Mechanics,* 1978. With permission).

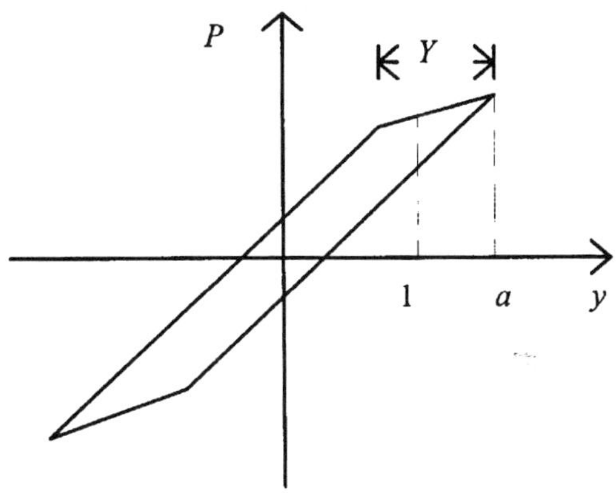

FIGURE 3.8
The relationship between the plastic yield displacement Y and the cycle amplitude a.

Furthermore, the probability that a yield will occur in any one cycle, P_Y is equal to the probability that a exceeds unity, i.e.,

$$P_Y = \int_1^\infty w(a)da \tag{3.122}$$

These relationships, in combination with equation (3.102), enable one to evaluate various yield statistics.

For the small levels of excitation assumed here, one can take $\omega_c = 1$, as shown earlier.

The rate of yielding, i.e., the average number of "upward" yields (yields with positive velocity) per unit time n^+ say, will equal the average number of downward yields per unit time. With

$$N^+ = 2\pi n^+/\omega_0 \tag{3.123}$$

defined as the average number of yields per cycle (and similarly for N^-), it is clear that this quantity is identical to the probability of a yield occurring in any one cycle, P_Y. Thus

$$N^+ = N^- = P_Y \tag{3.124}$$

where P_Y is given by equation (3.122).

The variation of P_Y with β can be found by straightforward numerical integration. In the particular case when $G(a) = 0$, it is found that

$$N^+ = \frac{2\,F_1(\beta)}{1 + 2\,F_1(\beta)} \tag{3.125}$$

where $F_1(\beta)$ is defined by equation (3.110) (here $n = 1$). The same result can be obtained by using Rice's formula for the rate of upcrossing and assuming that the yield levels have a negligibly small dispersion about the value of unity [38]. The present, quite different deviation is simpler and more direct.

When β is very small it is possible to obtain a useful asymptotic approximation for N^+. In the case where $h(y, \dot{y})$ is linear (see equation (3.76) the result is [38]

$$N^+ \rightarrow \frac{2\theta}{(1 - e^{-\theta})}\left[\frac{\pi\beta}{2(1+\lambda)}\right]^{\frac{1}{2}} e^{-\gamma^2 - \theta}(1 - \operatorname{erf}\gamma) \tag{3.126}$$

as $\beta \rightarrow 0$, where

$$\theta = \frac{\lambda}{2\beta}\,,\ \gamma = \frac{\lambda}{[2\beta(1+\lambda)]^{\frac{1}{2}}}\,,\ \lambda = \frac{\pi\zeta}{2(1-r)} \tag{3.127}$$

when the linear damping coefficient goes to zero ($\zeta \rightarrow 0$), equation (3.126) simplifies to

$$N^+ \rightarrow (2\pi\beta)^{\frac{1}{2}} \quad \text{as} \quad \beta \rightarrow 0 \tag{3.128}$$

Using the relationship between the plastic displacement Y, and the cycle amplitude, a, as given by equation (3.121), together with equation (3.99) for $w(a)$, the density function for Y, $p(Y)$, can be found. The result is

$$p(Y) = K(1 + \frac{Y}{2})^{1+1/\beta}\exp\left\{-\frac{1}{\beta}\left[\frac{Y}{2} + \frac{\pi G(1+\frac{Y}{2})}{2(1-r)}\right]\right\} \tag{3.129}$$

where K is another normalisation constant. From this result, the mean plastic displacement

$$m_Y = \int_0^\infty p(Y)Y\,dY \tag{3.130}$$

and the standard deviation of this displacement

$$\sigma_Y = \left\{ v - m_Y^2 \right\}^{\frac{1}{2}}, \quad v = \int_0^\infty p(Y) Y^2 dY \tag{3.131}$$

can be computed.

For very small values of β, a straightforward computation of m_Y and σ_Y can present numerical difficulties. These can be avoided by using an asymptotic form for $w(a)$ [38]. The corresponding asymptotic form for $p(Y)$ is

$$p(Y) = K \exp \left\{ -\frac{1}{\beta} \left[\frac{Y^2}{8} + \frac{\pi G(1 + \frac{Y}{2})}{2(1 - r)} \right] \right\} \tag{3.132}$$

For the special case of linear damping an explicit asymptotic approximation for m_Y, valid as $\beta \to 0$, may be found. Thus

$$m_Y \to \left[\frac{8\beta}{(1 + \lambda)} \right]^{\frac{1}{2}} \left\{ \frac{\exp(-\gamma^2)}{1 - \operatorname{erf}\gamma} - \gamma \right\} \tag{3.133}$$

When the damping coefficient $\zeta \to 0$, then $\gamma = 0$ and equation (3.133) reduces to

$$m_Y \to \left(\frac{8\beta}{\pi} \right)^{\frac{1}{2}} \quad \text{as } \beta \to 0 \tag{3.134}$$

Similarly, an asymptotic expression for σ_Y can be found. One obtains

$$v \to \frac{8\beta}{(1 + \lambda)} \left\{ \gamma^2 + \frac{1}{2} - \frac{\gamma e^{-\gamma^2}}{\pi^2 (1 - \operatorname{erf}\gamma)} \right\} \tag{3.135}$$

This, together with equations (3.131) and (3.133), gives the asymptotic form for σ_y. In the case where $\zeta \to 0$ the result reduces to

$$\sigma_Y \to 2 \left[\beta \left(1 - \frac{2}{\pi} \right) \right]^{\frac{1}{2}} \tag{3.136}$$

It has been shown by the author [38] that the above analytical results are in excellent agreement with corresponding simulation estimates.

It is interesting to note that the agreement with simulation also occurs in the case where $r = 0$ (zero secondary slope). As mentioned earlier, in this special case the response is no longer stationary. There will be a cumulative "drift" due to successive yields, the variance of which will increase continuously with elapsed time. Nonetheless, the statistics of the yielding behavior will still be time-invariant and can be successfully predicted by the averaging theory. Thus the amplitude distribution which is obtained can be interpreted, for the case $r = 0$, as relating to the "stationary component" of the process, $y(\tau)$, which is superimposed on a slowly varying drift component.

3.5 Stationary Response of Oscillators with Curvilinear Hysteresis

For hysteresis of the Bouc-Wen type, governed by equation (3.3) the area enclosed by a loop, of amplitude a, can be found by evaluating the integral in equation (3.23).

For the case $n = 1$ and β positive, an analytical expression for $A(a)$ can be found: thus

$$A(a) = 2A\left\{\left[\frac{\ln(1-\chi_0)+\chi_0}{\lambda_d^2}\right] - \left[\frac{\ln(1-\theta_0)+\theta_0}{\lambda_a^2}\right]\right\} \tag{3.137}$$

where

$$\chi_0 = \frac{\lambda_d z_m}{A}\ , \quad \theta_0 = \frac{\lambda_u z_m}{A} \tag{3.138}$$

$$\lambda_u = \beta + \gamma\ , \quad \lambda_d = \beta - \gamma \tag{3.139}$$

and z_m is the limiting value of the maximum of z as $a \to \infty$. On combining this result with (from equations (3.29), (3.30), and (3.21))

$$\varepsilon^2\eta(a) = \varepsilon^2\alpha(a) + \frac{(1-r)}{2\pi a}A(a) \tag{3.140}$$

and equation (3.86), the density function for $w(a)$ can be evaluated. The required integration must be performed numerically.

In the case where the excitation level is relatively high, the evaluation of $w(a)$ can be simplified very considerably. It is evident that $A(a)$ will grow linearly with a, as a becomes large, i.e.,

$$A(a) \to \mu + \lambda a \tag{3.141}$$

where μ and λ are constant, which depend on the loop parameters. This asymptotic form is true for both bilinear and curvilinear hysteretic loops. Referring to equation (3.97) for the loop area of bilinear hysteresis, and comparing with equation (3.141), it is seen that

$$\mu = -4 \ , \ \lambda = 4 \ \ (a > 1) \tag{3.142}$$

Thus, for bilinear hysteresis the asymptotic form for (3.141) is exact for all values of a greater than unity.

Returning to the curvilinear case, expressions for μ and λ can be found once $A(a)$ is evaluated. For $n = 1$ one finds, from equation (3.137), that

$$\mu = \frac{2A}{\lambda_d}\ln\left(1 - \frac{\lambda_d}{\lambda_u}\right)\left(\frac{1}{\lambda_d} + \frac{1}{\lambda_u}\right) - \frac{2A}{\lambda_u}\left(\frac{1}{\lambda_u} - \frac{1}{\lambda_d}\right) \tag{3.143}$$

$$\lambda = \frac{4A}{\lambda_u} \tag{3.144}$$

On substituting the asymptotic form for $A(a)$ into equation (3.140), and employing equation (3.86) again, the following result is obtained [40]:

$$w(a) = ca^{(1+\rho)}\exp(- \wedge a) \tag{3.145}$$

where

$$\rho = -\frac{\mu\omega_c(1-\alpha)}{\pi^2 S(\omega_c)} \tag{3.146}$$

$$\wedge = \frac{\lambda\omega_c(1-\alpha)}{\pi^2 S(\omega_c)} \tag{3.147}$$

When the inverse "strength parameter", ρ, becomes small (high level of excitation), equation (3.145) simplifies to

$$w(a) = ca\,\exp(- \wedge a) \tag{3.148}$$

This asymptotic result can be compared with the corresponding expression obtained for the use of bilinear hysteresis (see equation (3.106)). For bilinear hysteresis μ and λ are given by equation (3.142): hence

$$\wedge = -\rho = \frac{1}{\beta} \tag{3.149}$$

where β is defined by equation (3.100). Then equation (3.106) can be seen to correspond exactly with equation (3.148).

3.5.1 Response Statistics

From equation (3.145) some simple asymptotic results can be obtained for the statistics of y and $\dot{y}$. For example, the standard deviation of y, σ, is given by

$$\sigma^2 = \frac{(3+\rho)(2+\rho)}{2\wedge^2} \tag{3.150}$$

and for very high levels of excitation, this reduces to

$$\sigma^2 = \frac{3}{\wedge^2} \tag{3.151}$$

This corresponds to the asymptotic result obtained for the bilinear oscillator (see equation (3.113)).

The density function for y, $w(y)$, is found to be

$$w(y) = 2cy^{(1+\rho)}\omega_c \int_1^\infty \frac{\xi^{1+\rho}\exp(-\gamma\xi)}{\sqrt{\xi^2-1}}d\xi \tag{3.152}$$

where

$$\gamma = \wedge y^2 \tag{3.153}$$

For large γ the following asymptotic expressions can be used

$$w(y) \rightarrow 2cy^{(1+\rho)}\omega_c\sqrt{\frac{\pi}{2\gamma}}e^{-\gamma}\left[1 - \frac{(1+\rho)}{2\gamma} + \ldots\right] \tag{3.154}$$

Similarly, from equations (3.84) and (3.145) the following expressions for the upward crossing rate, ν^+, of a fixed level, b, may be obtained, using Rice's formula:

$$\nu^+ = cb^{\rho+2}\omega_c^2\gamma^{-(2+\rho)}\Gamma(2+\rho,\gamma) \tag{3.155}$$

For large γ one finds the simple asymptote

$$\nu^+ \rightarrow \frac{cb^{\rho+2}\omega_c^2\exp(-\gamma)}{\gamma} \tag{3.156}$$

3.5.2 Comparison with Experimental Results

The theoretical results presented in the last section have been compared with some experimental results obtained from dynamic tests on a simplified version of a Stockbridge damper, as shown in figure 3.9. Here the stranded cable acts as the hysteretic restoring element and the transverse vibration of the system, in the first "mode", can be modeled, to a good degree of approximation, as a single degree of freedom oscillator. Figure 3.10(a) shows the hysteretic behavior of the cable, obtained by applying a transverse fluctuating force at the tip of the cantilever, with the mass removed.

This behavior can be accurately modeled by the Bouc-Wen representation, applying parametric identification methods to estimate the loop parameters [51]. In the present case a very good fit can be obtained with $n = 1$. Figure 3.10(b) shows the theoretical hysteretic behavior predicted from the Bouc-Wen model, using the identified parameters.

Since the loops are slim, the Stockbridge averaging theory is appropriate. In fact, the experimental conditions relate to condition 3, in section 3.4.2, since most of the response relates to the near horizontal portions of the corresponding z loops.

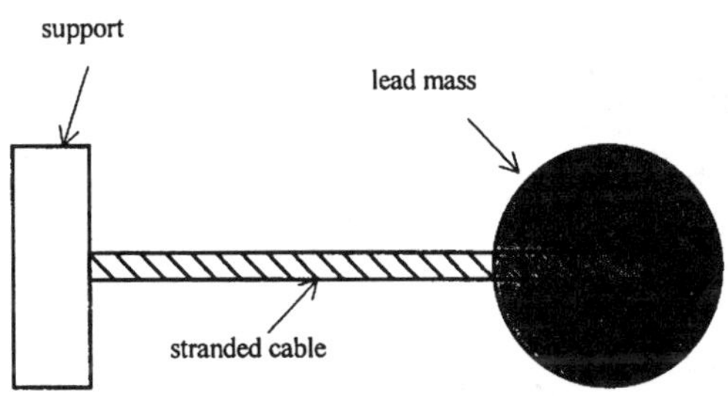

FIGURE 3.9
Mass-cable system.

A comparison between theoretical and experimental density functions for $a(\tau)$ is shown in figure 3.11 (for more detail see [40]). It can be seen that the stochastic averaging theory gives an excellent agreement with the experimental results, correctly predicting the effect of non-linearity on the distribution of the response. For comparison the "linear" Rayleigh density function is also shown in figure 3.11, based on the measured standard deviation. This comparison shows that linearisation will, in this case, lead to a serious under-prediction of the probability of reaching large amplitude levels.

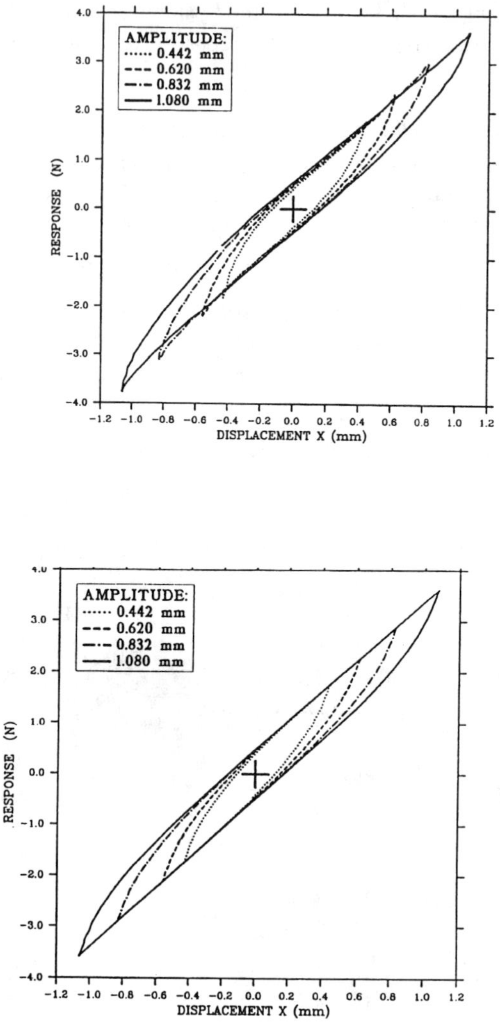

a)

b)

FIGURE 3.10

(a) Experimental hysteretic loops; (b) Reconstructed hysteretic loops using identified parameters. (From Roberts, J.B. and Sadeghi, A.H. in *Structural Safety*, 1990. With permission).

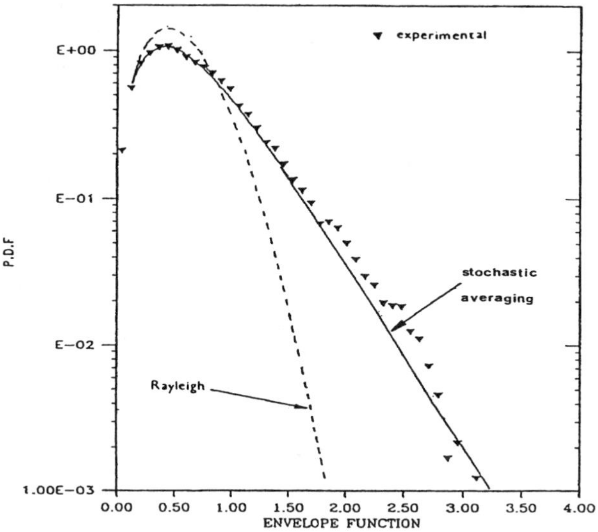

FIGURE 3.11
Probability density functions for the envelope process, a. (From Roberts, J.B. and Sadeghi, A.H., *Nonlinear Stochastic Mechanics*, 1991. With permission).

3.6 Non-stationary Excitation and Response

Equation (3.8) will be considered again, but now it will be assumed that the excitation, $f(\tau)$, is a non-stationary, wide-band process, with zero-mean. Such a process can be described in terms of an "evolutionary power spectrum", $S_f(\omega, t)$, where

$$S_f(\omega, t) = \mid A(t, \omega) \mid^2 \overline{S}_f(\omega) \tag{3.157}$$

Here $A(t, \omega)$ and $\overline{S}_f(\omega, t)$ are appropriate functions and the symbol $\mid \mid$ denotes the modulus of a complex function [55]. In the special case where the excitation is stationary

$$S_f(\omega, t) = S_f(\omega) \tag{3.158}$$

independent of t, where $S_f(\omega)$ is the usual power spectrum.

The analysis given in section 3.2 can be generalised to derive a Markov model for the joint process $[a, \phi]$, again. However, the use of a characteristic frequency, ω_c, is now not appropriate since ω_c cannot be evaluated as before, on the basis of a stationary response to a stationary input. This difficulty can be easily overcome by simply replacing ω_c with $\omega(a)$, where $\omega(a)$ can be found by the Krylov-Bogoliubov method, and is in fact given by equation (3.21).

As shown in [39], the appropriate FPK equation for the amplitude process $a(\tau)$ is given by

$$\frac{\partial p}{\partial \tau} = \frac{\partial}{\partial a}\left[\left\{\frac{\varepsilon^2\eta(a)}{\omega(a)} + \frac{\pi S_f[\omega(a),\tau]}{2a\omega^2(a)}\right\} p\right] + \frac{\pi S_f[\omega(a),\tau]}{2\omega^2(a)}\frac{\partial^2 p}{\partial a^2} \tag{3.159}$$

and, as in the stationary case, the amplitude process is uncoupled from the phase process, provided that $a = 0$ at $\tau = 0$. This initial condition will be assumed implicitly, henceforth. The true dependent joint density function $p(a, \phi : \tau)$ is thus given by (similar to equation (3.82))

$$p(a, \phi : \tau) = \frac{1}{2\pi} p(a : \tau) \tag{3.160}$$

where $p(a : \tau)$ is governed by equation (3.159). The joint density function for $y, \dot{y}$ is (similar to equation (3.84)), given by

$$p(y, \dot{y} : \tau) = \frac{1}{2\pi\omega(a)a} p(a : \tau) \tag{3.161}$$

An alternative to equation (3.159), which has an approximation error consistent with that equation, is

$$\frac{\partial p}{\partial \tau} = \frac{\partial}{\partial a}\left[\left\{\frac{\varepsilon^2\eta(a)}{\omega(a)} + \frac{\pi S_f[\omega(a),\tau]}{2a\omega^2(a)}\right\} p\right] + \frac{1}{2}\frac{\partial^2}{\partial a^2}\left[\frac{\pi S_f[\omega(a),\tau]}{\omega^2(a)} p\right] \tag{3.162}$$

As will be shown shortly, this equation can be readily solved using a physically based explicit finite difference scheme and will be used here in preference to equation (3.159).

Equation (3.162) can be expressed in the standard form

$$\frac{\partial p}{\partial \tau} = -\frac{\partial}{\partial a}[A(a,\tau)p] + \frac{1}{2}\frac{\partial^2}{\partial a^2}[B(a,\tau)p] \tag{3.163}$$

where

$$A(a,\tau) = \frac{\varepsilon^2\eta(a)}{\omega(a)} + \frac{\pi S_f[\omega(a),\tau]}{2a\omega^2(a)} \tag{3.164}$$

and

$$B(a,\tau) = \frac{\pi S_f[\omega(a),\tau]}{\omega^2(a)} \tag{3.165}$$

3.6.1 Numerical Solution of the FPK Equation

A diffusion equation, of the general form of equation (3.163), can be solved by a simple explicit finite difference scheme.

If τ is discretised into times $\tau_j = j\delta\tau$ and the amplitude a is discretised into the values $a_k = k\delta a$, where $\delta\tau$ and δa are (small) constant increments, then a suitable finite difference approximation to equation (3.163) is as follows:

$$\begin{aligned}\frac{p(k, j+1) - p(k, j)}{\delta\tau} &\\ = -\frac{1}{2\delta a}&\{p(k+1, j)A(k+1, j) - p(k-1, j)A(k-1, j)\}\\ +\frac{1}{2\delta a^2}&\{p(k+1, j)B(k+1, j) - 2p(k, j)B(k, j)\\ &+p(k-1, j)B(k-1, j)\}\end{aligned} \tag{3.166}$$

Here $p(k, j)$ denotes $p(a_k, \tau_j)$, and similarly for $A(k, j)$ and $B(k, j)$. After some re-arrangement equation (3.166) can be expressed as follows:

$$p(k, j+1) = p(k-1, j)\rho_{k-1,j} + p(k, j)\nu_{k,j} + p(k+1, j)\mu_{k+1,j} \tag{3.167}$$

where

$$\rho_{k,j} = \frac{\lambda}{2}[B(k, j) + A(k, j)\delta a] \tag{3.168}$$

$$\nu_{k,j} = 1 - \lambda B(k, j) \tag{3.169}$$

$$\mu_{k,j} = \frac{\lambda}{2}[B(k, j) - A(k, j)\delta a] \tag{3.170}$$

and

$$\lambda = \frac{\delta\tau}{\delta a^2} \tag{3.171}$$

Equation (3.167) provides a simple method of computing the values $p(k, j+1)$ at times τ_{j+1} from the values $p(k, j)$ at the preceding time τ_j. Thus the solution can be marched forward in time, from some initial condition.

A slightly different interpretation of equation (3.167) is obtained by multiplying throughout by δa. Then

$$P(k, j+1) = P(k-1, j)\rho_{k-1,j} + P(k, j)\nu_{k,j} + P(k+1, j)\mu_{k+1,j} \tag{3.172}$$

where

$$P(k, j) = p(k, j)\delta a \tag{3.173}$$

$P(k, j)$ can be interpreted as the (finite) probability of being in state (a_k, τ_k); thus the original, continuous probability "mass" is discretised into a set of "lumped masses" at the nodal points $a_k = k\delta a$. Clearly, for any j

$$\sum_k P(k, j) = 1 \tag{3.174}$$

Thus equation (3.172) represents an equation for diffusing finite probability "masses" over the discrete space a_k, τ_j.

The discrete process obeying equation (3.172) is, in fact, a random walk process. Thus, consider a random walk, $R(\tau_j)$, which can only assume the discrete amplitude a_k. If the process is at state (a_k, τ_j) then assume that it can do one of the following

(a) move to state (a_{k+1}, τ_{j+1}) with probability $\rho_{k,j}$

(b) move to state (a_k, τ_{j+1}) with probability $\nu_{k,j}$

(c) move to state (a_{k-1}, τ_{j+1}) with probability $\mu_{k,j}$.

This is represented graphically in figure 3.12(a).

Since only these three possibilities are allowed, it is evident that one must have

$$\rho_k + \mu_k + \nu_k = 1 \tag{3.175}$$

It follows that state a_k, τ_{j+1} can only be reached from (a_{k+1}, τ_j), (a_k, τ_j), or (a_{k-1}, τ_j), as illustrated in figure 3.12(b). Thus the probability $P(k, j+1)$ of being in state a_k, τ_{j+1} is, in fact, given by equation (3.172). Hence ρ_k, ν_k, and μ_k can be interpreted as transition probabilities, satisfying equation (3.175) at every j. This random walk scheme is a generalisation of that discussed in [56], where it was assumed that ν_k is zero, and that $A(a, \tau)$ and $B(a, \tau)$ are independent of τ (stationary excitation case). Clearly, from equation (3.169), if $\nu_k = 0$, then

$$\lambda = \frac{1}{B(k, j)} \tag{3.176}$$

Since λ must be constant, equation (3.176) can only be satisfied if $B(k, j)$ is independent of both k and j. The inclusion of a non-zero transition probability, ν_k, into the numerical scheme allows one to solve the more general type of diffusion equation where $B(a, \tau)$ is not a constant.

In the special case of stationary excitation the transition densities are independent of j and, therefore, need only be computed once, at the beginning of the

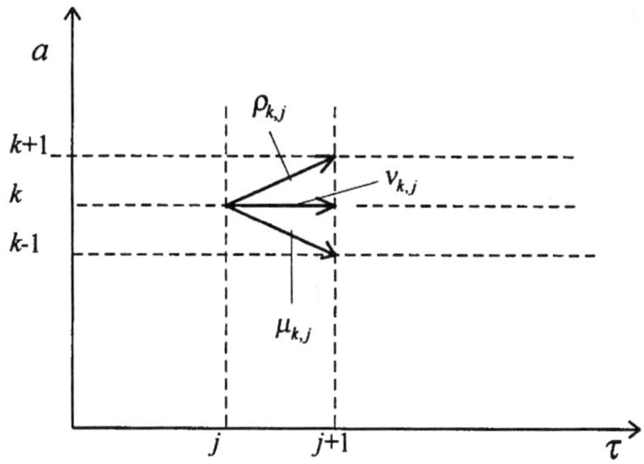

a)

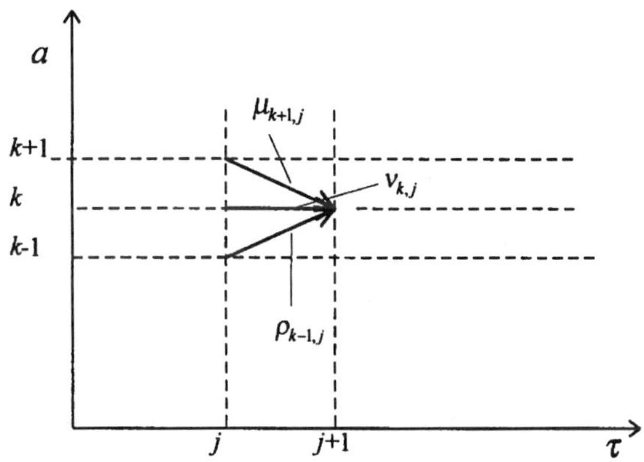

b)

FIGURE 3.12
(a) and (b) The random walk analogue for $a(\tau)$.

numerical diffusion process. In the more general, non-stationary case, they must be recalculated at every step.

In implementing the random walk scheme it is important to choose $\delta\tau$ and δa (and hence λ) such that the transition probabilities remain positive, otherwise there is a strong tendency towards numerical instability. Care must also be exercised in the vicinity of the lower amplitude limit $a = 0$, since $A(a, \tau) \to \infty$ as $a \to 0$. This difficulty can be overcome by treating $a = 0$ as a reflecting barrier and modifying the transition probabilities in the vicinity of the barrier accordingly.

3.6.2 Comparison with Simulation Results

Some comparisons have been made between predictions using the foregoing theory and digital simulation results [39]. For this purpose an excitation of the form

$$f(\tau) = \chi(\tau)w(\tau) \tag{3.177}$$

was assumed, where $\chi(\tau)$ is a deterministic modulating function and $w(\tau)$ is a stationary white noise.

Figure 3.13 shows a typical comparison, for the standard deviation of the response, for the case where

$$\begin{aligned} \chi(\tau) &= \exp(-\rho\tau) \qquad \tau \geq 0 \\ &= 0 \qquad\qquad\qquad \tau < 0 \end{aligned} \tag{3.178}$$

and ρ is a constant. Good agreement is obtained for both the chosen values of ρ (see [39] for details). For unit step modulation ($\rho = 0$), the standard deviation approaches a stationary value as time elapses; in the case of exponential modulation, the level of the response decays slowly as τ increases.

3.7 The Energy Envelope Method

As an alternative to introducing a characteristic frequency, ω_c, which depends on the level of excitation, or an amplitude dependent natural frequency, $\omega(a)$, one can employ an averaging method based on the energy envelope [36]. This approach will be illustrated by considering equation (3.8) again, in the case where the excitation is a zero-mean, stationary process. It will be assumed that z is a Bouc-Wen type hysteresis, governed by equation (3.3).

If the energy lost per cycle, through dissipation, is relatively small then the backbone model discussed in section 3.2.3 can be employed. Then the equation

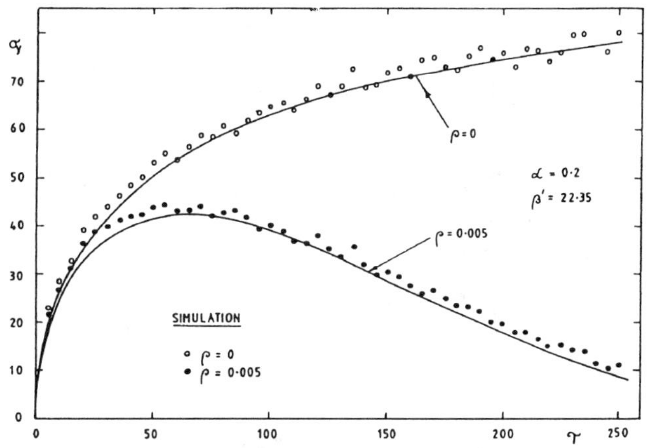

FIGURE 3.13
Variation of the standard deviation, $\sigma_y(\tau)$, with time $\alpha = 0.2, \beta = 22.5, \rho = 0$ and 0.005. Comparison with digital simulation estimates. (From Roberts, J.B. and Sadeghi, A.H., *Stochastic Structural Dynamics — Progress in Theory and Applications,* 1988. With permission).

of motion can be written as

$$\ddot{y} + \varepsilon^2 h(y, \dot{y}) + H(y) + (1 - r)q(y, \tau) = f(\tau) \tag{3.179}$$

where

$$H(y) = (1 - r)B(y) + ry \tag{3.180}$$

When the energy lost through viscous and hysteretic damping is small, the energy envelope

$$E(\tau) = \frac{\dot{y}^2}{2} + V(y) \tag{3.181}$$

where

$$V(y) = \int_0^y H(\xi)d\xi \tag{3.182}$$

will be slowly varying (in a macroscopic sense). In these circumstances, it is possible to obtain an approximate, first order stochastic equation for $E(t)$ [36].

On multiplying equation (3.179) throughout by $\dot{y}$ one finds, using equation (3.181), that

$$\dot{E} = -\dot{y}\left[\varepsilon^2 h(y, \dot{y}) + (1-r)q(y, \tau)\right] + \dot{y}f(\tau) \tag{3.183}$$

The first term on the right-hand side of this equation is the rate at which energy is lost due to damping. It can be approximated by averaging over one cycle of free, undamped oscillation. Thus

$$Q(E) = \frac{1}{T(E)} \int \dot{y}_0 \left[\varepsilon^2 h(y_0, \dot{y}_0) + (1-r)q(y_0, \tau)\right] d\tau \tag{3.184}$$

is the averaged rate of energy dissipation, where y_0 is the solution of

$$\ddot{y}_0 + H(y_0) = 0 \tag{3.185}$$

and $T(E)$ is the period of free, undamped oscillations, $y_0(\tau)$, given by

$$T(E) = 2\sqrt{2} \int_0^b \frac{d\xi}{[E - V(\xi)]^{\frac{1}{2}}} \tag{3.186}$$

where b is such that $V(b) = E$.

Equation (3.183) can now be approximated by

$$\dot{E} = -Q(E) + \dot{y}f(\tau) \tag{3.187}$$

Assuming now that $f(\tau)$ can be adequately modeled as a white noise, the second term on the right-hand side of equation (3.187) can also be averaged over the period $T(E)$. Thus, if the correlation function of $f(\tau)$ is

$$w_f(s) = E\{f(\tau)f(t+s)\} = I\delta(s) \tag{3.188}$$

where I is the "strength" of the input, then the fully averaged form of equation (3.187) is [36]

$$dE = \left[-Q(E) + \frac{I}{2}\right] dt + [IC(E)]^{\frac{1}{2}}\, dW \tag{3.189}$$

where W is a unit Wiener process and

$$C(E) = \frac{4\sqrt{2}}{T(E)} \int_0^b [E - V(\xi)]^{\frac{1}{2}} \, d\xi \tag{3.190}$$

As written, equation (3.189) should be interpreted in the Itô sense [36]. It follows that $E(\tau)$ is, approximately, a one-dimensional Markov process with a transition density function, $p(E \mid E_0; \tau)$ governed by the following FPK equation

$$\frac{\partial p}{\partial \tau} = \frac{\partial}{\partial E}\left[\left\{Q(E) - \frac{I}{2}\right\} p\right] + \frac{I}{2}\frac{\partial^2}{\partial E^2}[C(E)p] \tag{3.191}$$

The stationary density function, obtained by setting $\partial p/\partial \tau = 0$, is readily found to be

$$w(E) = kT(E)\exp\left\{-\frac{2}{I}\int_0^E \frac{Q(\xi)}{C(\xi)} d\xi\right\} \tag{3.192}$$

where k is a normalisation constant.

It is also possible to deduce an expression for the joint density function, $p(y, \dot{y})$ of $y(\tau)$ and $\dot{y}(\tau)$, in terms of $p(E)$. Thus [36]

$$p(y, \dot{y}) = \frac{p(E)}{T(E)} \tag{3.193}$$

where $E = \dot{y}^2/2 + V(y)$. Many useful statistics can be derived from equation (3.193). In particular, the mean square response, $\sigma^2 = E\left\{y^2(t)\right\}$, is given by

$$\sigma^2 = \int_0^\infty p(E) D_2(E) dE \tag{3.194}$$

where

$$D_2(E) = \frac{2\sqrt{2}}{T(E)} \int_0^b \frac{\xi^2 d\xi}{[E - V(\xi)]^{\frac{1}{2}}} \tag{3.195}$$

In the following, only the special case $r = 0$, $h(y, \dot{y}) = 0$ will be considered, i.e., the situation where all the damping and stiffness derives from the hysteretic characteristic $z(y, \tau)$. In this situation

$$Q(E) = \frac{A(E)}{T(E)} \tag{3.196}$$

where

$$A(E) = \oint z(y_0)dy_0 \tag{3.197}$$

is the area enclosed by the hysteretic loop, as defined earlier (see equation (3.23). Also, in the following, attention is focused on the case where $n = 1$.

3.7.1 Calculation of the Backbone

Here, as proposed in section 3.2.3, the backbone is defined as the locus of the extremities (y_m, z_m) of the loop. Thus $B(y)$ is defined by $B(y_m) = z_m$.

The relationship between y_m and z_m is easily obtained by integrating equation (3.3) and using a symmetry condition. One finds that

$$2y_m = -\frac{1}{\lambda_u}\ln\left[1 - \frac{\lambda_u z_m}{A}\right] - \frac{1}{\lambda_d}\ln\left[1 - \frac{\lambda_d z_m}{A}\right] \tag{3.198}$$

where λ_u and λ_d are defined by equations (3.139). This non-linear algebraic relationship must, in general, be solved numerically. However, if γ is small the hysteresis loop is slim and one can use the following simple asymptotic approximation:

$$B(y) = \frac{A}{\beta}\left[1 - \exp(-\beta y)\right] \tag{3.199}$$

Irrespective of the value of γ the backbone is linear, as $y_m \to 0$; thus $B(y) \to Ay$.

3.7.2 Calculation of the Area Enclosed by a Loop

An exact expression for the loop area can be found (see equation (3.137). This can be transformed into a function of E by using the expressions

$$z_m = B(y_m)\ ;\ E = V(y_m) \tag{3.200}$$

Asymptotic expressions for $A(E)$ can be found in certain special cases. For example

$$A(E) \to \frac{8}{3}\left(\frac{2}{A}\right)^{\frac{1}{2}} \gamma E^{3/2} \quad E \to 0 \tag{3.201}$$

irrespective of the magnitudes of β and γ. The above expression for $A(E)$ also holds, asymptotically, for any E, if $\beta = 0$ and $\gamma \to 0$.

For $\beta > 0$ the asymptotic expression for $A(E)$ given by equations (3.141), (3.143) and (3.144) are applicable. This expression is useful for the purposes of numerical evaluation since the computation of $A(E)$ from equation (3.137) can lead to numerical difficulties when E is large.

3.7.3 Calculation of $T(E)$, $C(E)$, and $D_2(E)$

The integral in equation (3.186) becomes infinite as $\xi \to b$. To remove this singularity it is convenient to use a transformation of variables, as follows:

$$V(y) = E \sin^2 \theta \tag{3.202}$$

Equation (3.186) can, therefore, be expressed as

$$T(E) = 4\sqrt{2E} \int_0^{\pi/2} \frac{\sin\theta}{g(y)} d\theta \tag{3.203}$$

and the integral can now be evaluated by straightforward methods. $C(E)$ and $D_2(E)$ can be evaluated using the same transformation [42].

Asymptotic expressions for the above functions of E are useful for the purpose of evaluating $p(E)$ and σ^2. As $E \to 0$ the backbone function $B(y) \to Ay$, as previously mentioned. Hence one finds that

$$\left.\begin{aligned} T(E) &\to 2\pi/A^{\frac{1}{2}} \\ C(E) &\to E \\ D_2(E) &\to E/A \end{aligned}\right\} \text{ as } E \to 0 \tag{3.204}$$

For $\beta > 0$ the following asymptotic expressions for the case $E \to \infty$ can also be found

$$\left.\begin{aligned} T(E) &\to 4\sqrt{2E}\beta/A \\ C(E) &\to 2E/3 \\ D_2(E) &\to \frac{8}{15}\left[\frac{E\beta}{A}\right]^2 \end{aligned}\right\} \text{ as } E \to \infty \tag{3.205}$$

3.7.4 The Loss Factor

The function $Q(E)$, defined by equation (3.196), represents the average rate of loss of energy, due to hysteretic damping. It is convenient to introduce a non-dimensional loss factor $L(E)$, defined as

$$L(E) = \frac{Q(E)}{2E} = \frac{A(E)}{2T(E)E} \tag{3.206}$$

which is such that, in the case of a linear, viscously damped oscillator, $L(E) = \zeta$, where ζ is the usual critical damping factor.

It has been shown that, in the linear case, the stochastic averaging approximation gives very good results if $\zeta < 0 \cdot 1$ [42]. Accordingly, it is reasonable to assume that in the present non-linear case the energy envelope averaging method will give good results when $L(E) < 0 \cdot 1$.

3.7.5 The Case $\beta = 0$

In the special case where $\beta = 0$ the backbone is linear. If furthermore, γ is small, so that the asymptotic area expression given by equation (3.201) is relevant, then it is possible to obtain a complete closed-form analytical solution for $p(E)$, as follows [42]:

$$p(E) = k \exp\left\{-\rho \frac{\gamma}{I} E^{3/2}\right\} \tag{3.207}$$

where

$$\rho = \frac{16\sqrt{2}}{9\pi} \sim 0.800 \tag{3.208}$$

and

$$k = \frac{3}{2}\left(\frac{\rho\gamma}{I}\right)^{2/3} \frac{1}{\Gamma(^2/_3)} \tag{3.209}$$

An explicit expression for the mean square displacement can also be found. The result is

$$\sigma^2 = \frac{\lambda}{A}\left[\frac{I}{\rho\gamma}\right]^{2/3} \qquad \gamma \to 0 \tag{3.210}$$

where

$$\lambda = \frac{\Gamma(^1/_3)}{3\Gamma(^2/_3)} \sim 0.66 \tag{3.211}$$

It is noted that the results given in this section could also be obtained by the standard stochastic averaging method, described earlier in this Chapter, since the backbone function is linear.

3.7.6 Comparison with Simulation

In the case of a stiffening backbone the present theory gives excellent agreement with simulation results, provided that the loss factor $L(E)$ is small (< 0.1). For the case of a softening backbone the present theory gives a good model for the non-drifting component of the response [42].

3.8 Concluding Remarks and References

It has been shown that the stochastic averaging method can be applied successfully to study the probability distribution of the response of oscillators, with various kinds of hysteretic restoring force, to broad-band random excitation. The standard method has been applied to oscillators with both bilinear and curvilinear hysteresis and a very good degree of agreement has been obtained with both simulation results and real experimental data, in cases where the response is narrow-band in character. Moreover, it has been shown that, when the backbone of the hysteresis is significantly non-linear and the loops are slim, a generalised version of the method, based on the energy envelope of the response is appropriate and gives good results.

Further work needs to be done in establishing whether, or not, the generalised stochastic averaging method leads to improvements in accuracy in all cases where the response is not narrow-band.

A disadvantage of the generalised method is that, unlike the standard method it is very difficult, if not impossible, to apply to multi-degree of freedom (MDOF) systems. Some interesting work remains to be done on the applicability of stochastic averaging methods to MDOF systems.

References

[1] Bouc, R., Modele Mathematique d'Hysteresis (in French), *Acustica*, 24, 1961, pp. 16–25.

[2] Wen, Y-K., Equivalent Linearisation for Hysteretic Systems under Random Excitation, *Journal of Applied Mechanics*, ASME, 47, 1980, pp. 150–154.

[3] Wen, Y-K., Methods of Random Vibration for Inelastic Structures, *Applied Mechanics Reviews*, ASME, 42, 1989, pp. 39–52.

[4] Roberts, J.B. and Spanos, P.D., *Random Vibration and Statistical Linearisation*, J. Wiley and Sons, Chichester, U.K., 1990.

[5] Caughey, T.K., Random Excitation of a System with Bilinear Hysteresis, *Journal of Applied Mechanics*, ASME, 27, 1960, pp. 649–652.

[6] Baber, T.T. and Wen, Y-K., Random Vibration of Hysteretic, Degrading Systems, *Journal of Engineering Mechanics*, ASCE, 107, 1981, pp. 1069–1087.

[7] Baber, T.T and Wen, Y.K., Stochastic Response of Multi-storey Yielding Frames, *Journal of Earthquake Engineering and Structural Dynamics*, 10, 1982, pp. 403–416.

[8] Baber, T.T., Nonzero Mean Random Vibration of Hysteretic Systems, *Journal of Engineering Mechanics*, ASCE, 110, 1984, pp. 1036–1049.

[9] Baber, T.T. and Noori, M.N., Random Vibration of Degrading, Pinching Systems, *Journal of Engineering Mechanics*, ASCE, 111, 1985, 1010–1026.

[10] Casciati, F. and Faravelli, L., Methods of Nonlinear Stochastic Dynamics for the Assessment of Structural Fragility, *Nuclear Engineering and Design*, 90, 1985, pp. 341–345.

[11] Baber, T.T., Nonzero Mean Random Vibration of Hysteretic Frames, *Computers and Structures*, 23, 1986, pp. 265–277.

[12] Noori, M., Choi, J-D., and Davoodi, H., Zero and Nonzero Mean Random Vibration Analysis of a New General Hysteresis Model, *Probabilistic Engineering Mechanics*, 1, 1986, pp. 192–201.

[13] Baber, T.T. and Noori, M., Modeling General Hysteresis Behavior and Random Vibration Application, *Journal of Vibration, Acoustics, Structures and Reliability in Design*, ASME, 108, 1986, pp. 411–420.

[14] Park, Y.J., Wen, Y-K., and Ang, A. H-S., Random Vibration of Hysteretic Systems under Bi-directional Ground Motions, *Earthquake Engineering and Structural Dynamics*, 14, 1986, pp. 543–557.

[15] Wen, Y-K., Stochastic Response and Damage Analysis of Inelastic Structures, *Probabilistic Engineering Mechanics*, 1, 1986, pp. 49–57.

[16] Faravelli, L., Casciati, F., and Singh, M.P., Stochastic Equivalent Linearisation Algorithms, *Meccanica*, 22, 1988, pp. 2–6.

[17] Casciati, F. and Faravelli, L., Hysteretic 3-dimensional Frames under Stochastic Excitation, *Res Mechanica*, 26, 1989, pp. 193–213.

[18] Casciati, F. and Faravelli, L., Stochastic Equivalent Linearisation of Hysteretic Behavior, in *Proceedings of the 8th International Conference on Offshore Mechanics and Arctic Engineering*, ASME, 1989, pp. 383–385.

[19] Pradlwarter, H.J. and Schuëller, G.I., Equivalent Linearisation — a Suitable Tool for Analysing MDOF Systems, *Probabilistic Engineering Mechanics*, 8, 1992, pp. 115–126.

[20] Casciati, F., Faravelli, L., and Venini, P., Frequency Analysis in Stochastic Linearisation, *Journal of Engineering Mechanics*, ASCE, 120, 1994, pp. 2498–2518.

[21] Chang, T.-P., Random Vibration Analysis of Nonlinear Hysteretic Plates, *Journal of Sound and Vibration*, 172, 1994, pp. 539–547.

[22] Naess, A., Prediction of Extreme Response of Nonlinear Structures by Extended Stochastic Linearisation, *Probabilistic Engineering Mechanics*, 10, 1995, pp. 153–160.

[23] Suzuki, Y. and Minai, R., Application of Stochastic Differential Equations to Seismic Reliability Analysis of Hysteretic Structures, *Probabilistic Engineering Mechanics*, 3, 1988, pp. 43–52.

[24] Suzuki, Y. and Minai, R., Stochastic Seismic Damage and Reliability Analysis of Hysteretic Structures, in *Proceedings of the IUTAM Symposium on Nonlinear Stochastic Dynamic Engineering Systems*, Ziegler F., Ed., Springer-Verlag, Berlin, 1988, pp. 407–418.

[25] Suzuki, Y. and Minai, R., Seismic Damage and Reliability Analysis of Hysteretic Multi-degree of Freedom Structures, in *Proceedings of the 9th World Conference on Earthquake Engineering*, 8, Tokyo-Kyoto, Japan, 1988, pp. 773–778.

[26] Noori, M., Davoodi, H., and Saffar, A., An Itô-based General Approximation Method for Random Vibration of Hysteretic Systems, I: Gaussian Analysis, *Journal of Sound and Vibration*, 127, 1988, pp. 331–342.

[27] Noori, M. and Davoodi, H., Comparison between Equivalent Linearisation and Gaussian Closure for Random Vibration of Several Nonlinear Systems, *International Journal of Engineering Science*, 28, 1990, pp. 897–905.

[28] Davoodi, H. and Noori, N.N., Extension of an Itô-based Approximation Technique for Random Vibration of a BBW General Hysteresis Model, II: Non-Gaussian Analysis, *Journal of Sound and Vibration*, 140, 1990, pp. 319–339.

[29] Liu, Q., and Davies, H.G., Response Analysis of a Hysteretic System under Random Excitation, *Journal of Sound and Vibration*, 166, 1993, pp. 477–492.

[30] Liu, Q. and Afagh, F.F., Random Response and Energy Dissipation of a Hysteretic Structure subjected to White Noise Excitation, *Transactions of the Canadian Society of Mechanical Engineering*, 18, 1994, pp. 333–352.

[31] Kimura, K., Yasumuro, H., and Sakata, M., Non-Gaussian Equivalent Linearisation for Nonstationary Random Vibration of Hysteretic System, *Probabilistic Engineering Mechanics*, 9, 1994, pp. 15–22.

[32] Mochio, T., Random Vibration Analysis of Multi-degree of Freedom Structures with Hysteretic Characteristics subjected to Combined Random Loads, *JSME International Journal Series C - Dynamics Control Robotics Design and Manufacturing*, 38, 1995, pp. 17–22.

[33] Wen, Y.K., Method for Random Vibration of Hysteretic Systems, *Journal of the Engineering Mechanics Division*, ASCE, 102, 1976, pp. 249–263.

[34] Spencer, B.F. and Bergman, L.A., On the Reliability of a Simple Hysteretic System, *Journal of Engineering Mechanics*, ASCE, 111, 1985, pp. 1502–1514.

[35] Spencer, B.F., *Reliability of Randomly Excited Hysteretic Structures*, Lecture Notes in Engineering, Springer-Verlag, Berlin, 1986.

[36] Roberts, J.B. and Spanos, P.D., Stochastic Averaging: an Approximate Method for Solving Random Vibration Problems, *International Journal of Non-linear Mechanics*, 21, 1986, pp. 111–134.

[37] Roberts, J.B., The Response of an Oscillator with Bilinear Hysteresis to Stationary Random Excitation, *Journal of Applied Mechanics*, ASME, 45, 1978, pp. 923–928.

[38] Roberts, J.B., The Yielding Behavior of a Randomly Excited Elasto-plastic Structure, *Journal of Sound and Vibration*, 72, 1980, pp. 71–85.

[39] Roberts, J.B. and Sadeghi, A.H., Nonstationary Response of Oscillators with Bilinear Hysteresis to Random Excitation, in *Stochastic Structural Dynamics*, Ariaratnam, S.T. et al., Eds., Elsevier, Amsterdam, 1988, pp. 197–218.

[40] Roberts, J.B., Distribution of the Response of Hysteretic Oscillators with Wide-band Random Excitation, in *Proceedings of the IUTAM Symposium on Nonlinear Mechanics*, Bellomo, N. and Casciati, F., Eds., Springer-Verlag, Berlin, 1992, pp. 439–452.

[41] Yar, M. and Hammond, J.K., Modeling and Response of Bilinear Hysteretic Systems, *Journal of Engineering Mechanics*, ASCE, 113, 1987, pp. 1000–1013.

[42] Roberts, J.B., Application of Averaging Methods to Randomly Excited Hysteretic Systems, in *Proceedings of the IUTAM Symposium on Nonlinear Stochastic Dynamic Engineering Systems*, Ziegler F., Ed., Springer-Verlag, Berlin, 1988, pp. 361–379.

[43] Zhu, W-Q. and Lei, Y., Stochastic Averaging of Energy Envelope of Bilinear Hysteretic Systems, in *Proceedings of the IUTAM Symposium on Nonlinear Stochastic Dynamic Engineering Systems*, Ziegler F., Ed., Springer-Verlag, Berlin, 1988, pp. 381–392.

[44] Cai, G.Q. and Lin, Y.K., On Randomly Excited Hysteretic Structures, *Journal of Applied Mechanics*, ASME, 57, 1990, pp. 442–448.

[45] Roberts, J.B., Energy Method for Non-linear Systems with Non-white Excitation, in *Proceedings of the IUTAM Symposium on Random Vibrations and Reliability*, Hennig, K. Ed., Akademie Verlag, Berlin, 1982, pp. 285–294.

[46] Cai, G.Q. and Lin, Y.K., Response of a Hysteretic System under Nonstationary Earthquake Excitations, in *Proceedings of the IUTAM Symposium on Advances in Nonlinear Stochastic Mechanics*, Trondheim, Norway (to be published), 1995.

[47] Noori, M., Dimentberg, M., Hou, Z., Christodoulidou, R., and Alexandrou, A., First-passage Study and Stationary Response Analysis of a BWB Hysteresis Model using Quasi-conservative Stochastic Averaging Method, *Probabilistic Engineering Mechanics*, 10, 1985, pp. 161–170.

[48] Ditlevsen, O. and Bognar, L., Plastic Displacement Distributions of the Gaussian White Noise Excited Elasto-plastic Oscillator, *Probabilistic Engineering Mechanics*, 8, 1993, pp. 209–231.

[49] Ditlevsen, O. and Randrup-Thomsen, S., Gaussian White Noise Excited Elasto-plastic Oscillator of Several Degrees of Freedom, in *Proceedings of the IUTAM Symposium on Advances in Nonlinear Stochastic Mechanics*, Trondheim, Norway, July 1995 (to be published).

[50] Randrup, S. and Ditlevsen, O., One-floor Building as Elasto-plastic Oscillator subject to and Interacting with Gaussian Base Motion, in *Proceedings of the 3rd International Conference on Stochastic Structural Dynamics*, San Juan, Puerto Rico, 1995.

[51] Roberts, J.B. and Sadeghi, A.H., Sequential Parametric Identification and Response of Hysteretic Oscillators with Random Excitation, *Journal of Structural Safety*, 8, 1990, pp. 45–68.

[52] Noori, M., Davoodi, H., and Baber, T.T., A Comprehensive Stationary Non-Gaussian Analysis of BWB Hysteresis, in *Proceedings of the IUTAM Symposium on Nonlinear Mechanics*, Bellomo, N. and Casciati, F., Eds., Springer-Verlag, Berlin, 1992, pp. 415–425.

[53] Stratonovich, R.L., *Topics in the Theory of Random Noise*, Vol.1, Gordon and Breach, New York, 1964.

[54] Khasminskii, R.Z., A Limit Theorem for the Solutions of Differential Equations with Random Right-hand Sides, *Theory of Probability and its Applications*, 11, 1966, pp. 390–405.

[55] Priestley, M.B., Power Spectral Analysis of Nonstationary Random Processes, *Journal of Sound and Vibration*, 6, 1967, pp. 86–97.

[56] Roberts, J.B., First Passage Time for Oscillators with Nonlinear Damping, *Journal of Applied Mechanics*, ASME, 45, 1978, pp. 175–180.

4

Non-parametric Estimation of Failure Probabilities

A.M. Hasofer
Victoria University, Melbourne, Australia

4.1 Introduction

In recent years, there has been much work done on the calculation of small probabilities of failure associated with structural systems specified by a p-dimensional vector of physical variables and a failure region in p-dimensional space. However, the bulk of the work assumed that the distribution of the physical variables was multivariate normal. When the joint distribution is not multivariate normal, Hohenbichler and Rackwitz [14] advocate the use of the Rosenblatt transformation to transform the variables to joint normality. It is then necessary to specify a family of conditional distributions to carry out the transformation. As long as one stays in the realm of theory, this can be done, at least in principle. But if the vector of physical variables is known only through a sample, either by simulation or from field measurements, it is well-nigh impossible to evaluate the required conditional probability distributions.

It is intuitively reasonable to look for an approach based on the sample points that fall either in the failure region or near it. The novelty of the proposed method lies essentially in the fact that the failure probability can then be accurately estimated from the sample, provided the sample size is large enough, without making any assumptions about the underlying joint distribution. This can be achieved by using recent results of extreme value theory. The calculation of failure probability is first reduced to a one-dimensional problem. The domain of attraction is then determined. Finally, the appropriate asymptotic estimators are applied.

There have been two main approaches to the selection of the appropriate sample points in the one-dimensional case: the "threshold" method and the k top order statistics method. In the threshold method some high level u is chosen and only the sample values above u are considered. In the alternative method it is the number k of top order statistics which is chosen. In this chapter attention will be focused on the k top order statistics method for which the author and his collaborators have developed a fairly complete and easy to use theory. The threshold method will be outlined in Section 4.12 with adequate references to enable readers to use it if they so wish.

4.2 The Single Dimensional Case

Let the upper ϵ-quantile $q(\epsilon)$ of a random variable X be defined by $P[X > q(\epsilon)] = \epsilon$, where ϵ is some small number.

Suppose that we are interested in evaluating the quantile $q(\epsilon)$, but that the probability distribution of X is not known explicitly. The only information we have about X is either:

(a) a random sample of size n, or

(b) some algorithm for simulating such a random sample.

In classical statistics it is customary to postulate a parametric form for the distribution of X depending on one or more unknown parameters. The validity of the form of the distribution would be tested by applying some goodness of fit test such as the chi-squared test. The parameters are then estimated from the sample and the estimators used to obtain an estimator $\hat{q}(\epsilon)$ of the required quantile.

A typical example of such a procedure would be to assume that X is normally distributed with mean μ and standard deviation σ. The parameters μ and σ are estimated by $\hat{\mu} = \sum X_i/n$ and $\hat{\sigma}^2 = \sum(X_i - \bar{X})^2/(n-1)$. The quantile $q(\epsilon)$ is then estimated by $\hat{q}(\epsilon) = \hat{\mu} + k(\epsilon)\hat{\sigma}$ where $k(\epsilon)$ is the corresponding quantile of the standard normal distribution.

The above approach suffers from two grave defects:

1. As pointed out by Ofverbeck [21] and later by Lind [19], it may happen that an increase in some of the lower values of the sample will actually decrease $\hat{q}(\epsilon)$, which is highly counter-intuitive in the context of reliability analysis. In fact, Lind [19] has proposed the following *Principle of Reliability Consistency:* "The probability of failure should be a monotonically non-decreasing function of any sample element of a load variable" (and conversely for a resistance variable). This implies that high quantiles of load variables should be non-decreasing functions of the sample elements. Lind shows that this consistency condition is often violated when

the procedure outlined above for the normal distribution is used.

2. It may well be that the high values of X are generated by a different mechanism than that which generates average and/or low values of X. In that case, the latter values are essentially irrelevant to the determination of the high quantiles of X. Estimating the probability distribution from all values of the sample may lead to a totally wrong value of the high quantiles. This will be illustrated later in the chapter.

To remedy the above defects, we shall follow up Ofverbeck's suggestion to base the high quantile estimation on the top k order statistics of the sample only. The novelty of our approach will be the use of recent results in extreme value theory which will enable us to propose essentially non-parametric estimators for the high quantiles.

4.3 A Short Statement of Extreme Value Theory

Suppose $X_1, X_2, \ldots, X_n$ is a random sample from the distribution function $F(x)$. Let $X_{1n}, X_{2n}, \ldots, X_{nn}$ be the corresponding descending order statistics, satisfying

$$X_{1n} \geq X_{2n} \geq \ldots \geq X_{nn}.$$

Then, of course, X_{1n} is the maximum of $X_1, X_2, \ldots, X_n$, and

$$\lim_{n\to\infty} P(X_{1n} \leq x) = \lim_{n\to\infty} F^n(x) = \begin{cases} 0, & \text{if } F(x) < 1 \\ 1 & \text{if } F(x) = 1. \end{cases} \tag{4.1}$$

We are interested in finding sequences of real constants $\{a_n\}$ and $\{b_n\}$, $n = 1, 2, \ldots$ such that $F^n(a_n + b_n x)$ converges (weakly) to a non-degenerate distribution function $H(x)$. If such sequences exist, the distribution function $F(x)$ is said to belong to the domain of attraction of the extreme value distribution $H(x)$. The sequences a_n and b_n are called the coefficients of attraction of $F(x)$.

The central result of extreme value theory is that the distribution $H(x)$ must be, up to a change of scale and origin, of the form

$$H_\alpha(x) = \exp[-(1 + \alpha x)^{-1/\alpha}]. \tag{4.2}$$

For $\alpha = 0$, $H_\alpha(x)$ is defined as

$$H_0(x) = \lim_{\alpha\to 0} H_\alpha(x) \tag{4.3}$$

$$= \exp[-\exp(-x)] \quad -\infty < x < \infty. \tag{4.4}$$

This distribution is known as Type I (or *Gumbel*) extreme value distribution.

For $\alpha > 0$,

$$H_\alpha(x) = \begin{cases} \exp[-(1+\alpha x)^{-1/\alpha}], & \text{for } x > -1/\alpha \\ 0 & \text{for } x < -1/\alpha. \end{cases} \tag{4.5}$$

This distribution is known as Type II extreme value distribution.

For $\alpha < 0$,

$$H_\alpha(x) = \begin{cases} 1 & \text{for } x > -1/\alpha \\ \exp[-(1+\alpha x)^{1/\alpha}], & \text{for } x < -1/\alpha. \end{cases} \tag{4.6}$$

This distribution is known as Type III extreme value distribution.

It is easy to see that if X has distribution $H_0(x)$, then

$$Y = [\exp(\alpha X) - 1]/\alpha$$

has distribution $H_\alpha(x)$ and conversely, if Y has distribution $H_\alpha(x)$, then

$$X = \ln[(1+\alpha Y)^{1/\alpha}]$$

has distribution $H_0(x)$.

Type II and Type III extreme value distributions are usually expressed somewhat differently to the scheme outlined above, which emphasizes the fact that extreme value distributions form a continuum indexed by the parameter α which varies from $-\infty$ to $+\infty$.

For Type II distribution ($\alpha > 0$) let $Y = 1 + \alpha X$ and $\gamma = 1/\alpha$. Then

$$P(Y \le x) = H_{II,\gamma}(x) = \begin{cases} \exp(-x^{-\gamma}), & \text{for } x > 0 \\ 0 & \text{for } x < 0. \end{cases} \tag{4.7}$$

For Type III distribution ($\alpha < 0$) let $Y = -(1+\alpha X)$ and $\gamma = -1/\alpha > 0$. Then

$$P(Y \le x) = H_{III,\gamma}(x) = \begin{cases} 1 & \text{for } x > 0 \\ \exp[-(-x)^\gamma], & \text{for } x < 0. \end{cases} \tag{4.8}$$

We note that if X has distribution $H_0(x)$ (Type I), then $Y = \exp(X/\gamma)$, $\gamma > 0$, has distribution $H_{II,\gamma}(x)$ and $Y' = -1/Y$ has distribution $H_{III,\gamma}(x)$.

Necessary and sufficient conditions for a distribution function to belong to the domain of attraction of each of the three types of extreme value distribution are known. See e.g., Galambos [5]. There are distribution functions that do not belong to any domain of attraction. However, they are essentially pathological distributions devised by mathematicians and are of no concern in the practical applications we shall consider here.

We shall only give a set of sufficient conditions due to Von Mises and de Haan. Under certain conditions, they are sufficient as well.

Let $F(x)$ have a density function $f(x)$ and ω be the upper end point of $F(x)$, i.e., $F(\omega) = 1$ and $F(x) < 1$ for $x < \omega$. If $F(x) < 1$ for all x then $\omega = +\infty$. For all $x < \omega$ define the hazard function of $F(x)$, $h(x)$, by

$$h(x) = \frac{f(x)}{1 - F(x)},$$

i.e.,

$$h(x)dx = P(x < X \leq x + dx | X > x).$$

Let u_n be the unique solution of the equation $F(u_n) = 1 - 1/n$. Then

Theorem 2.1:

1. A sufficient condition for $F(x)$ to belong to the domain of attraction of Type I is that $f'(x)$ be negative in some interval (x_0, ω) and that

$$\lim_{x \to \omega} \frac{d}{dx}\left(\frac{1}{h(x)}\right) = 0.$$

   Then

$$\lim_{n \to \infty} F^n\left(u_n + \frac{y}{h(u_n)}\right) = \exp[-\exp(-y)].$$

   Thus, in this case, $a_n = u_n$ and $b_n = 1/h(u_n)$.
   Moreover, if $f'(x)$ is non-decreasing in (x_0, ω), the above condition is also necessary.

2. A sufficient condition for $F(x)$ to belong to the domain of attraction of Type II is that $F(x) < 1$ for all x and that

$$\lim_{x \to \infty} xh(x) = \gamma, \quad \gamma > 0.$$

   Then

$$\lim_{n \to \infty} F^n(u_n y) = \exp(-y^{-\gamma}), \quad y \geq 0.$$

Thus $a_n = 0$ and $b_n = u_n$.
Moreover, if $f(x)$ is non-increasing for large x, the above condition is also necessary.

3. A sufficient condition for $F(x)$ to belong to the domain of attraction of Type III is that ω, the end point of $F(x)$, be finite, and that

$$\lim_{x \to \omega} (\omega - x)h(x) = \gamma, \quad \gamma > 0.$$

Then

$$\lim_{n \to \infty} F^n[\omega + (\omega - u_n)y] = \exp[-(-y)^\gamma], \quad y \leq 0.$$

Thus $a_n = \omega$ and $b_n = \omega - u_n$.
Moreover, if $f(x)$ is non-increasing in $(\omega - \epsilon, \omega)$ for some $\epsilon > 0$, then the above condition is also necessary.

It should be noted that the choice of the coefficients of attraction is not unique. The following theorem provides conditions for a valid choice.

Theorem 2.2:

If a_n and b_n are coefficients of attraction for the distribution function $F(x)$, then a'_n and b'_n will qualify if and only if

$$\frac{b'_n}{b_n} \to 1 \text{ and } \frac{a'_n - a_n}{b'_n} \to 1.$$

Therefore, for instance, in (2) of Theorem 2.1, the coefficient a_n may be taken as any bounded sequence, since $b_n \to \infty$.

Theorem 2.1 enables us to verify that the Gamma (including chi-square), Exponential, Weibull, Normal, Lognormal, and Rayleigh distributions belong to the domain of attraction of Type I, the Cauchy, Pareto, and t-distributions belong to the domain of attraction of Type II and the Uniform and Negative Gamma (i.e., the distribution of $-Y$ where Y is Gamma distributed) belong to the domain of attraction of Type III.

The above theory has been couched in terms of the distribution of the maximum of a sample. If we are interested in the asymptotic properties of the minimum of the sample, we use the fact that $\min(X_1, X_2, \ldots, X_n) = -\max(-X_1, -X_2, \ldots, -X_n)$ and modify the equations appropriately.

It should be emphasized that although the Type I distribution belongs to the continuum of extreme value distributions, as shown above, it differs fundamentally from Type II and Type III and has to be treated separately. In essence, a Type I variable basically behaves like the *logarithm* of a Type II or Type III variable.

4.4 Asymptotics of the Top Order Statistics

Let as before $X_1, X_2, \ldots, X_n$ be a random sample from the distribution function $F(x)$ in the domain of attraction of $H(x)$ and let $X_{1n}, X_{2n}, \ldots, X_{nn}$ be the corresponding descending order statistics, satisfying

$$X_{1n} \geq X_{2n} \geq \ldots \geq X_{nn}.$$

Let us normalize these order statistics by writing

$$Z_{in} = (X_{in} - a_n)/b_n$$

where a_n and b_n are the coefficients of attraction of $F(x)$. Consider now the variables $Z_{1n}, \ldots, Z_{kn}$, where k is a fixed number and let n tend to infinity. The following results are proved in Weissman [28] (or follow easily, see Hasofer and Wang [12]).

1. As $n \to \infty$ the variables Z_{in}, $(n = 1, \ldots, k)$ tend to limit random variables U_i, $(i = 1, \ldots, k)$. The points defined on the line by these variables constitute a non-stationary Poisson process with intensity $\lambda(x) = -\ln H(x)$.
   Suppose now that $H(x) = H_0(x)$.

2. Let $V_i = U_i - U_{i+1}$, $(i = 1, \ldots, k-1)$. Then for each k the random variables

   $$\{U_k, V_1, \ldots, V_{k-1}\}$$

   are independent, each V_i is a negative exponential random variable with mean i^{-1}, and U_1 has the distribution $H_0(x)$.

3. Let $T_i = U_i - U_k$, $(i = 1, \ldots, k-1)$. Then the random variables T_i are distributed like the order statistics of a sample of size $k-1$ from the negative exponential distribution.

4. Let $M_i = \theta + \delta U_i$, $(i = 1, \ldots, k)$ and let

   $$\bar{M} = \frac{1}{k}\sum_{j=1}^{k} M_j.$$

Then the pair $(M_k, \bar{M})$ is a sufficient statistic for (θ, δ) and the unique minimum variance unbiased estimators $\hat{\theta}, \hat{\delta}$ of θ, and δ are given by

$$\hat{\delta} = \left(\frac{k}{k-1}\right)(\bar{M} - M_k)$$

and

$$\hat{\theta} = \hat{\delta}(S_k - \gamma) + M_k$$

where

$$S_k = \sum_{j=1}^{k-1} j^{-1}$$

and $\gamma = .5722...$ is Euler's constant. Since

$$\gamma = \lim_{k\to\infty} S_k - \ln k$$

we see that $\hat{\delta}$ is approximately equal to $\bar{M} - M_k$ and that $\hat{\theta}$ is approximately equal to $\hat{\delta} \ln k + M_k$ for sufficiently large k.

4.5 Estimation of High Quantiles for Type I

Using the above results, Weissman has obtained the asymptotically minimum variance estimator $\hat{q}(\epsilon)$ for the quantile $q(\epsilon)$ based on the top k order statistics $X_{1n}, X_{2n}, \ldots, X_{kn}$ when $F(x)$ belongs to the domain of attraction of Type I. It is given by

$$\hat{q}(\epsilon) = \hat{a}\ln(k/n\epsilon) + X_{kn}$$

where

$$\hat{a} = \left(\sum_{i=1}^{k} X_{in}\right)/k - X_{kn}.$$

Because it is linear in the order statistics, this estimator is also the Best Linear Unbiased Estimator ("BLUE") of $q(\epsilon)$.

Boos [2] has carried out large-scale simulations to test Weissman's estimator and has confirmed its high efficiency.

4.6 A Test for Extreme Value Domain of Attraction

Before Weissman's estimator can be applied, it must be determined that the hypothesis that $F(x)$ belongs to the domain of attraction of Type I is consistent with the relevant data, namely the high order statistics of the given sample. If that hypothesis is rejected, a suitable transformation of the variable studied must be carried out to fulfill the above requirement.

Hasofer and Wang [12] have developed a simple but effective statistic (based on the top k order statistics) to test the hypothesis that $F(x)$ belongs to the domain of attraction of Type I. It is denoted by W and is given by the formula

$$W = \frac{k(\bar{X} - X_{kn})^2}{(k-1)\left[\sum_{i=1}^{k}(\bar{X} - X_{in})^2\right]}$$

where

$$\bar{X} = \left(\sum_{j=1}^{k} X_{jn}\right)/k.$$

The W-test can be considered as a generalisation of the Shapiro-Wilk test of normality (Shapiro and Wilk [23]). It is easy to see that W is invariant under a linear transformation of X (which will apply the same linear transformation to each order statistic).

The null hypothesis is that $F(x)$ belongs to the domain of attraction of Type I, while the alternatives are that $F(x)$ belongs to the domain of attraction of Type II or of Type III.

The critical regions of the test are as follows: Let W_L and W_U be the lower and upper chosen percentage points of W. Then if $W < W_L$ we accept $H_2 : X$ belongs to the domain of attraction of Type II, while if $W > W_U$ we accept $H_3 : X$ belongs to the domain of attraction of Type III. Otherwise, we accept that X belongs to the domain of attraction of Type I and use Weissman's estimator as is.

The rationale for the above choice of critical regions is as follows: We recollect that the extreme value domains of attraction can be indexed by a parameter α varying from $-\infty$ to $+\infty$. $\alpha > 0$ corresponds to Type II while $\alpha < 0$ corresponds to Type III. But it can be shown (see Hasofer and Wang [12]) that as α increases the distribution of W shifts continuously towards the left. Thus large values of W correspond to Type III while low values correspond to Type II (and intermediate values to Type I).

The upper and lower percentage points of W (asymptotic for large n) are given in Table 4.1 as functions of the number k of high-order statistics used. For larger values of k, W may be taken to be normally distributed with mean $1/k$ and standard deviation $2/k^{3/2}$.

Table 4.1 Upper and lower percentage points for $W \times 10^4$

	Percentage level							
	Lower Tail				Upper Tail			
k	1	2.5	5	10	10	5	2.5	1
13	333.8	395.6	457.7	538.4	1599.5	1827.5	1998.5	2204.6
14	305.6	361.2	416.3	489.6	1406.3	1589.4	1749.6	1934.4
15	288.1	339.4	389.3	456.5	1275.6	1450.4	1596.2	1768.1
16	272.8	321.3	368.8	431.5	1183.3	1334.9	1469.4	1629.5
17	258.6	306.0	349.6	407.7	1095.0	1234.7	1357.9	1508.3
18	247.4	290.2	332.0	386.5	1017.1	1143.9	1259.7	1397.7
19	237.1	277.6	316.6	368.5	950.4	1064.5	1171.1	1301.4
20	226.9	265.4	302.5	350.7	888.8	997.3	1096.5	1220.9
21	217.7	255.0	289.1	333.6	836.0	938.1	1027.2	1145.8
22	210.1	244.9	278.3	321.0	787.3	881.9	968.7	1077.2
25	189.4	219.3	248.8	285.4	672.1	747.4	821.1	910.8
30	165.0	189.2	212.6	241.7	535.5	593.6	650.7	719.9
40	131.7	149.5	166.1	186.4	377.1	413.6	451.5	497.0
50	110.4	124.1	136.6	151.8	289.5	316.5	340.2	371.1
60	96.4	106.9	116.6	128.8	232.9	253.8	270.8	293.3
80	76.5	84.2	91.0	98.9	168.4	179.9	190.7	205.5
100	63.6	69.2	74.3	80.2	129.6	138.4	146.4	155.5
200	35.4	37.6	39.9	41.9	59.4	62.2	64.8	68.1
500	16.3	16.9	17.4	18.0	22.6	23.2	23.8	24.5

4.7 Estimating Quantiles for Type III

The most serious departure from Type I occurs when the W test indicates a Type III domain of attraction. This is because in that case the variable X must have an upper bound. Of course all physical variables are theoretically bounded, but often the bound is so high that it is of no practical significance (e.g., wind velocity) and the variable may be assumed to be unbounded. However, some load variables have a practically significant upper bound, e.g., rainfall and earthquake magnitude. On the other hand, practically all resistance variables have a significant lower bound, simply because they cannot be negative. That lower bound can be of great significance in reliability analysis. See e.g., the paper of Dalgliesh and Taylor [3] on the strength and testing of window glass.

Let X have a finite upper bound ω_0, and let as before $X_{1n} \geq X_{2n} \geq \ldots \geq X_{kn}$ be the k top order statistics from a sample of size n. Clearly we must have $\omega_0 \geq X_{1n}$. It follows from the results of Section 4.3 that the limiting distribution of the $Y_{in} = -\ln(\omega_0 - X_{in})$, $i = 1, \ldots, k$, is, after a transformation of scale and origin by a suitable pair of sequences, the limiting distribution of the top k order statistics corresponding to Type I. We recollect that the statistic W is invariant with respect to a linear transformation of the order statistics, and is thus only a function of the X_{in} and of the additional parameter ω_0.

To estimate ω_0 we use a recently developed method called the "estimating equation" method. (See [6].) This consists in finding a function G of the sample and of the required parameters that has zero mean (when the parameters take the correct value) and then solving the equation $G = 0$ for the parameters. Special cases of the "estimating equation" method are the method of moments, the method of least squares and the method of maximum likelihood. One great advantage of the method is that it is often possible to find a G that excludes some nuisance parameters, thus considerably simplifying the form of the estimators.

Let $\mathbf{Y}(\omega)$ be the vector $\{-\ln(\omega - X_{1n}), \ldots, -\ln(\omega - X_{kn})\}$. We shall take as an estimating function for ω_0

$$G = W[\mathbf{Y}(\omega)] - \mathrm{E}(W[\mathbf{Y}(\omega)]|\omega = \omega_0)$$

Clearly $\mathrm{E}(G) = 0$ when $\omega = \omega_0$. Also, for large k, we have approximately $\mathrm{E}(W|\omega = \omega_0) = 1/k$. Simulation work indicates that this value can be used in the estimating equation for k as low as 10.

The important fact that gives the proposed method the edge over other methods which have been proposed to estimate ω_0 is that if the W-test rejects the null hypothesis that X belongs to the domain of attraction of Type I in favour of the alternative that X belongs to the domain of attraction of Type III with a significance level of less than 20%, then our estimating equation has always a unique solution. This follows from the following properties of W, whose proof is given in Hasofer and Wang [13]:

1.
$$\lim_{\omega \to X_{1n}} W(\mathbf{Y}) = 1/(k-1)^2.$$

2.
$$\lim_{\omega \to \infty} W(\mathbf{Y}) = W(X_{1n}, \ldots, X_{kn}).$$

3. $W[\mathbf{Y}(\omega)]$ is a strictly monotone increasing function of ω for all $\omega > X_{1n}$.

Indeed, it is then clear that $W[\mathbf{Y}(\omega)]$ increases monotonically from $1/(k-1)^2$ which is less than $\mathrm{E}(W[\mathbf{Y}(\omega)]|\omega = \omega_0)$ to $W(X_{1n}, \ldots, X_{kn})$ which is greater than $W_U > \mathrm{E}(W[\mathbf{Y}(\omega)]|\omega = \omega_0)$ for $k \geq 3$, so that our equation has a unique solution. Moreover, it can be shown that the obtained estimator is consistent as k tends to infinity (see [13]). In the reference just quoted will be found numerous simulation results which indicate that the proposed estimation method for ω_0 as well as for the high quantiles of the variable under study is efficient and easy to use.

High quantiles in the situation under consideration are obtained by carrying out the transformation

$$\hat{Y}_{in} = -\ln(\hat{\omega}_0 - X_{in}),$$

where $\hat{\omega}_0$ is the obtained estimator of ω_0, evaluating the required quantile of $\hat{Y}$ by using the Weissman estimator, and then carrying out the reverse transformation to obtain the estimator of the quantile of X.

4.8 Estimating Quantiles for Type II

The situation when the W test indicates that X is in the domain of attraction of Type II is not as critical as in the case of the domain of attraction of Type III. As pointed out in Section (4.3), the sequence a_n may be taken to be null without affecting the asymptotic results. Thus, the limiting distribution of $Y_{in} = \ln(X_{in})$ will be of Type I and the Weissman estimator may be applied to it. This is in fact the approach advocated by Weissman himself in [28].

However, in the practical application of the method, where the asymptotic theory is applied to finite samples, this approach is unsatisfactory on two counts:

1. A quantile estimator should be invariant under translation, in the sense that if the underlying variable is increased by some amount X_0 then the quantile estimator should be increased by the same amount. This is clearly not the case with the algorithm just described.

2. There is no good reason to believe that all sample values will be necessarily positive. For example, our sample may measure sea levels below some reference level. If some sample values are negative, taking logarithms will not be possible.

For the above reasons, we advocate the use of the transformation $Y_{in} = \ln(X_{in} - \omega_0)$, with $\omega_0 \leq X_{kn}$, when the W-test rejects the null hypothesis that X belongs to the domain of attraction of Type I in favour of the alternative that X belongs to the domain of attraction of Type II. We use again the estimating function method to obtain an estimator of ω_0. Let $\mathbf{Y}(\omega)$ be the vector $\{\ln(X_{1n} - \omega), \ldots, \ln(X_{kn}\} - \omega)$. As previously, we shall take as an estimating function for ω

$$G = W[\mathbf{Y}(\omega)] - \mathrm{E}(W[\mathbf{Y}(\omega)] | \omega = \omega_0).$$

Clearly $\mathrm{E}(G) = 0$ when $\omega = \omega_0$. Also, for $k > 10$, we still have approximately $\mathrm{E}(W | \omega = \omega_0) = 1/k$. Here W has the following properties (Hasofer and Li [9]):

1.

$$\lim_{\omega \to X_{kn}} W(\mathbf{Y}) = 1.$$

2.

$$\lim_{\omega \to -\infty} W(\mathbf{Y}) = W(X_{1n}, \ldots, X_{kn}).$$

3. $W[\mathbf{Y}(\omega)]$ is a strictly monotone increasing function of ω for all $\omega < X_{kn}$.

Here $W(X_{1n}, \ldots, X_{kn})$ is less than $\mathrm{E}(W[\mathbf{Y}(\omega)] | \omega = \omega_0)$, since it is less than W_L at a significance level of less than 20%. Also, $\mathrm{E}(W[\mathbf{Y}(\omega)] | \omega = \omega_0)$, which is approximately $1/k$ for large k, is less than unity and this ensures the existence of a unique solution for the estimating equation. As for Type III, it can be shown that the obtained estimator is consistent as k tends to infinity (See [9]). In that reference will be found numerous simulation results which indicate that the proposed estimation method for ω_0 as well as for the high quantiles of the variable under study is efficient and easy to use. Evidence is also brought out to show that the assumption that $\omega_0 = 0$ may lead to serious error in finite samples.

High quantiles in the situation under consideration are obtained by carrying out the transformation

$$\hat{Y}_{in} = \ln(X_{in} - \hat{\omega}_0),$$

where $\hat{\omega}_0$ is the obtained estimator of ω_0, evaluating the required quantile of $\hat{Y}$ by using the Weissman estimator, and then carrying out the reverse transformation to obtain the estimator of the quantile of X.

4.9 The Choice of *k*

The choice of an optimal value for the number k of high order statistics to be used involves conflicting considerations. On the one hand, the standard deviation of the quantile estimator $\hat{q}$ is approximately equal to $1/\sqrt{k}$. More precisely

$$\sigma(\hat{q}) = c_n \left[\zeta_n^2 \frac{(k-1)}{k^2} + \frac{\pi^2}{6} - R_k \right]^{1/2}$$

where $\zeta_n = \ln(k/n\epsilon)$, $c_n = q(1/ne) - q(1/n)$, $(e = 2.718\ldots)$, and $R_k = \sum_{n=1}^{k-1}(1/n^2)$.

In the formulae just given $q(\epsilon)$ denotes the true quantile corresponding to ϵ, which is, of course, unknown. However, the standard deviation of $\hat{q}$ can be estimated from the above formula by replacing $q(1/ne)$ and $q(1/n)$ by their Weissman estimators.

Thus, to enhance the precision of the estimation, k should be chosen as large as possible. However, when k/n is taken too large, bias is introduced by departure from the asymptotic distribution. To balance the effects of variance decrease against bias increase, Boos [2] recommends, for $\epsilon \leq 0.05$, $k/n = 0.02$ for $50 \leq n \leq 500$, and $k/n = 0.1$ for $500 < n \leq 5,000$. My own investigations suggest that k can be taken to equal approximately $1.5\sqrt{n}$.

It is worth noting that as long as $k > n\epsilon$, $\hat{q}$ is a monotonically increasing function of the order statistics, except X_{kn}. To ensure that the coefficient of X_{kn} be positive as well, the additional condition $k > 2.718n\epsilon$ should be fulfilled. This will ensure that Lind's Principle of Reliability Consistency (see Section 4.2) will be satisfied.

A more sophisticated approach to the determination of the optimal k has been proposed by Wang [27]. The proposed procedure treats as a whole the two problems of determination of the domain of attraction and selection of k. It consists in calculating for each value of $k = 3, 4, \ldots$ the value of W based on $X_{1n} \geq X_{2n} \geq \ldots \geq X_{kn}$. The values are then plotted against k on a graph, together with the corresponding values of W_L and W_U for some chosen significance level. If the graph of W shows an early downward or upward trend, this is taken as an indication that the domain of attraction is Type II or Type III, respectively. An appropriate logarithmic transformation is then carried out (see Sections 4.8 and 4.7) and a new set of W values is computed. These transformations, by their very nature, will bring back W to the region between W_L and W_U. We then take the optimal k to be equal to the first value of k for which W leaves the region between W_L and W_U minus one. A further improvement would be to then recalculate the parameter of the logarithmic transformation and repeat the procedure.

Example 1: Heat load generated by a fire in a compartment.

Hasofer and Beck [10] have developed a stochastic model for compartment fires derived from physical laws which consist of just three time-varying variables forming a Markov vector which satisfies a stochastic differential equation.

The variables are the temperature T, the rate of burning of the combustible material R, and the oxygen deficiency (below 0.23) D. They satisfy the equations

$$dT - \beta R dt + q(T)dt = f_1(T)dW_1$$

$$dD - \delta(k_1 - D)Rdt + \mu D dt = f_2(T)dW_2$$

$$dR - \alpha(k - D)Z(T)dt = f_3(T)dW_3$$

where $q(T)$ is a scaled-down measure of the rate of heat generation and $Z(T)$ is some increasing function of T derived from experiments. For details see the reference given above. When the combustible material has been burned out, the fire decays exponentially.

To illustrate the capacity of the model to simulate the intrinsic variability of a fire, a Monte-Carlo simulation of a fire was carried out. The functions f_i were chosen as follows:

$$f_1(T) = \phi T/1000$$

$$f_2(T) = 15\phi T/1000$$

$$f_3(T) = 12\phi T/10^6.$$

By varying the parameter ϕ, various degrees of stochasticity may be achieved.

As an example of the kind of result obtainable, one particular measure of fire severity, namely the "Normalized Heat Load" proposed by Harmathy and Mehaffy [7] was studied. We ignored the normalizing parameters and simply evaluated the variable H defined by

$$H = \int_0^{t_{max}} q[T(u)]du \tag{4.9}$$

where q is, as above, the (scaled) rate of heat generated and t_{max} is the time when the burning rate is reduced to a negligible value.

In their paper, Harmathy and Mehaffy argue that the heat load is a measure of the maximum temperature that some key load-bearing elements of the enclosure boundaries will attain as a result of fire exposure and constitutes thereby a measure of the destructive potential of the fire. Since the performance of these key elements depends sensitively on the maximum temperature reached, it is of great importance to be able to calculate the upper quantiles of the heat load, as the heat load is a measure of the maximum temperature attained.

Figure 4.1 shows a histogram of the values of H based on $1,500$ simulations of the fire with $\phi = 100$. It is interesting to note that for some simulations the fire was extinguished at the outset, and for some others the heat load remained low. However, the main body of the data is bell shaped, as expected, and there are no particular anomalies at the upper end, which is the significant part of the data as far as reliability calculations are concerned. The mean of H was 8063.5 and the standard deviation 849.8.

A test of normality (Kolmogorov-Smirnov) as applied to the data and the normality hypothesis was rejected, even when the outliers at the lower end were removed. This was not unexpected, in view of the non-linearities in the model equations. The data thus appeared to be an ideal application of the methods presented in this paper.

The first task was to determine an appropriate value for k. The W-test was applied to the data with various values of k. The results, together with the 5% confidence limits for W, are plotted in Figure 4.2. We see that the W-values show an early upward trend, indicating possibly a domain of attraction of Type

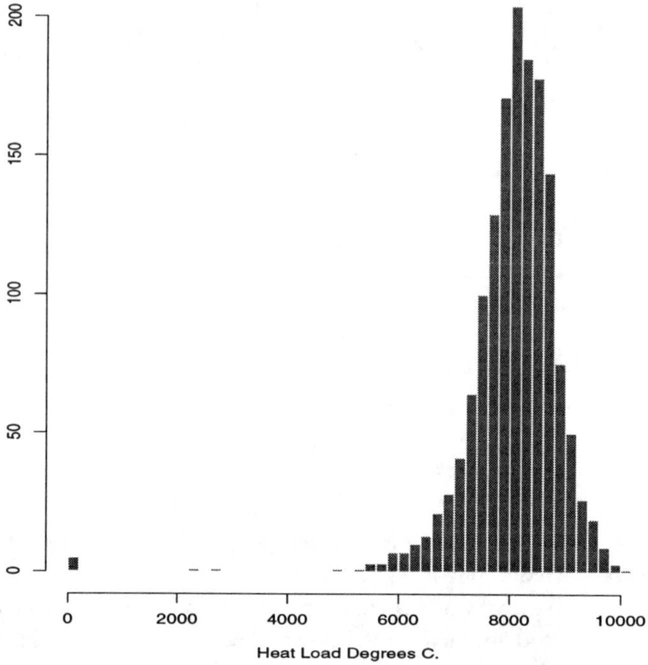

FIGURE 4.1
Histogram of heat load.

III. We chose $k = 39$ and estimated the finite upper bound ω_0, using the method described above, to be $\hat{\omega}_0 = 10,236$. We then carried out the transformation $H^* = -\log(\hat{\omega}_0 - H)$ and recalculated W with the new data H^* for various values of k. The results are shown in Figure 4.3 from which we now conclude that H^* may assumed to be in the domain of attraction of Type I with $k = 199$. Finally some quantiles of H^* were calculated, using the Weissman estimator and the corresponding values of H found. For comparison purposes the quantiles were also calculated on the assumption that H belonged to the domain of attraction of Type I, and also on the assumption that H was normally distributed with the mean and standard deviation given above. The results are given in Table 4.2.

It can be seen that quantile estimation in this case is insensitive to the choice of domain of attraction as well as to the precise value of k. This has been our experience in other applications as well. On the other hand, it is quite clear that quantile evaluation under the normality assumption is grossly in error. For $\epsilon = 0.01$ the quantile is overestimated by more than half a standard deviation! Of course, as is to be expected, the error diminishes as the tail probability increases, but since

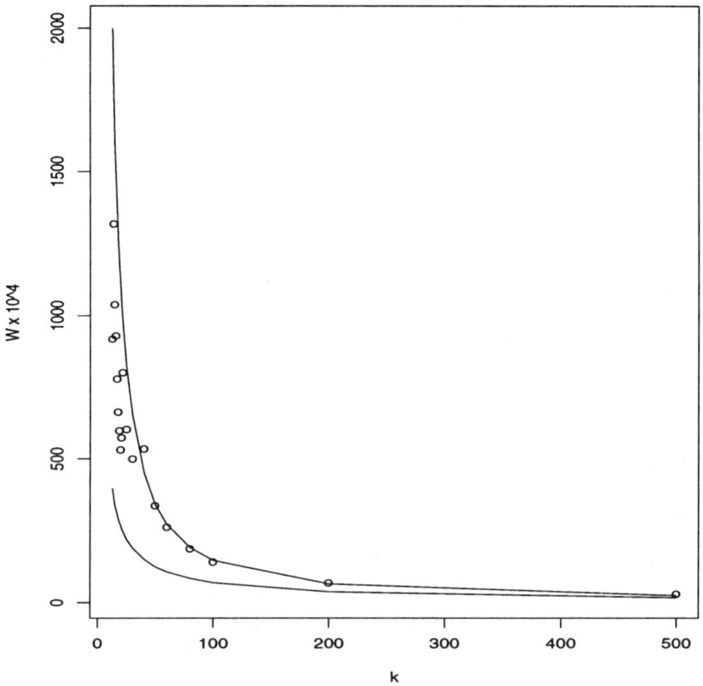

FIGURE 4.2
Original values of W for various values of k.

Table 4.2 Estimated high quantiles of H

ϵ	$k = 100$		$k = 199$		Normal
	Type I	Type III	Type I	Type III	
0.01	9,551	9,556	9,599	9,539	10,040
0.02	9,360	9,397	9,378	9,386	9,808
0.05	9,108	9,130	9,087	9,132	9,461

in reliability applications it is small tail probabilities which are of interest, the example illustrates dramatically the superiority of the method advocated in this Chapter over methods based on the entire sample.

Example 2: Sizes of bushfires (Wang [27])

The sizes (in hectares) of the 30 largest bushfires that burned one hectare or more in Australia during 1986-87 are listed below. There were 75 such fires.

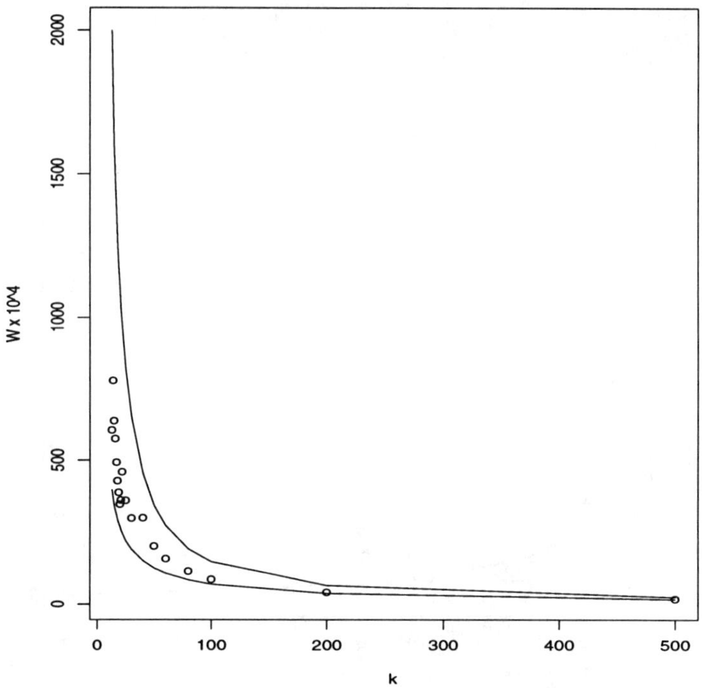

FIGURE 4.3
Transformed values of *W* for various values of *k*.

5800	3500	2300	2160	1000	900	800	700	450	400
360	300	300	300	200	200	200	200	150	150
120	110	100	99	82	80	80	55	50	40

The plot of W against k for the data indicates that $k = 12$ for a significance level $\alpha = 10\%$. However, the downward trend of the graph points to the assumption of Type II domain of attraction. After a logarithmic transformation $Y = \ln(X)$, we find that the optimal k is 27, and that the graph of the transformed W then outcrosses the W_U curve.

4.10 Extension to More than One Dimension

Consider a structural system represented by a vector of physical variables $\mathbf{x} = (x_1, \ldots, x_p)$ and a safe region S in R_p, where R_p is p-dimensional Euclidean space. Suppose that S contains the origin and is star-shaped, i.e., if $\mathbf{x} \in S$ then all points $\lambda\mathbf{x}$, $0 \leq \lambda \leq 1$, also belong to S.

A function of $\mathbf{x}$ which takes values smaller than 1 in the interior of S, values larger than 1 in the exterior of S (the interior of S^c) and is equal to 1 on the boundary of S will be called a failure function. A particularly simple form of failure function, denoted by $\Phi(\mathbf{x})$, can be defined as follows:

1. If all points on the half line through the origin and $\mathbf{x}$ are in S set $\Phi(\mathbf{x}) = 0$.
2. Otherwise, let $\mathbf{x}_0$ be the first point where the half line meets the boundary of S. Put $\Phi(\mathbf{x}_0) = 1$ and let $\Phi(\mathbf{x})$ increase linearly from zero at the origin along the half line. Thus, if $\mathbf{x} = \lambda\mathbf{x}_0$, let $\Phi(\mathbf{x}) = \lambda$. It is easy to verify that the function $\Phi(\mathbf{x})$ has the following two properties:
   (a) For all $\mathbf{x} \in R_p$ and all $\lambda \geq 0$ $\Phi(\lambda\mathbf{x}) = \lambda\Phi(\mathbf{x})$, i.e., Φ is homogeneous of degree one.
   (b) $\mathbf{x} \in S^c$ if and only if $\Phi(\mathbf{x}) \geq 1$.

$\Phi(\mathbf{x})$ will be called the canonical failure function.

Let now the physical variables become random and be denoted by $\mathbf{X}$. Then the probability of failure p_f will be given by

$$p_f = P[\Phi(\mathbf{X}) \geq 1].$$

If the joint distribution of $(X_1, \ldots, X_p)$ is not known explicitly but either a sample of size n is given or there exists an algorithm for simulating $\mathbf{X}$, the p-dimensional problem can be reduced to a one-dimensional problem by means of the following device: Consider the family of functions $\Phi(\mathbf{x}) = \lambda$ and let λ decrease from infinity to zero. Assuming that the distribution of $X_1, \ldots, X_p$ is absolutely continuous, we will obtain a sequence of values $\Lambda_1 \geq \Lambda_2 \geq \ldots \geq \Lambda_k$ at which the boundary $\Phi(\mathbf{X}) = \Lambda$ contains a sample value of $\mathbf{X}$. On the average, np_f of these values will be larger than one.

Choose now the first k values of Λ. They will be a set of upper order statistics. As in the one-dimensional case, we first test whether the underlying distribution can be assumed to belong to the domain of attraction of Type I. If the Type I hypothesis is acceptable then p_f can be estimated by solving for p_f the quantile equation

$$1 = \hat{a}\ln(k/np_f) + \Lambda_k$$

where $\hat{a} = (\sum_{i=1}^{k} \Lambda_i)/k - \Lambda_k$. Otherwise, the appropriate logarithmic transformation, as detailed in Sections 4.7 and 4.8 should be carried out.

4.10.1 An Illustration

Hasofer and Wang [11] describe a bivariate problem where the joint distribution of the variables is a mixture of two bivariate normal distributions. This can occur when the sample values are produced by two different mechanisms, but the observer is unaware of them.

Let $\mathbf{X} = (X_1, X_2)$ be a pair of jointly normal variables with zero mean, unit variance and correlation coefficient +0.5. Let $\mathbf{Y} = (Y_1, Y_2)$ be similarly a pair of jointly normal variables with zero mean, unit variance and correlation −0.5, independent of $\mathbf{X}$. Let A be a Bernoulli variable with $P(A = 1) = 0.5$ and $P(A = 0) = 0.5$, independent of $\mathbf{X}$ and $\mathbf{Y}$. Let $\mathbf{Z} = (Z_1, Z_2) = A\mathbf{X} + (1 - A)\mathbf{Y}$. It is easy to see that Z_1 and Z_2 have marginal standard normal distributions and zero correlation. However, their joint distribution is not normal.

Consider a failure region defined by $Z_1 + Z_2 > M$. For a failure probability $p_f = 0.005$ the true value of M is 4.34. Suppose however that on the basis of a test of the marginal distributions and a finding of zero correlation we treat Z_1 and Z_2 as independent standard normal. The incorrect value of M obtained is $M = 3.64$, corresponding to a failure probability $p_f = 0.0009$, 80% too high!

Applying the W test with $k = 50$ and $n = 5,000$ to M, it turns out that M belongs to the domain of attraction of Type I. The Weissman estimator, also based on $k = 50$ and $n = 5000$, then yields $\mathrm{E}(\hat{M}) = 4.12$ with standard deviation 0.10.

4.11 Introducing Importance Sampling

4.11.1 A Primer of Importance Sampling

Since Importance Sampling is a powerful and increasingly popular method of reducing estimator variance, it is natural to seek to use it in the context of non-parametric estimation.

The basic idea of Importance Sampling is very simple. Let D be the failure domain when the system under study is represented by the p-dimensional vector of physical variables $\mathbf{X}$. Then the probability of failure p_f is given by $p_f = \int_D f(\mathbf{x})d\mathbf{x}$ where $f(\mathbf{x})$ is the joint probability density function of $\mathbf{X}$. The raw Monte-Carlo estimator of p_f is obtained by drawing a sample of size n from $f(\mathbf{x})$ and calculating

$$\hat{p}_f = \frac{1}{n} \sum_{i=1}^{n} \chi_D(\mathbf{x}_i)$$

where $\chi_D(\mathbf{x})$ is the indicator function of D.

Unfortunately, since the probability of the failure domain is small, most of the sample values are wasted. This is what makes the raw Monte-Carlo method inefficient in reliability problems.

To overcome this difficulty, we draw the sample from a different probability density function $h(\mathbf{x})$, called the "Importance Sampling Density", which samples heavily from the failure domain. Now since p_f can be written in the form

$$p_f = \int_D \frac{f(\mathbf{x})}{h(\mathbf{x})} h(\mathbf{x}) d\mathbf{x}$$

an unbiased estimator of p_f is given by

$$\hat{p}_f = \frac{1}{n} \sum_{i=1}^{n} \chi_D(\mathbf{x}_i) \frac{f(\mathbf{x})}{h(\mathbf{x})}$$

where $\mathbf{x}_1, \ldots, \mathbf{x}_n$ is a sample from $h(\mathbf{x})$. The variance of this latter estimator is given by

$$\mathbf{Var}(\hat{p}_f) = \frac{1}{n} \left[\int_D \frac{f^2(\mathbf{x})}{h(\mathbf{x})} d\mathbf{x} - p_f^2 \right].$$

It is clear that the choice of the Importance Sampling Density is crucial to the precision of the estimator. Indeed, if $h(\mathbf{x})$ could be chosen to be $\chi_D(\mathbf{x}) f(\mathbf{x})/p_f$, then the variance would be zero. Of course this choice is not practical, since it requires knowing exactly the failure probability p_f, which is precisely what we are trying to evaluate. However, it provides a goal, an ideal sampling density, which we should try to approximate.

4.11.2 The Weissman Estimator Revisited

We recall that the Weissman estimator $\hat{q}(\epsilon)$ of the quantile $q(\epsilon)$ based on the top k order statistics $X_{1n}, X_{2n}, \ldots, X_{kn}$ of a sample of size n from the distribution $F(x)$ when $F(x)$ belongs to the domain of attraction of Type I is given by

$$\hat{q}(\epsilon) = \hat{a} \ln(k/n\epsilon) + X_{kn}$$

where

$$\hat{a} = \left(\sum_{i=1}^{k} X_{in} \right) / k - X_{kn}. \tag{4.10}$$

Let now $F_n(x)$ be the empirical distribution function of our sample, namely a step function with jumps of size $1/n$ at each sample value. Equation (4.10) can be rewritten in the form

$$\hat{a} = \frac{n}{k}\left[\frac{1}{n}\sum_{i=1}^{k}(X_{in} - X_{kn})\right] = \frac{\int_{X_{kn}}^{\infty}(x - X_{kn})dF_n(x)}{1 - F_n(X_{kn})}. \tag{4.11}$$

Let now n and k tend to infinity in such a way that X_{kn} tends to x_0. Since $F_n(x) \to F(x)$ as $n \to \infty$, the right-hand side of (4.11) tends to

$$a_L = \left[\int_{x_0}^{\infty}(x - x_0)dF(x)\right]\left[\int_{x_0}^{\infty}dF(x)\right]^{-1} \tag{4.12}$$

and furthermore $k/n \to 1 - F(x_0)$ so that

$$\hat{q}(\epsilon) \to x_0 + a_L\{\ln[1 - F(x_0)] - \ln \epsilon\}. \tag{4.13}$$

It should be noted that the upper tail of the Type I distribution is asymptotic to that of the negative exponential:

$$1 - e^{-e^{-x}} \sim 1 - e^{-x}$$

as $x \to \infty$. This is also true of the respective density functions. What is interesting is that if $X = a_L W + b$ where W has the density function e^{-x}, $x > 0$, then formulae (4.12) and (4.13) hold exactly for any x_0. Moreover, we have

$$b = x_0 + a_L \ln[1 - F(x_0)]$$

and b is estimated by

$$\hat{b} = X_{kn} + \hat{a}\ln(k/n). \tag{4.14}$$

4.11.3 A Modified Weissman Estimator

Assume that $F(x)$ has a density function $f(x)$ and let $g(x)$ be an Importance Sampling density function. Then we can rewrite (4.12) in the form

$$a_L = \left[\int_{x_0}^{\infty}(x - x_0)\frac{f(x)}{g(x)}g(x)dx\right]\left[\int_{x_0}^{\infty}\frac{f(x)}{g(x)}g(x)dx\right]^{-1}.$$

Suppose now that we generate the k upper order statistics of a sample of size n from the density function $g(x)$ and denote them by $Y_{1n} \geq Y_{2n} \geq \ldots \geq Y_{kn}$. Then, by the usual method of Importance Sampling, we set $x_0 = Y_{kn}$ and estimate a_L by

$$\hat{a} = \left[\frac{1}{n}\sum_{i=1}^{k}(Y_{in} - Y_{kn})\frac{f(Y_{in})}{g(Y_{in})}\right]\left[\frac{1}{n}\sum_{i=1}^{k}\frac{f(Y_{in})}{g(Y_{in})}\right]^{-1}$$

and $q(\epsilon)$ by

$$\hat{q}(\epsilon) = Y_{kn} + \hat{a}\left[\ln\left\{\frac{1}{n}\sum_{i=1}^{k}\frac{f(Y_{in})}{g(Y_{in})}\right\} - \ln(\epsilon)\right]. \tag{4.15}$$

The proposed scheme in the univariate case will therefore consist of the following steps:

1. Generate a pilot sample of size n_1 from the distribution of X. Select the k_1 upper order statistics and test whether X belongs to the domain of attraction of Type I (see Section (4.6)). If it does not, carry out an appropriate logarithmic transformation (see Sections (4.7) and (4.8)).
2. From the same k_1 order statistics obtain rough estimates $\tilde{a}$ and $\tilde{b}$, using equations (4.11) and (4.14).
3. Simulate the top k order statistics of a sample of size $n \gg n_1$ from the shifted negative exponential distribution with density function $f(x) = \exp[-(x-\tilde{b})/\tilde{a}]/\tilde{a}$. The required algorithm is described in the next section.
4. Estimate $q(\epsilon)$ accurately from equation (4.15).

Since it is only necessary to generate k random variables from a simple distribution, followed by the calculation of $f(x)$ for these k values, one can afford to choose k and n much larger than for raw simulation. And since the standard deviation of $q(\epsilon)$ is approximately proportional to $1/\sqrt{k}$ (see Section 4.9) the estimator obtained will have a much higher precision than otherwise attainable with the same amount of information. Large-scale simulation confirms this conclusion.

4.11.4 Direct Simulation of k Upper Order Statistics

One of the simplest ways of simulating the k top order statistics from the shifted negative exponential distribution is to use the "sequential" method (see Kennedy and Gentle [16]). First we generate k independent uniform (0, 1) variables $A_1, \ldots, A_k$. Then we calculate sequentially $B_1 = A_1^{1/n}$, $B_{r+1} = B_r A_{r+1}^{1/(n-r)}$, $(r = 1, \ldots, k-1)$. Finally we set $Y_r = \tilde{b} - \tilde{a}\ln(1 - B_r)$.

4.11.5 Extension to the Multidimensional Case

In the single-dimensional case, the importance sampling distribution was taken to be the upper tail of a Type I distribution (provided the W test supported such a premise). It turned out that this was asymptotically equivalent to the tail of a shifted negative exponential distribution.

To extend our method to the multivariate case, we follow the asymptotic results given by Maes, Breitung, and Dupuis [20] (see also the earlier results given in Dupuis and Maes [4]).

Our setup is the same as in Section 4.10. But the approach is based on the remark that the joint density function $f(\mathbf{x})$ of $\mathbf{X}$ (also called the "likelihood" of $\mathbf{X}$) has, under normal circumstances, a maximum on the failure hypersurface $\Phi(\mathbf{x}) = \lambda$, and that maximum is reached at just one point, the "point of maximum likelihood" or PML for short. If $\mathbf{X}$ is a standard normal vector, the PML is just the well-known beta-point used in second order reliability analysis.

We further assume that the failure hypersurface is suitably smooth in the neighborhood of the PML and that the likelihood function can be approximated by a second-order Taylor expansion near the PML. Let us carry out a translation and rotation transformation $\mathbf{X} \to \mathbf{Y}$ to a coordinate system with origin at the PML, the first coordinate along the outward normal to the failure hypersurface and the other $p - 1$ coordinates in the hyperplane tangent to the failure hypersurface. Under such circumstances, it is shown in [20] that the random vector $\mathbf{Y}$ has asymptotically (as $\lambda \to \infty$) a distribution which is the product of shifted negative exponential distribution (for Y_1) and an independent zero-mean multivariate normal along the other $p - 1$ coordinates.

The proposed procedure to estimate p_f is as follows:

1. Obtain a pilot sample of size n_1 and select the sample values that correspond to the k_1 top order statistics of Λ. Test that they belong to the domain of attraction of type I. If not, carry out an appropriate logarithmic transformation.

2. Transform the chosen sample values to the $\mathbf{Y}$ coordinate system.

3. Estimate the parameters of the distribution of $Y_2, \ldots, Y_p$ from the chosen sample with the usual estimators for variances and covariances.

4. Let $V_1 \geq \ldots \geq V_{k_1}$ be the ordered values of the Y_1 variable. Asymptotically $V = aW + b$ where W is negative exponential. Estimate a and b by the usual Weissman estimators

$$\tilde{a} = \left[\frac{1}{k_1} \sum_{i=1}^{k_1} V_i \right] - V_{k_1}$$

and

$$\tilde{b} = V_{k_1} + \tilde{a} \ln(k_1/n).$$

Let the joint density function of $Y_1, \ldots, Y_p$ be denoted by $h(\mathbf{y})$.

5. Generate the top k order statistics from a sample of size $n \gg n_1$ from the negative exponential variable V, following the procedure given in Section 4.11.4.

6. Generate a random sample of size k from the multivariate normal distribution $h(\mathbf{y})$. This is a legitimate procedure because the $Y_2, \ldots, Y_p$ are asymptotically independent of Y_1.

7. Calculate the order statistics $\Lambda_i, i = 1, \ldots, k$ of the canonical variable Λ corresponding to the sample values of $\mathbf{Y}$.

8. Estimate the upper quantile $q(\epsilon)$ of Λ by

$$\hat{q}(\epsilon) = \Lambda_k + \hat{a}\left\{\ln\left[\frac{1}{n}\sum_{i=1}^{k}\frac{f(\mathbf{y}_i)}{h(\mathbf{y}_i)}\right] - \ln\epsilon\right\}$$

where

$$\hat{a} = \left[\frac{1}{k}\sum_{i=1}^{k}(\Lambda_i - \Lambda_k)\frac{f(\mathbf{y}_i)}{h(\mathbf{y}_i)}\right]\left[\frac{1}{k}\sum_{i=1}^{k}\frac{f(\mathbf{y}_i)}{h(\mathbf{y}_i)}\right]^{-1}.$$

Alternatively, the probability $\hat{\epsilon}$, which estimates the failure probability corresponding to the failure surface $\Phi(\mathbf{x}) = q(\epsilon)$, is given by

$$\hat{\epsilon} = \frac{1}{n}\sum_{i=1}^{k}\frac{f(\mathbf{y}_i)}{h(\mathbf{y}_i)}\exp[\hat{a}^{-1}(\Lambda_k - q(\epsilon)].$$

4.11.6 Simulation Examples

The results given below are all based on 1, 000 replications.

1. A two-dimensional example.
Let X_1 be $N(0, 1)$ along the x-axis and X_2 be $N(17, 1)$ along the y-axis and suppose they are independent. We are interested in finding $P(X_1^2 + X_2^2 \geq 20^2$. The true value of the probability is 0.001476, calculated by a noncentral chi-square distribution, since the joint density function of X_1 and X_2 is

$$f(x, y) = (2\pi)^{-1}\exp\{-[x^2 + (y-17)^2]/2\}, \quad -\infty < x, y < \infty$$

We estimate the probability by the following steps:

(a) A sample of size 500 is obtained from X_2 and the top 80 values are selected. From them we estimate the scale and location parameters

$\tilde{a}$ and $\tilde{b}$. The Importance Sampling density is then given by

$$h(x, y) = (2\pi)^{-1/2} \exp(x^2/2) \exp[-\tilde{a}^{-1}(y - \tilde{b})]$$

for $-\infty < x < \infty$, $y > \tilde{b}$.

(b) The top 400 order statistics from a sample X of size 40,000 are generated from the negative exponential distribution and 400 values are generated from a sample Y from the distribution $N(0, 1)$.

(c) Let $\Lambda = (X^2 + Y^2)^{1/2}$ and $\Lambda_1 \geq \Lambda_2 \geq \ldots \geq \Lambda_{400}$ be the corresponding order statistics.

(d) For given $q(\epsilon) = 20$, we estimate the probability $\epsilon = P[\Lambda > q(\epsilon)]$.

The results are tabulated in Table 4.3.

Table 4.3 Simulation results: example 1

True value	0.001476
Estimated mean value	0.001487
Standard deviation	0.000115
Mean square error	1.322×10^{-8}
Minimum	0.001216
Maximum	0.001943

2. Effect of increasing the curvature at the PML.
   Suppose that in the previous example we reduce the mean of X_2 from 17 to 7 and change the failure curve to $X_1^2 + X_2^2 = 10^2$. This has the effect of increasing the curvature of the failure curve with respect to the joint distribution of X_1 and X_2. The results of the simulation are given in Table 4.4.
   It is apparent that increased curvature of the failure hypersurface adversely affects the precision of the estimates.

3. A three-dimensional example.
   Let X_i be $N(0, 1)$ along the x_i-axes, $i = 1, 2$, and X_3 be $N(17, 1)$ along the y-axis, and suppose they are independent. The true value of the probability $P(X_1^2 + X_2^2 + X_3^2 \geq 20^2)$ is 0.001610. Here we have for the original joint density

$$f(x_1, x_2, y) = (2\pi)^{-3/2} \exp\{-[x_1^2 + x_2^2 + (y - 17)^2]/2\}$$

for $-\infty < x_1, x_2, y < \infty$ and for the Importance Sampling density

Table 4.4 Simulation results: example 2

True value	0.001650
Estimated mean value	0.001622
Standard deviation	0.000137
Mean square error	1.967×10^{-8}
Minimum	0.001320
Maximum	0.002201

$$h(x_1, x_2, y) = (2\pi)^{-1} \exp[(x_1^2 + x_2^2)/2] \exp[-\tilde{a}^{-1}(y - \tilde{b})]$$

for $-\infty < x_1, x_2 < \infty$, $y > \tilde{b}$.
The results are given in Table 4.5.

Table 4.5 Simulation results: example 3

True value	0.001610
Estimated mean value	0.001628
Standard deviation	0.000141
Mean square error	2.006×10^{-8}
Minimum	0.001288
Maximum	0.002250

4. A four-dimensional example.
   Let X_i be $N(0, 1)$ along the x_i-axes, $i = 1, 2, 3$, and X_4 be $N(17, 1)$ along the y-axis, and suppose they are independent. The true value of the probability $P(X_1^2 + X_2^2 + X_3^2 + X_4^2 \geq 20^2)$ is 0.001760. The results of the simulation are given in Table 4.6.

4.12 The Threshold Method

As mentioned in the Introduction, an alternative method for estimating high quantiles of a distribution, based on a sample, is to chose some high level u and

Table 4.6 Simulation results: example 4

True value	0.001760
Estimated mean value	0.001744
Standard deviation	0.000149
Mean square error	2.249×10^{-8}
Minimum	0.001420
Maximum	0.002564

to base the estimation on the sample points above the level u, known as the "exceedances".

The basic asymptotic result used in the threshold method is due to Pickands [22]. Let $F(x)$ be the distribution function of X, let ω denote the upper end point of $F(x)$ and let

$$F_u(y) = \frac{F(u+y) - F(u)}{1 - F(u)}, \quad u < \omega, 0 < y < \omega, \tag{4.16}$$

be the conditional distribution function of $X - u$ given $X > u$. Define the Generalised Pareto Distribution (GPD) by

$$G(y;\sigma,\alpha) = \begin{cases} 1 - \left(1 + \frac{\alpha y}{\sigma}\right)^{-1/\alpha}, & \text{for } \alpha \neq 0,\ \sigma > 0, \\ 1 - \exp(-y/\sigma), & \text{for } \alpha = 0,\ \sigma > 0, \end{cases} \tag{4.17}$$

for $0 < y < \infty$ when $\alpha > 0$ or $0 < y < -\sigma/\alpha$ when $\alpha < 0$.

It is seen that, just as in the case of the extreme value distributions (Section 4.3), the case $\alpha = 0$ is the limit of the case $\alpha \neq 0$ as $\alpha \to 0$. The case $\alpha > 0$ is just a reparametrization of the ordinary Pareto distribution. Pickands showed that the Generalised Pareto Distribution is a good approximation of $F_u(y)$ in the sense that

$$\lim_{u \to \omega} \sup_{0<y<\omega-u} |F_u(y) - G(y;\sigma(u),\alpha)| = 0 \tag{4.18}$$

for some fixed α and some positive function $\sigma(u)$ if and only if $F(x)$ is in the domain of attraction of one of the three limit laws. The value of α is the same as in equation (4.2) of Section 4.3 while $\sigma(u)$ may be taken to be

$$\sigma(u) = -\frac{F(u)\log F(u)}{F'(u)} \tag{4.19}$$

where $F'(u) = dF(u)/du$ whenever $F(x)$ is differentiable.

Using this result, Pickands derived an estimate of the upper tail of $F(x)$ which is consistent for any of the three domains of attraction.

Using the same approach, Smith [25] gave another procedure, in which maximum likelihood estimation was applied. Let $X_1, X_2, \ldots, X_n$ be a large sample from the distribution $F(x)$ and let the corresponding order statistics be $X_{1n} \geq X_{2n} \geq \ldots \geq X_{nn}$. Fix a high threshold, say u, and let N denote the number of exceedances of u. Let $Y_1, Y_2, \ldots, Y_N$ denote the excesses, i.e., $Y_i = X_{in} - u, \quad i = 1, 2.\ldots, N$. Conditionally on N, the excesses are independent identically distributed random variables with distribution function $F_u(y)$. To obtain a tail estimate, Smith then proposes to approximate $F_u(y)$ by the Generalised Pareto Distribution and to estimate σ and α by maximum likelihood based on $Y_1, Y_2, \ldots, Y_N$. These estimators are asymptotically efficient and normally distributed provided $\alpha > -0.5$. When that condition is violated, a variety of other results are known (see Smith [24]).

A useful graphical device both to assess the overall fit of the model and more specifically as an aid to choosing the threshold is a 'mean residual life plot' in which the mean excess over a threshold u is plotted against u for a wide range of values of u. The name comes from the use of a similar idea in survival analysis. The device is based on the fact that, if Y has a Generalised Pareto Distribution with parameters σ and α, then, provided $\alpha < 1$ and $u \geq 0$, we have

$$\mathrm{E}(Y - u | Y \geq u) = \frac{\sigma + \alpha u}{1 - \alpha}. \tag{4.20}$$

Consequently, we expect the plot to be roughly linear with intercept $\sigma/(1 - \alpha)$ and slope $\alpha/(1-\alpha)$. In practice, the plot tends to be very irregular at its right-hand end, when u is large and the number of exceedances of u is small, but away from there it can be a useful way of checking upon the appropriateness of the generalised Pareto Distribution.

4.13 Serial Dependence and Seasonality

When the data considered arise from successive observations in a time series, the assumption that the observations are independent must be questioned. When dependence exists, observations tend to appear in clusters rather than at random. The treatment of serial dependence in modern statistical analysis has been much influenced by developments in the theory of extremes of stationary stochastic processes. See for instance Leadbetter, Lindgren, and Rootzén [17], Leadbetter and Rootzén [18], and Aldous [1]. In discrete time processes, provided dependence between observations decreases to zero fast enough when the distance between the observations goes to infinity ("mixing" conditions), it can be shown that the extremes are attracted to the same limit laws as extremes of random samples. However, instead of the formula $P(M_n \leq x) = F^n(x)$ for the maximum M_n of a random sample of size n from $F(x)$, which applies when the observations are independent, we have a relation of the form

$$P(M_n \leq x) \approx F^{n\theta}(x) \tag{4.21}$$

where the parameter θ may take any value between 0 and 1 (including 0 and 1 themselves) and has the rough interpretation that $1/\theta$ is the mean cluster size. Thus $\theta = 1$ is the case of no clustering (as with independent data) and $\theta = 0$ is the opposite extreme when the mean cluster size is infinite. In other words, dependence can be dealt with approximately by simply reducing the sample size in the formulae from n to $n\theta$, treating each cluster as one observation. Extensions to non-stationary series are available (Hüsler [15]).

It is also possible that the observations vary seasonally. This can be dealt with by breaking up the year into independent seasons, and treating each season separately. It is a critical decision how many seasons to use: too few and the seasonal effect will not be properly allowed for, but if too many are used there will not be enough data in each season to allow good estimation of the parameters.

4.14 Conclusions and References

When the vector of physical variables is known only through a sample, either by simulation or from field measurements, it is well-nigh impossible to evaluate the conditional probability distributions required by a standard procedure of reliability assessment.

The novelty of the proposed method lies essentially in the fact that the failure probability can then be accurately estimated from the sample, provided the sample size is large enough, without making any assumptions about the underlying joint distribution. This can be achieved by using recent results of extreme value theory. The calculation of failure probability is first reduced to a one-dimensional problem. The domain of attraction is then determined. Finally, the appropriate asymptotic estimators are applied.

References

[1] Aldous, D.J., *Probability Approximations via the Poisson Clumping Heuristic,* Applied Mathematical Sciences Series No. 77, Springer Verlag, New York, 1989.

[2] Boos, D.D., Using Extreme Value Theory to Estimate Large Percentiles. *Technometrics*, **26**, 1, 1984, pp. 33–39.

[3] Dalgliesh, W.A. and Taylor, D.A., The Strength and Testing of Window Glass. *Can. J. Civ. Eng.*, **17**, 1990, pp. 752–762.

[4] Dupuis, D.J. and Maes, M.A., A Computationally Effective, Self-adjusting Technique for Determining Failure Probability. *Proc. CERRA-ICASP 6*, Mexico, 1991, Vol. I, pp. 240–249.

[5] Galambos, J., *The Asymptotic Theory of Order Statistics.* John Wiley & Sons, New York, 1978.

[6] Godambe, V.P., The Foundations of Finite Sample Estimation in Stochastic Processes. *Biometrika*, **72**, 1985, pp. 419–428.

[7] Harmathy,T.Z. and Mehaffy, J.R., Normalized Heat Load: A Key Parameter in Fire Safety Design, *Fire and Materials*, **6**, 1, 1982, pp. 27–31.

[8] Hasofer, A.M., High Order Statistics in Structural Reliability. *Proc. Specialty Conference on Probabilistic Mechanics and Structural and Geotechnical Reliability*, Denver, Colorado, ASCE, 1992.

[9] Hasofer, A.M. and Li, G.S., Estimation for Type II Domain of Attraction Based on the *W* Statistic (in preparation), 1995.

[10] Hasofer, A.M. and Beck, V.R., A Stochastic Model for Compartment Fires. *Proc. CERRA-ICASP 7*, Paris, 1995, Vol. I, pp. 525–528.

[11] Hasofer, A.M. and Wang, J.Z., System Reliability Calculations Using Extreme Value Theory. *Proc. CERRA-ICASP 6*, Mexico, 1991, Vol. I, pp. 41–47.

[12] Hasofer, A.M. and Wang, J.Z., A Test for Extreme Value Domain of Attraction. *J. Am. Stat. Assoc.*, **87**, 417, 1992, pp. 171–177.

[13] Hasofer, A.M. and Wang, J.Z., Estimating Quantiles for a Type III Domain of Attraction, Based on the *k* Largest Observations. *Proceedings of the Conference on Extreme Value Theory and Applications*, Editors: J. Galambos, J. Lechner, and E. Simiu, NIST Special Publication 866, National Institute of Standards and Technology, U.S.A., 1993, Vol. 3, pp. 149–155.

[14] Hohenbichler, M. and Rackwitz, R., Non-normal Vectors in Structural Safety. *J. Eng. Mech. Div., ASCE,* **107**, No EM6, 1981, pp. 1227–1238.

[15] Hüsler, J., Extreme Values of Non-stationary Random Sequences, *J. Appl. Prob.*, **23**, 1986, pp. 937–950.

[16] Kennedy, W.J. and Gentle, J.E., *Statistical Computing*. Marcel Dekker, New York, 1980.

[17] Leadbetter, M.R., Lindgren, G., and Rootzén, H., *Extremes and Related Properties of Random Sequences and Series*, Springer Verlag, New York, 1993.

[18] Leadbetter, M.R. and Rootzén, H., Extremal Theory for Stochastic Processes, *Ann. Probab.*, **16**, 1988, pp. 431–478.

[19] Lind, N.C., Consistent Distribution Parameter Estimation for Reliability Analysis. *Structural Safety*, **4**, 2, 1987, pp. 141–149.

[20] Maes, M.A., Breitung, K., and Dupuis, D.J., Asymptotic Importance Sampling. *Structural Safety*, **12**, 3, 1993, pp. 167–186.

[21] Ofverbeck, P., Small Sample Control and Structural Safety. Report TVBK-3009, Lund Institute of Technology, Dept. of Structural Engineering, Lund, Sweden, 1980.

[22] Pickands, J., Statistical Inference using Extreme Order Statistics. *Ann. Statist.*, **3**, 1975, pp. 119–131.

[23] Shapiro, S.S. and Wilk, M.B., An Analysis of Variance Test for Normality (Complete samples). *Biometrika*, **52**, 1965, pp. 591–611.

[24] Smith, R.L., Maximum Likelihood Estimation in a Class of Nonregular Cases. *Biometrika*, **72**, 1985, pp. 67–92.

[25] Smith, R.L., Estimating Tails of Probability Distributions. *Ann. Statist.*, **15**, 3, 1987, pp. 1174–1207.

[26] Wang, J.Z., Determination of Domain of Attraction for Annual Extremes. *Australian Statistical Society Conference*, Melbourne, July 1994.

[27] Wang, J.Z., Selection of the k Largest Order Statistics for the Domain of Attraction of the Gumbel Distribution. To appear in *J. Am. Stat. Assoc.*, 1995.

[28] Weissman, I., Estimation of Parameters and Large Quantiles Based on the k Largest Observations. *J. Am. Stat. Assoc.*, **73**, 364, 1978, pp. 812–815.

5

Response Surface Methods and Asymptotic Approximations

K. Breitung
University of Pavia, Italy

L. Faravelli
University of Pavia, Italy

5.1 Introduction

In problems of structural reliability we study structures in the presence of random influences, which are described by stochastic models. Such models are random variables, random vectors, stochastic processes, and random fields. Often also the properties of the structure itself are modeled by random quantities. The structure fails if certain functions (limit state functions) of these quantities exceed some limits (see [31], [39], and [15]).

There are several levels of modeling the behavior of technical structures in reliability analysis. The simplest model is to describe all random quantities by a random vector $\boldsymbol{X}$ which does not depend on time. Such models are called time-invariant.

More advanced models incorporate random processes or random fields; they are called time-variant models. Approximations for time-variant problems can be obtained by introducing time as an additional component into a time-invariant model. If the state of the structure at a fixed time t is described by an n-dimensional random vector $\boldsymbol{X}(t)$, we can consider this vector at different time points $t_1, \dots, t_k$. To include time-variant behavior in such a random vector, we study the $n \times k$ random vector $(\boldsymbol{X}(t_1), \dots, \boldsymbol{X}(t_k))$.

Here we will consider the case of the description by a random vector, but we will assume that its distribution depends on a parameter vector and give some tools for optimization, i.e., methods for calculating derivatives with respect to the parameters.

The basic reliability model, which will be studied in the following, is the parameter-dependent integral that defines the probability of failure $I\!P(\boldsymbol{\tau})$ by:

$$I\!P(\boldsymbol{\tau}) = \int_{g(\boldsymbol{x},\boldsymbol{\tau})\leq 0} f(\boldsymbol{x},\boldsymbol{\tau})\, d\boldsymbol{x}. \tag{5.1}$$

Here $\boldsymbol{\tau}$ is a parameter, vector $\boldsymbol{X}$ assumes its values, and the lower case $\boldsymbol{x}$ denotes a vector in the n-dimensional space where the random vector for fixed $\boldsymbol{\tau}$ the function $f(\boldsymbol{x},\boldsymbol{\tau})$ is an n-dimensional p.d.f. (probability density function) of a random vector $\boldsymbol{X}(\boldsymbol{\tau})$ and $g(\boldsymbol{x},\boldsymbol{\tau})$ is a limit state function. The parameter vector may include design parameters and/or distribution parameters of the random influences.

Three possible situations are the following:

1. $X_1 =$ load, $X_2 =$ resistance, $\tau_1 = I\!E[X_1] =$ mean value of X_1. It follows that $g(\boldsymbol{x}) = X_2 - X_1$ does not depend on τ_1,
2. $X_1 =$ load, $X_2 =$ local resistance, $\tau_1 =$ load/load effect. It follows that $g(\boldsymbol{x},\tau_1) = X_2 - X_1/\tau_1$ and $f(\boldsymbol{x})$ does not depend on τ_1,
3. $X_1 =$ load, $X_2 =$ local resistance, $\tau_1 = I\!E[X_1]$, $\tau_2 =$ load/load effect. It follows that both g and f depend on $\boldsymbol{\tau}$.

For the given parameter vector $\boldsymbol{\tau}_0$, the random vector $\boldsymbol{X}(\boldsymbol{\tau}_0)$ having the p.d.f. $f(\boldsymbol{x},\boldsymbol{\tau}_0)$ describes the random influences on the structure under consideration. There are four steps in the probabilistic analysis of such a problem:

1. the determination of the distribution of $\boldsymbol{X}(\boldsymbol{\tau})$, i.e., to find the functional form of $f(\boldsymbol{x},\boldsymbol{\tau})$,
2. the determination of the functional form of $g(\boldsymbol{x},\boldsymbol{\tau})$,
3. the computation of the integral $I\!P(\boldsymbol{\tau})$ given in equation (5.1),
4. the optimization of the structure with respect to the failure probability, i.e., to find a parameter vector, which minimizes $I\!P(\boldsymbol{\tau})$ under design restrictions.

If the parameter vector includes distribution parameters whose values are not known exactly, a further problem is to find confidence intervals or Bayesian posterior distributions for the failure probabilities.

The first point is a standard problem of statistical inference; it is treated in all statistical textbooks. The two main approaches are based on the classical frequentist concept and on the Bayesian idea. In this context we will use, if necessary, the Bayesian approach as outlined for example in [1] and [5].

We will assume in the following that the distributional form is known, i.e., the function $f(\boldsymbol{x}, \boldsymbol{\tau})$ and that any uncertainty about the values of the parameter is described in the Bayesian concept by a p.d.f. $p(\boldsymbol{\tau})$ over the parameter space of admissible values.

The second problem is a more specific problem of reliability theory. In many cases it is assumed that the limit state function $g(\boldsymbol{x}, \boldsymbol{\tau})$ is known in an analytic form. But often this is not the case and the form has to be estimated from data. Since the dimension of the random vector is usually large and the shape of the limit state function complicated, a straightforward application of the statistical techniques for such a problem could lead one to unsatisfactory results. Both the specific requirements of a reliability analysis and the known distributional form of the random vector $\boldsymbol{X}(\boldsymbol{\tau})$ have to be considered to find an adequate solution. A method for determining the limit state function will be outlined in section 5.2.

The third problem, the computation of failure probabilities, is a mathematical problem. Exploiting the special structure of the integral in equation (5.1) helps to find efficient methods for this task. Methods for computing such integrals are developed in section 5.3.

For the optimization of structures we derive in section 5.4 approximations for the derivatives of the failure probability integral with respect to the components of the parameter vector $\boldsymbol{\tau}$. The uncertainty of these parameters is also discussed.

Numerical examples are developed in section 5.5 to illustrate the main aspects of the reliability analysis.

5.2 Response Surface Methods

In studying the structure of limit state functions, we often encounter the problem that these functions are not known explicitly or have a complicated functional form. One deals with two different problems, of a mathematical and statistical nature, respectively, and it is necessary to distinguish carefully between them.

If there is a limit state function g given, which assigns to every point $\boldsymbol{x}$, in the state space, exactly one value $g(\boldsymbol{x})$ of the limit state function, there are no statistical problems. If we attempt to replace this function by a simpler function $\hat{g}$, which is easier to handle, we have an interpolation problem. The quality of the interpolation can be judged if there are estimates for the function g and its derivatives. For example, near a point a function can be approximated by its Taylor expansion to the k degree. Then upper bounds for the approximation error can be obtained from estimates of the maxima of derivatives of degree $k + 1$.

Another problem appears if the limit state function is not known beforehand and has to be estimated from data and/or experiments. Then we have now the problem that

$$y = g(x_1, \dots, x_n). \tag{5.2}$$

but the functional form of g is unknown. Here the $x_1, \dots, x_n$ are called the independent or regressor variables and y the dependent or response variable. The usual method of statistical inference to find such a relationship is the response surface method ([43], [6], [45], [7], [25], and [15]).

Response surface methods consist of the following steps:

1. Choice of one of several families of functions, which appear to be suitable to approximate the unknown function $g(x_1, \dots, x_n)$,
2. Design of experiments, which give optimal estimators of the parameters of the functions chosen in the last step,
3. Validation of the derived approximation model by statistical tests or other methods.

A possible further step in response surface methodology is to test the significance of the contribution of terms in the derived functional form, for example, in a quadratic polynomial, the significance of the square terms. If they are not significant, a simpler model without these terms might be used instead of the whole polynomial expression.

If experiments are made, we have in general some random variability. If the random experiment is run m times for the same values of $x_1, \dots, x_n$, the resulting values $y_1, \dots, y_m$ of the response variable will be different. This means that there is inherently a pure random error. Therefore the model should be put in the form

$$y = g(x_1, \dots, x_n) + \epsilon(x_1, \dots, x_n) \tag{5.3}$$

with $\epsilon(x_1, \dots, x_n)$ representing the error term. To judge the quality of a model, estimates of the magnitude of this error term are important. Since in such an experimental design we have different responses for the identical set of regressor variables, it will not be possible in general to fit a model without any error term.

If the experiments are of a numerical nature, contribution to this pure error arises from the impossibility of working in the space of all the influencing variables. Usually, a projection in a space of reduced dimension is introduced, where then the influence of the neglected variables results in a random effect.

Let us assume that a complete description of the state of a structural system can be given by an n-dimensional random vector $\boldsymbol{X}$. As we will see soon, this is not possible, in general, since a stochastic process or a random field can only be described by an infinite expansion. But if we make for the moment this simplifying assumption, we can write the failure probability in the following form

$$\mathbb{P}(\mathbb{F}) = \int_{\mathbb{R}^n} \mathbf{1}_F(x_1, \ldots, x_n) f(\boldsymbol{x})\, d\boldsymbol{x}. \tag{5.4}$$

Here $\mathbf{1}_F(\boldsymbol{x})$ is the characteristic function of the failure set $\mathbb{F}$, i.e., $\mathbf{1}_F(\boldsymbol{x}) = 1$ if $\boldsymbol{x} \in \mathbb{F}$ and zero elsewhere. $f(\boldsymbol{x})$ is again the p.d.f. of the random vector $\boldsymbol{X}$.

Now, in reality, we know only a subset of these variables. Without loss of generality we can assume that these are the first k $(k < n)$ variables. We have then

$$\mathbb{P}(\mathbb{F}) = \int_{\mathbb{R}^k} \mathbb{E}[\mathbf{1}_F(X_1, \ldots, X_k, X_{k+1}, \ldots, X_n) | \boldsymbol{X}^k = \boldsymbol{x}^k] f^k(\boldsymbol{x}^k)\, d\boldsymbol{x}^k \tag{5.5}$$

Here $\boldsymbol{x}^k = (x_1, \ldots, x_k)$ and $f^k(\boldsymbol{x}^k)$ is the p.d.f. of the random vector $\boldsymbol{X}^k = (X_1, \ldots, X_k)$. $\mathbb{E}[\mathbf{1}_F(X_1, \ldots, X_k, X_{k+1}, \ldots, X_n) | \boldsymbol{X}^k = \boldsymbol{x}^k]$ is the conditional expectation of the function $\mathbf{1}_F(\boldsymbol{X})$ under the condition that the first k components are equal to $x_1, \ldots, x_k$.

If these conditional expected values were known, the probability of failure could be calculated exactly as before. But if the conditional distribution of the remaining variables $X_{k+1}, \ldots, X_n$ under the condition $\boldsymbol{X}^k = \boldsymbol{x}^k$ is unknown and is replaced by some approximating function $h_F(\boldsymbol{x}^k)$, we get an error term. Our approximation $\hat{\mathbb{P}}(\mathbb{F})$ is now

$$\hat{\mathbb{P}}(\mathbb{F}) = \int_{\mathbb{R}^k} h_F(\boldsymbol{x}^k) f(\boldsymbol{x}^k)\, d\boldsymbol{x}^k \tag{5.6}$$

and for the difference, we have

$$\mathbb{P}(\mathbb{F}) - \hat{\mathbb{P}}(\mathbb{F}) = \int_{\mathbb{R}^k} \left(\mathbb{E}[\mathbf{1}_F(\boldsymbol{X}) | \boldsymbol{X}^k = \boldsymbol{x}^k] - h_F(\boldsymbol{x}^k) \right) f^k(\boldsymbol{x}^k)\, d\boldsymbol{x}^k \tag{5.7}$$

An extended response surface model from (5.3) which accounts for this space reduction can be formulated as follows:

$$g(\boldsymbol{x}|\boldsymbol{x}_{ss}) = F_R(\boldsymbol{x}_v, \boldsymbol{x}_{sv}, \vartheta) + \epsilon(\boldsymbol{x}_{ss}|\boldsymbol{x}_v, \boldsymbol{x}_{sv}) \tag{5.8}$$

with $\boldsymbol{x}_v$ a vector of random variables, $\boldsymbol{x}_{sv}$ a vector of random central values, and $\boldsymbol{x}_{ss}$ vectors of deviations from the corresponding central values. ϵ takes into account here the pure error and the effect of the space reduction.

5.2.1 Response Surface Model of Limit State Functions

In the case that the form of the limit state function $g(\boldsymbol{x}, \boldsymbol{\tau})$ is not known beforehand, its functional form has to be estimated. Since in general the form of such a function is complex, it is not very promising to try to fit a simple approximating function for the whole state space of the random influences. There are two alternatives:

1. to model the structure of the limit state function by a set of different approximating functions; each of these models expresses the limit state function in a different domain of the state space;
2. to use a model restricted to the part (or parts) of the state space important for reliability analysis.

The second approach is certainly numerically less demanding, but here we have to identify first the region, which is important for reliability analysis. In section 5.3 it is shown that in such problems the value of the failure probability is determined mainly through the position of the point on the limit state function where the p.d.f. with respect to the failure domain is maximal, the PML (point of maximum likelihood), and the shape of the limit state function around this point. Therefore it is sufficient to model the limit state function near this point and not in the whole space.

It can be shown (see section 5.3) that the asymptotic behavior of the failure estimation only depends on the quadratic Taylor expansion of the limit state function in the PML. Therefore the fitting of a quadratic response surface is appropriate.

Consider equation (5.8) written in the central region of all the variables in order to identify the actual location of the surface $G = \{\boldsymbol{z};\, g_z(\boldsymbol{z}) = 0\}$ in the standardized $\boldsymbol{z}$ space. Here the standardization of $\boldsymbol{x}_v$ and $\boldsymbol{x}_{vs}$ (i.e., reduction to zero mean, unit variance variables) is still introduced in order to preserve some useful features of the experiment planning. A linear polynomial form is preliminarily selected for the response surface. An appropriate fractional replicate of a factorial design centered at the origin of the space $\boldsymbol{z}$ defines the experiment plan [12] (see figure 5.1a). This first response surface application (where the contribution of the quantities $\boldsymbol{x}_{ss}$ is neglected) leads to an estimation of the design point (i.e., the point of G with minimum distance from the origin) $\mathbf{z}^0_{(1)}$ and of the maximum likelihood point $\boldsymbol{z}^*_{(1)}$

for the linear model. The log-likelihood function $l(\boldsymbol{z})$ is defined as $\ln(f(\boldsymbol{z}))$, $f(\boldsymbol{z})$ being the joint p.d.f. of the variables $\boldsymbol{z}$.

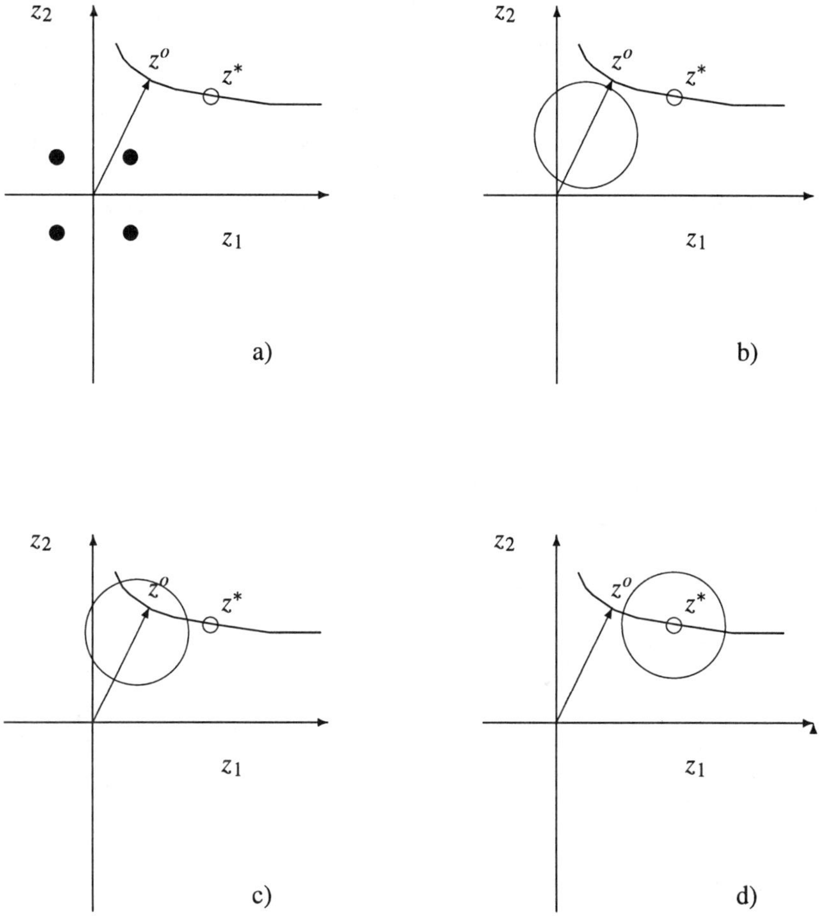

FIGURE 5.1
a) Factorial design for fitting a preliminary linear approximation of the response surface; b) and c) subsequent shifted composite central designs during the iterative procedure of reliability assessment; d) final composite central design in the estimated point of maximum likelihood.

The sensitivity factors of each variable z_j can be estimated from the gradient of the log-likelihood function at the point $\mathbf{z}^*_{(1)}$ due to the fact that the probability of failure satisfies in first approximation the relation:

$$\partial I\!P(I\!F)/\partial z_i \sim (\partial l(\mathbf{z}^*)/\partial z_i) I\!P(I\!F)$$

No substantial error is generally introduced by considering the random variables with low sensitivity factors as fixed and equal to their means in the further calculations. Alternatively, one can account for their randomness in the error terms ϵ.

A better estimation is then pursued by a sequence of repeated response surface applications. Each of them consists of several deterministic structural analyses for different input data (see figure 5.1b and c). They form an experiment design conceived for both fitting second-order polynomial response surface and incorporating the terms $\boldsymbol{x}_{ss}$ (and the $\boldsymbol{x}_v$ and $\boldsymbol{x}_{vs}$ with low sensitivity factors).

At the i-th step, the resulting approximation for the response function is:

$$g_{\boldsymbol{v}}^{(i)}(\boldsymbol{v}_{(i)}) = F_R(\boldsymbol{v}_{(i)}, \boldsymbol{\vartheta}_{(i)}) + \epsilon_{(i)}(\boldsymbol{x}_{ss})$$

where the dependence of ϵ on $\boldsymbol{v}_{(i)}$ is neglected and

$$\boldsymbol{v}_{(i)} = \boldsymbol{z} - \boldsymbol{z}_{(i)}^{x} \tag{5.9}$$

By equation (5.9), the origin is shifted along the vector from the past origin to the relevant design point $\boldsymbol{z}_{(i-1)}^{0}$. The convergence to the actual value is pursued by selecting a point on the line between this point and the origin: this defines $\boldsymbol{z}_{(i)}^{x}$.

Practical reasons suggest that one should work, rather than in the space $\boldsymbol{v}$, in a particular space $\boldsymbol{u}$, obtained from it by rotation. The new first axis u_1 coincides with the normal to the function $g(\,) = \text{const.}$ at the selected point. This iterative procedure stops when convergence on the distance of the maximum-likelihood point (that coincides with the design point in the Gaussian standardized variables, see section 5.3) from the origin is achieved. At each step the current estimate of the maximum-likelihood $\boldsymbol{z}_{(i)}^{*}$ is computed by the algorithm of constrained nonlinear optimization [12].

5.2.2 The Regression Model in a Projection Framework

In section 5.2.1 we considered as model a quadratic function of the observed variables. Let n_e be the number of available experiments. The model without error terms is then

$$y = \vartheta_0 + \boldsymbol{x}^T \boldsymbol{\vartheta}_1 + \boldsymbol{x}^T \Theta_2 \boldsymbol{x} \tag{5.10}$$

with $\boldsymbol{x} = (x_1, \ldots, x_n)^T$ the vector of regressor variables, ϑ_0 an unknown constant, $\boldsymbol{\vartheta}_1$ an unknown n-dimensional vector, and Θ_2 an unknown symmetric $n \times n$ matrix. The number of parameters in the model is then

$$n_c = 1 + n + \frac{n(n+1)}{2}. \tag{5.11}$$

Then we assume as a model with errors the following (see [25], [26], and [4])

$$y = A\vartheta + \epsilon_L + \epsilon. \tag{5.12}$$

Here A is the observation matrix of the regressor variables including the constant, i.e., an $n_e \times n_c$ matrix in the form

$$A = \begin{pmatrix} 1 & x_1^{(1)} & \dots & x_n^{(1)} & (x_1^{(1)})^2 & x_1^{(1)}x_2^{(1)} & \dots & (x_n^{(1)})^2 \\ 1 & x_1^{(2)} & \dots & x_n^{(2)} & (x_1^{(2)})^2 & x_1^{(2)}x_2^{(2)} & \dots & (x_n^{(2)})^2 \\ \vdots & \vdots & \vdots & \vdots & \vdots & \vdots & \vdots & \vdots \\ 1 & x_1^{(n_e)} & \dots & x_n^{(n_e)} & (x_1^{(n_e)})^2 & x_1^{(n_e)}x_2^{(n_e)} & \dots & (x_n^{(n_e)})^2 \end{pmatrix} \tag{5.13}$$

and ϑ is an n_c-dimensional vector of the regression coefficients composed of ϑ_0, ϑ_1 and Θ_2, which has to be estimated, i.e.,

$$\vartheta = \begin{pmatrix} \vartheta_0 \\ \vartheta_1 \\ \Theta_{21} \\ \Theta_{22} \\ \vdots \\ \Theta_{2n} \end{pmatrix}. \tag{5.14}$$

where Θ_{21} is the first column of the lower triangular part of Θ_2 and Θ_{2n} is its last column. Further $y = (y^{(1)}, \dots, y^{(n_e)})^T$ is the vector of the observed responses. The components of $\epsilon_L = (\epsilon_L^{(1)}, \dots, \epsilon_L^{(n_e)})^T$ are the systematic errors due to lack of fit and the components of $\epsilon = (\epsilon_1, \dots, \epsilon_{n_e})^T$, the random errors introduced in (5.3). It will be assumed that the random errors have mean zero, are uncorrelated, and have the same variance σ^2. This gives for the matrix of the variances of the random errors:

$$\mathbb{E}[\epsilon\epsilon^T] = \sigma^2 I_{n_e}. \tag{5.15}$$

If the covariance structure of these random variables is different, we have to use weighted least square estimation techniques (see for example [33]).

The estimated vector of the regression coefficients is then

$$\hat{\vartheta} = (A^T A)^{-1} A^T y. \tag{5.16}$$

The estimated value of the response vector y results

$$\hat{y} = A(A^T A)^{-1} A^T y = P_A y. \tag{5.17}$$

This is the projection of the vector $\boldsymbol{y}$ on the subspace spanned by the columns of $\boldsymbol{A}$, which has dimension n_c, and $\boldsymbol{P}_A$ is the projection matrix of this subspace (see appendix 5.7.1).

If we look on the residual sum of squares, we find

$$SS_R = (\boldsymbol{y} - \hat{\boldsymbol{y}})^T(\boldsymbol{y} - \hat{\boldsymbol{y}}). \tag{5.18}$$

This gives then, using equation (5.17),

$$SS_R = ((\boldsymbol{I}_{n_e} - \boldsymbol{P}_A)\boldsymbol{y})^T((\boldsymbol{I}_{n_e} - \boldsymbol{P}_A)\boldsymbol{y}) \tag{5.19}$$

Using equations (5.17) and (5.12) we get for the residual sum of squares the expression

$$SS_R = (\epsilon_L + \epsilon)^T \boldsymbol{P}^T \boldsymbol{P}(\epsilon_L + \epsilon) = |\boldsymbol{P}(\epsilon_L + \epsilon)|^2 \tag{5.20}$$

with

$$\boldsymbol{P} = \boldsymbol{I}_{n_e} - \boldsymbol{A}(\boldsymbol{A}^T\boldsymbol{A})^{-1}\boldsymbol{A}^T, \tag{5.21}$$

i.e., $\boldsymbol{P}$ is a projection matrix. It projects the error vector onto the space orthogonal to the space spanned by the columns of the matrix $\boldsymbol{A}$, since $\boldsymbol{P} = \boldsymbol{I}_{n_e} - \boldsymbol{P}_{\boldsymbol{A}}$ (see appendix 5.7.1). Since the dimension of this space is n_c, the dimension of its orthogonal complement is $n_e - n_c$.

So we have

$$|\boldsymbol{y}|^2 = |\hat{\boldsymbol{y}}|^2 + |\boldsymbol{P}(\epsilon_L + \epsilon)|^2. \tag{5.22}$$

The sum of the squared $\boldsymbol{y}$'s is decomposed into the sum of the $\hat{\boldsymbol{y}}$'s, the projection on the space spanned by the rows of $\boldsymbol{A}$ and the remaining error terms, which are orthogonal to this subspace.

From the model above we also estimate the error variance of the model by repeating m times the experiment at the central point. We assume that these are the first m observations. Since there is no variation of the regressor variables, the value $\epsilon_L^{(1)} = \ldots = \epsilon_L^{(m)}$ of the systematic error is always the same. We have for the pure error sum of squares the form

$$SS_E = \sum_{j=1}^{m}(y^{(j)} - \bar{y})^2 = \sum_{j=1}^{m}(y^{(j)})^2 - m\bar{y}^2 \tag{5.23}$$

with $\bar{y} = m^{-1} \sum_{j=1}^{m} y^{(j)}$. This can be written as

$$SS_E = \epsilon_0^T A_E \epsilon_0 \tag{5.24}$$

with $\epsilon_0 = (\epsilon^{(1)}, \ldots, \epsilon^{(m)})^T$. The matrix A_E is defined by $A_E = I_m - E_m$, where $E_m = m^{-1}[(1)_{i,j=1,\ldots,m}]$.

We can also write $E_m = m^{-1}(1, \ldots, 1)^T(1, \ldots, 1)$ and this is the projection matrix on the one-dimensional subspace spanned by the vector $(1, \ldots, 1)^T$. Therefore A_E is the projection on the orthogonal complement of the one-dimensional subspace spanned by the vector $(1, \ldots, 1)^T$, i.e., on an $(m-1)$-dimensional subspace.

We consider then the lack of fit sum of squares SS_L. It is defined as the residual sum of squares minus the pure error sum of squares SS_E

$$SS_R = SS_L + SS_E. \tag{5.25}$$

We get the form

$$\begin{aligned} SS_L &= (\epsilon_L + \epsilon)^T (P - A^*)^T (P - A^*)(\epsilon_L + \epsilon) \\ &= |(P - A^*)(\epsilon_L + \epsilon)|^2 \end{aligned} \tag{5.26}$$

with

$$A^* = \begin{pmatrix} I_m - E_m & o_{m,n_e-m} \\ o_{n_e-m,m} & o_{n_e-m,n_e-m} \end{pmatrix}. \tag{5.27}$$

where $o_{h,k}$ denotes a matrix of zeroes of size $h \times k$. The matrix A^* applied to a vector gives the projection on an $m-1$ dimensional subspace. We get then that $P - A^*$ is the projection on an $n_e - n_c - (m-1)$-dimensional subspace.

5.2.3 A Test for Lack of Fit

Now there are two possibilities to make an F-test. The first one is to compare the amount of variation explained by the model, i.e., the variance of $|\hat{y}|^2$ with the remaining variance, which is due to pure random error and lack of fit. The second is to compare the error variance with the pure random error variance to find out if there is a systematic error component. In this case we check if the value of $|P(\epsilon_L + \epsilon)|^2/\sigma^2$ is too large to come from a central χ^2-distribution with $n_e - n_c$ degrees of freedom. In structural reliability applications the error terms are relatively large and therefore F-tests for the first case might give no significant results (see [25]).

We consider the second possibility and construct a test. We have the SS_L sum of squares and an estimator for the variance of the error variables $\hat{\sigma}^2 = SS_E/(m-1)$. The ϵ_i/σ are then standard normal random variables and the sum of the squares $((\epsilon_i + \epsilon_L^{(i)})/\sigma)^2$ has a non-central χ^2-distribution with n_e degrees of freedom and non-centrality parameter $|\epsilon_L|^2/\sigma^2$ (see [38], p. 130). Therefore the squared length of the random vector obtained by projecting this random vector $\epsilon + \epsilon_L$ on a subspace with dimension $n_e - n_c$ is a non-central χ^2-square distribution with non-centrality parameter defined by

$$\Delta = \sigma^{-2}|\boldsymbol{P}\epsilon_L|^2. \tag{5.28}$$

In the case that there is no lack of fit, i.e., $\Delta = 0$, the SS_L and SS_E sum of squares are independent and both have a central χ^2- distribution, with $\nu_L = n_e - n_c - m - 1$ and $\nu_E = m - 1$ degrees of freedom, respectively. Then the ratio

$$\lambda^2 = \frac{SS_L \cdot \nu_E}{\nu_L \cdot SS_E} \tag{5.29}$$

has an F-distribution with ν_L and ν_E degrees of freedom.

We write $f_\alpha(\nu_1, \nu_2)$ for the upper α fractile of an F-distribution with ν_1 and ν_2 degrees of freedom, i.e., $\text{Prob}[F > f_\alpha(\nu_1, \nu_2)] = \alpha$ for $F > 0$. We have for the lower alpha fractiles

$$f_{1-\alpha}(\nu_1, \nu_2) = \frac{1}{f_\alpha(\nu_2, \nu_1)} \tag{5.30}$$

(note that ν_1 and ν_2 are interchanged in the denominator).

The further properties of the F-distribution can be found in [38] (chapter 26). In particular we have for the moments

$$\mathbb{E}[F] = \frac{\nu_2}{\nu_2 - 2} \tag{5.31}$$

$$\text{Var}[F] = \frac{2\nu_2^2(\nu_1 + \nu_2 - 2)}{\nu_1(\nu_2 - 2)^2(\nu_2 - 4)}. \tag{5.32}$$

For the computation of the distribution, it is important to know the relation with the incomplete beta function, given in [38] (p. 78, equation (8))

$$\text{Prob}[F \le C] = I_K(\nu_1/2, \nu_2/2) \tag{5.33}$$

with $K = C\nu_1/(\nu_2 + C\nu_1)$. Here $I_x(p,q)$ is the incomplete beta function divided by $B(p,q)$:

$$I_x(p,q) = (B(p,q))^{-1} \int_0^x t^{p-1}(1-t)^{q-1}\, dt \tag{5.34}$$

with $0 < x < 1$. The term $B(p,q)$ represents the beta function (i.e., the function defined by the integral in the r.h.s. of the previous equation, when it spans between 0 and 1).

A test about the hypothesis "the model is correct" is now made. There are two reasons for an incorrect model. Firstly that there might be a lack of fit and secondly that the model might be misspecified such that the pure error variances are different at different points.

We make a test in the following form:

1. First we choose a significance level α. Then we compute the fractile $f_\alpha(\nu_L, \nu_E)$ with $I\!P(F > f_\alpha(\nu_L, \nu_E)) = \alpha$.
2. Then the value of λ^2 is compared with these numbers.
3. If we have that

$$\lambda^2 \leq f_\alpha(\nu_L, \nu_E), \tag{5.35}$$

   the hypothesis is not rejected.
4. If we have

$$f_\alpha(\nu_L, \nu_E) < \lambda^2, \tag{5.36}$$

   the hypothesis is rejected, since the value is too large. It must be assumed that there is either a lack of fit $\Delta > 0$ or the pure error variances at the other points are larger than at the midpoint.

If the model passes this test, it might be adopted. But there is still the problem of selecting the best model among all those which are not rejected by the test. In [25] the use of the statistic λ, i.e., the square root of the statistic defined in equation (5.29) is proposed. It is defined by

$$\lambda = \sqrt{\frac{SS_L \cdot \nu_E}{SS_E \cdot \nu_L}}. \tag{5.37}$$

It will be used not to reject the model, but to find the best one among the not-rejected models. If there is no lack of fit, the expected value of λ is given by

$$\mathbb{E}[\lambda] = \sqrt{\frac{\nu_E}{\nu_E - 2}} = \sqrt{\frac{m-1}{m-3}} \approx 1. \tag{5.38}$$

For the last approximation we assume that m is not too small. The models, whose λ-values are near to unity, are therefore taken as the best fitting.

We show now that the λ-criterion is closely related to another widely used criterion, the C_p statistic, defined by

$$C_p = \frac{SS_R}{\hat{\sigma}^2} - (n_e - 2n_c). \tag{5.39}$$

Here $\hat{\sigma}^2$ is an estimate for the error variance. This statistic was proposed for comparing different regression models by [42] (see also [22], p. 299 and [49], p. 447-9). It is a weighted sum of the variances of the regression estimates $\hat{y}_i$ and their biases $(y_i - \hat{y}_i)^2$. If there is no lack of fit, the residual sum of squares shows a central χ^2-distribution with $n_e - n_c$ degrees of freedom.

In the used model the variance is estimated within the model, therefore the assumptions for the derivation of the C_p statistic are not fulfilled precisely, but we will show that we obtain an analogous result. Neglecting that the estimator SS_E/ν_E is not independent of SS_R, we write

$$\begin{aligned} C_p &= \frac{SS_R}{SS_E/\nu_E} - (n_e - 2n_c) = \frac{SS_L + SS_E}{SS_E/\nu_E} - (n_e - 2n_c) \\ &= \frac{SS_L \nu_E}{SS_E} + \nu_E - (n_e - 2n_c). \end{aligned} \tag{5.40}$$

If there is no lack of fit, we have for $(SS_L \cdot \nu_E)/(SS_E \cdot \nu_L)$ the expected value $\nu_E/(\nu_E - 2)$. This gives for the expected value of C_p

$$\mathbb{E}[C_p] = (\nu_L \nu_E)/(\nu_E - 2) + \nu_E - (n_e - 2n_c) \tag{5.41}$$

$$\approx n_e - n_c - (m-1) + (m-1) - n_e + 2n_c = n_c \tag{5.42}$$

i.e., under the assumptions made, the expected value of C_p is n_c. For a model with a good fit the actual value should be near n_c; for models with a bad fit the values are larger.

The relation between λ and C_p becomes clear from equation (5.40). Since $\lambda^2 = (SS_L \cdot \nu_E)/(SS_E \cdot \nu_L)$ we find

$$C_P = \nu_L \lambda^2 + \nu_E - (n_e - 2n_c). \tag{5.43}$$

and for λ we get

$$\lambda = \sqrt{\frac{C_p - \nu_E + (n_e - 2n_c)}{\nu_L}}. \tag{5.44}$$

Since the expected value of C_p is approximately n_c, we get then for the expected value of λ^2 that

$$\begin{aligned} I\!E[\lambda^2] &\approx \nu_L^{-1}(n_c - (m-1) + n_e - 2n_c) \qquad (5.45) \\ &= \frac{n_e - n_c - (m-1)}{n_e - n_c - (m-1)} = 1 \end{aligned}$$

and, hence,

$$I\!E[\lambda] \approx 1. \tag{5.46}$$

The expected value of λ is, for all models, larger than or equal to unity. Only for the correct model is it approximately equal to unity.

In conclusion, in order to fit a quadratic regression model, we estimate the regression coefficients and the λ statistic. The latter one should be used in two ways. Firstly we check if there is a systematic error by performing an F-test and then we identify the best model among those that were not rejected by the test.

5.3 The Calculation of Failure Probabilities

In this section we outline the concepts of approximation methods for the failure probability integrals in equation (5.1).

5.3.1 The Basic Problem

The failure probability $I\!P(\boldsymbol{\tau})$ is given by an n-dimensional integral over the failure domain $I\!F$ defined by $I\!F = \{\boldsymbol{x}; g(\boldsymbol{x}, \boldsymbol{\tau}) \le 0\}$. If the usual methods for calculating such integrals, i.e., numerical or Monte Carlo integration are applied, problems appear. There are three reasons, which make these methods inefficient.

1. The domain $\mathbb{F}$ is given only implicitly by the limit state function $g(\boldsymbol{x}, \boldsymbol{\tau})$,
2. the domain $\mathbb{F}$ is usually far off from the mean values of the random vector $\boldsymbol{X}$,
3. the magnitude of the integrand $f(\boldsymbol{x}, \boldsymbol{\tau})$ is changing rapidly in the failure domain.

The first point makes it difficult to obtain bounds for the integration domains or the random number generation. The second point also inhibits an efficient random number generation and the third point requires a careful choice of the integration scheme that no peaks of the integrand are missed.

Due to these problems, attempts were made to develop alternative methods for this problem, which give better results. Such ideas were pursued starting from 1956, when Freudenthal proposed in [31] to linearize the limit state function.

5.3.2 Analytic Approximation Methods

In the sixties several authors proposed to estimate the reliability by the so-called reliability indices. The problem with these indices was that they were depending on the specific formulation of the reliability problem. Different mathematical formulations of the same limit state condition lead to different reliability indices for the same failure domain (see [21]).

In 1974 then Hasofer and Lind [34] considered the case of a standardized n-dimensional normal random vector $\boldsymbol{X} = (X_1, \dots, X_n)^T$ with independent components, i.e., a random vector with p.d.f.

$$\begin{aligned} f(\boldsymbol{x}) &= (2\pi)^{-n/2} \exp\left(-\frac{1}{2}\sum_{i=1}^{n} x_i^2\right) \\ &= (2\pi)^{-n/2} \exp\left(-\frac{1}{2}|\boldsymbol{x}|^2\right). \end{aligned} \tag{5.47}$$

For a failure domain $\mathbb{F}$ they proposed as reliability index $\beta(\mathbb{F})$ the distance of the failure domain to the origin, i.e., to the point representing the mean value of the vector $\boldsymbol{X}$. The index is defined therefore by

$$\beta(\mathbb{F}) = \min_{\boldsymbol{x} \in \mathbb{F}} |\boldsymbol{x}|. \tag{5.48}$$

Then $\beta(\mathbb{F})$ is the minimal distance of $\mathbb{F}$ to the origin.

This index does not depend on the formulation of the problem, since the failure domain will be the same under all equivalent mathematical descriptions. Moreover, since the index $\beta(\mathbb{F})$ depends only on the shape of the domain $\mathbb{F}$, it will be always the same, no matter by which function the domain is defined.

Another property of this index is that for all failure sets with the origin in its interior, its value is zero. But, in reliability problems, failure generally occurs if some extreme values, far from the mean, are observed. So the failure domain has generally a large distance from the origin.

The relation between the index and the probability content of the failure domains is not clear at the first glance. Only in the case of a linear limit state function we obtain straightforward that the probability content of the failure domain $\mathbb{F}$ is given by

$$P(\mathbb{F}) = \Phi(-\beta(\mathbb{F})). \tag{5.49}$$

where $\Phi(\ )$ is the standard Gaussian distribution. So in the case of linear limit state functions there is a one-to-one correspondence between the reliability index and the failure probability. But this does not hold in general. A simple example is given in [39] (p. 57).

A heuristic attempt to find approximations for the probability contents of failure domains with non-linear boundaries was to use equation (5.49) to obtain an approximation, i.e.,

$$P(\mathbb{F}) \approx \Phi(-\beta(\mathbb{F})). \tag{5.50}$$

Even if no precise mathematical justification was given for this concept, it turned out that often the obtained results were quite satisfactory.

The computational method for finding this approximation evaluates the point of minimal distance to the origin. Its distance to the origin was then the reliability index for the failure domain $\mathbb{F}$. Another interpretation for this approximation was that the limit state surface is replaced by its first order Taylor expansion at the minimal distance point $\boldsymbol{x}^0$. Due to the Lagrange multiplier theorem at $\boldsymbol{x}^0$, the gradient of g is parallel to $\boldsymbol{x}^0$, since here there is minimum of the function $|\boldsymbol{x}|$ with respect to the surface $G = \{\boldsymbol{x}; g(\boldsymbol{x}) = 0\}$. So this linearized limit state function g_L has the form

$$g_L(\boldsymbol{x}) = (\nabla g(\boldsymbol{x}^0))^T(\boldsymbol{x} - \boldsymbol{x}^0) = -\frac{|\nabla g(\boldsymbol{x}^0)|}{|\boldsymbol{x}^0|}(\boldsymbol{x}^0)^T(\boldsymbol{x} - \boldsymbol{x}^0). \tag{5.51}$$

The minus sign appears since the gradient points towards the origin and the vector $\boldsymbol{x}^0$ in the opposite direction. Then the random function $g_L(\boldsymbol{X})$ is a linear function of a standard normal random vector and therefore again a normal random variable.

We get

$$\mathbb{E}[g_L(\boldsymbol{X})] = |\nabla g(\boldsymbol{x}^0)||\boldsymbol{x}^0| \tag{5.52}$$

$$\mathrm{Var}[g_L(\boldsymbol{X})] = |\nabla g(\boldsymbol{x}^0)|^2. \tag{5.53}$$

Therefore we get for the probability $\mathbb{P}(g_L(\boldsymbol{X}) \leq 0)$ simply

$$\mathbb{P}(g_L(\boldsymbol{X}) \leq 0) = \Phi(-|\boldsymbol{x}^0|). \tag{5.54}$$

This means that the approximation is equivalent to linearize the limit state function at $\boldsymbol{x}^0$ and calculating the failure probability of the failure domain bounded by this function. The methods based on linearizing the limit state function are nowadays known as FORM (First Order Reliability Methods).

If a random vector $\boldsymbol{X} = (X_1, \ldots, X_n)^T$ with independent components has a joint p.d.f. $f(\boldsymbol{x}) = \prod_{i=1}^n f_i(x_i)$, it can be easily transformed into a standard normal random vector $\boldsymbol{Z} = (Z_1, \ldots, Z_n)$. The transformation $\boldsymbol{x} = (x_1, \ldots, x_n) \mapsto \boldsymbol{z} = (z_1, \ldots, z_n)$ is given by

$$z_i = \mathcal{T}(x_i) = \Phi^{-1}[F_i(x_i)] \tag{5.55}$$

with $i = 1, \ldots, n$. Here $F_i(x_i)$ is the cumulative distribution function (c.d.f.) of the random variable X_i, i.e., $F_i(x_i) = \int_{-\infty}^{x_i} f_i(y)dy$. Written in vector form we have the equation

$$\boldsymbol{z} = \mathcal{T}(\boldsymbol{x}) = \left(\Phi^{-1}[F_1(x_1)], \ldots, \Phi^{-1}[F_n(x_n)]\right)^T. \tag{5.56}$$

Even for a random vector $\boldsymbol{X}$ with dependent components, such a transformation can be constructed theoretically (see [36]), but since it requires the iterative calculation of condition distributions, numerical problems will appear if these distributions are not known analytically.

Following these ideas, approximations for failure probabilities were derived by first transforming the random vector into a standard normal random vector, calculating for the transformed failure domain $\mathcal{T}(\mathbb{F})$ the reliability index $\beta(\mathcal{T}(\mathbb{F}))$ and then using $\Phi(-\beta(\mathcal{T}(\mathbb{F})))$ as approximation for $\mathbb{P}(\mathbb{F})$.

Several questions remained unclear or problematic in this concept:

1. What if there are several minimal distance points? How can an approximation for the failure probability be found in this case?

2. In the case of a sphere with radius β the approximation above $\Phi(-\beta)$ for the probability content of the domain outside is clearly very bad. Indeed,

the FORM approximation for the probability content of the sphere with radius β would be $\Phi(-\beta)$, but the exact probability $P_s(\beta)$ is (see [11], p. 29)

$$P_s(\beta) \sim \frac{2}{2^{\frac{n}{2}}\Gamma(\frac{n}{2})}\beta^{n-2}\exp(-\frac{\beta^2}{2}).$$

For $n > 1$ the asymptotic order of this probability is different from $\Phi(-\beta)$. What can be said about the quality of such approximations?

A number of improvements were proposed, but the principal problems remained unsolved. One method was to use higher order approximations for the limit state surface, for example a second order Taylor expansion at the minimal distance point (see for example [30]).

To obtain a general solution for all these problems, a different approach based on asymptotic concepts is needed. In appendix 5.7.7 the basic ideas of asymptotic analysis are discussed briefly.

We consider now a failure probability integral in the form

$$\mathbb{P}(\mathbb{F}) = (2\pi)^{-n/2} \int\limits_{g(\boldsymbol{x})\leq 0} \exp\left(-\frac{1}{2}|\boldsymbol{x}|^2\right)\, d\boldsymbol{x} \tag{5.57}$$

i.e., we are back to the case of a vector $\boldsymbol{X}$ of standardized Gaussian variables. We assume that there is only one point $\boldsymbol{x}^0$ on the limit state surface G with minimal distance to the origin, i.e., we have $\beta(\mathbb{F}) = |\boldsymbol{x}^0|$. If we look at the integral in equation (5.57), we see that the integrand is maximal at $\boldsymbol{x}^0$, since at this point the value of $|\boldsymbol{x}|$ is minimal throughout $\mathbb{F}$.

To use now asymptotic concepts, we imbed the surface G into a manifold of surfaces. The distance of G to the origin is $\beta(\mathbb{F})$. G is defined by $G = \{\boldsymbol{x};\, g(\boldsymbol{x}) = 0\}$. If we consider the surfaces $G(\beta)$ defined by $G(\beta) = \{\boldsymbol{x};\, g((\beta/\beta(\mathbb{F}))\boldsymbol{x}) = 0\}$, then we have $G(\beta(\mathbb{F})) = G$ and for each surface $G(\beta)$ we have that its distance to the origin is β. This can be seen from the fact that, since the point $\boldsymbol{x}^0$ is the point with minimal distance to the origin on G and all other surfaces are created from G by a linear transformation, the minimal distance point on $G(\beta)$ is $(\beta/\beta(\mathbb{F}))\boldsymbol{x}^0$ and the distance of this point to the origin is $|\beta/\beta(\mathbb{F})||\boldsymbol{x}^0| = \beta$.

With all these surfaces $G(\beta)$ are associated failure domains $\mathbb{F}(\beta) = \{\boldsymbol{x};\, g((\beta/\beta(\mathbb{F})))\boldsymbol{x} \leq 0\}$. We have for the probabilities of failure

$$\mathbb{P}(\mathbb{F}(\beta)) = (2\pi)^{-n/2} \int\limits_{g((\beta/\beta(\mathbb{F}))\boldsymbol{x})\leq 0} \exp\left(-\frac{1}{2}|\boldsymbol{x}|^2\right)\, d\boldsymbol{x}. \tag{5.58}$$

The trick is now to make a transformation $\boldsymbol{x} \to \boldsymbol{y} = \beta^{-1}\boldsymbol{x}$. Then we have for the probabilities

$$\mathbb{P}(\mathbb{F}(\beta)) = \beta^n (2\pi)^{-n/2} \int\limits_{g(\beta(\mathbb{F})^{-1}\boldsymbol{y})\leq 0} \exp\left(-\frac{\beta^2}{2}|y|^2\right) d\boldsymbol{y}. \tag{5.59}$$

Now we can use the theory of Laplace type integrals, since we have a fixed domain $\{\boldsymbol{y};\, g((\beta(\mathbb{F})^{-1})\boldsymbol{y}) \leq 0\}$ and in the argument of the exponential function a parameter β approaching infinity. We have in particular that $\mathbb{P}(\mathbb{F}(\beta(\mathbb{F})))$ is equal to $\mathbb{P}(\mathbb{F})$. If we now derive an asymptotic approximation $I_a(\beta)$ for the integral in the last equation, then the relative error of approximating the function value $\mathbb{P}(\mathbb{F}(\beta))$ by the value $\beta^n(2\pi)^{-n/2}I_a(\beta)$ approaches zero as $\beta \to \infty$. Therefore, if $\beta(\mathbb{F})$ is large, the error of approximating $\mathbb{P}(\beta(\mathbb{F})) = \mathbb{P}(\mathbb{F})$ by $\beta^n(2\pi)^{-n/2}I_a(\beta(\mathbb{F}))$ should be small.

The asymptotic approximation for the integrals $I_a(\beta)$ is derived in appendix 5.7.8. We have here

$$I_a(\beta) = (2\pi)^{(n-1)/2} \exp(-\beta^2/2) \prod_{i=1}^{n-1} (1 - \beta(\mathbb{F})\kappa_i)^{-1/2}. \tag{5.60}$$

The κ_i's are the main curvatures of the surface G at the minimal distance point $\boldsymbol{x}^0$. For the failure probability we get the approximation

$$\mathbb{P}(\mathbb{F}) \sim \frac{1}{(2\pi)^{1/2}\beta(\mathbb{F})} \exp(-\beta(\mathbb{F})^2/2) \prod_{i=1}^{n-1} (1 - \beta(\mathbb{F})\kappa_i)^{-1/2}. \tag{5.61}$$

Using Mill's ratio $\exp(-x^2/2)/(\sqrt{2\pi}x) \sim \Phi(-x)$ as $x \to \infty$ gives the basic formula for asymptotic approximations of multinormal integrals

$$\mathbb{P}(\mathbb{F}) \sim \Phi(-\beta) \prod_{i=1}^{n-1} (1 - \beta(\mathbb{F})\kappa_i)^{-1/2}. \tag{5.62}$$

This shows that asymptotically the failure probability is determined by a quadratic expansion at the minimal distance point.

Methods using quadratic expansions at the minimal distance points are called SORM (Second Order Reliability Methods). The methods using the asymptotic concept are called asymptotic SORM. For these methods the mathematical justification outlined above holds. In the case of several minimal distance points it follows from the asymptotic theory that for each point separately the asymptotic approximation based on a second order expansion around the point has to be calculated and then all these approximations are added to give the approximation for the whole failure domain.

Similar results are found if there are no single minimal distance points, but whole surface parts of G have minimal distance. This will be shown by the following classification of the possible types of asymptotic behavior. We summarize the possible forms of asymptotic behavior of failure probabilities by considering a limit state function g defining a failure domain $\mathbb{F}$ and a limit state surface G with $\min_{\boldsymbol{x}\in\mathbb{F}} |\boldsymbol{x}| = \min_{\boldsymbol{x}\in G} |\boldsymbol{x}| = 1$. For the failure domains $\beta\mathbb{F}$ with $\beta > 1$ we have then $\min_{\boldsymbol{x}\in\beta\mathbb{F}} |\boldsymbol{x}| = \beta$. For the asymptotic behavior of the failure probabilities $\mathbb{P}(\beta\mathbb{F})$ there are then three possibilities:

1. The function g is a linear function, i.e., $g(x_1,\ldots,x_n) = \sum_{i=1}^{n} \alpha_i x_i - b$. Then we have $\mathbb{P}(\beta\mathbb{F}) = \Phi(-\beta)$.

2. The function g is a non-linear function and there is a finite number k of points $\boldsymbol{x}_1,\ldots,\boldsymbol{x}_k$ with $g(\boldsymbol{x}_i) = 0$ and $|\boldsymbol{x}_i| = 1$. Then

$$\mathbb{P}(\beta\mathbb{F}) \sim \Phi(-\beta)\left[\sum_{i=1}^{k}\prod_{j=1}^{n-1}(1-\kappa_{ij})^{-1/2}\right],\ \beta\to\infty \tag{5.63}$$

with the κ_{ij}'s ($j = 1,\ldots,n-1$) the main curvatures of G at $\boldsymbol{x}_i$.

3. There is a submanifold M of dimension $k \leq n-1$ of the surface G with $|\boldsymbol{y}| = 1$ for all $\boldsymbol{y} \in M$. Then

$$\mathbb{P}(\beta\mathbb{F}) \sim \tilde{Q}(\beta|k+1) \times \frac{\Gamma((k+1)/2)}{2\pi^{(k+1)/2}} \int_M \frac{ds_M(\boldsymbol{x})}{\sqrt{|\det(\boldsymbol{H}^*(\boldsymbol{x}))|}},$$

$$\beta\to\infty. \tag{5.64}$$

Here $ds_M(\boldsymbol{x})$ denotes surface integration over M and

$$\tilde{Q}(x|k) = \left[2^{(k-1)/2}\Gamma(k/2)\right]^{-1}\int_x^\infty y^{k-1}e^{-y^2/2}dy$$

is the complementary distribution function of the χ_k-distribution (χ_k not χ_k^2!). The $(n-k-1)\times(n-k-1)$-matrix $\boldsymbol{H}^*(\boldsymbol{x})$ is defined by $\boldsymbol{H}^*(\boldsymbol{x}) = \boldsymbol{A}^T(\boldsymbol{x})\boldsymbol{H}(\boldsymbol{x})\boldsymbol{A}(\boldsymbol{x})$. Here the matrix $\boldsymbol{H}(\boldsymbol{x})$ is given by $\boldsymbol{H}(\boldsymbol{x}) = (-\delta_{ij} - |\nabla g(\boldsymbol{x})|^{-1}g^{ij}(\boldsymbol{x}))_{i,j=1,\ldots,n}$ and $\boldsymbol{A}(\boldsymbol{x}) = (\boldsymbol{a}_{n-k+1}(\boldsymbol{x}),\ldots,\boldsymbol{a}_{n-1}(\boldsymbol{x}))$. The $\boldsymbol{a}_{n-k+1}(\boldsymbol{x}),\ldots,\boldsymbol{a}_{n-1}(\boldsymbol{x})$ form an orthonormal basis of the $(n-k-1)$ dimensional orthogonal complement of the k-dimensional tangential space of M at $\boldsymbol{x}$ in the tangential space of G at $\boldsymbol{x}$. $\Gamma(\,)$ is the gamma function.

A proof for these results is given in [11]. If there are several limit state functions, there can appear some more special cases of minimal distance points, where they lie on the intersection of two or more limit state surfaces. Such cases are treated in [35] and [13].

SORM approximations require to find the structure of a quadratic form on a subspace towards the solution of two problems:

1. to identify a local extremum under the constraint $g(\boldsymbol{x}) = 0$;
2. to calculate the curvature factor in equation (5.62).

In appendix 5.7.3 we derive a simple method for approaching these two problems.

The approximation for the failure probability given in equation (5.62) is an asymptotic approximation, since the asymptotic analysis shows that the cross terms g_{in}, which express the second derivatives with respect to the i-th and n-th variables in the Taylor expansion of the limit state function, are asymptotically negligible and therefore the limit state function can be approximated near the minimal distance point $\boldsymbol{x}^0$ under a suitable linear transformation (i.e., $\boldsymbol{x}^0 = (0, \ldots, 0, 1)$ and the $x_1, \ldots, x_{n-1}$-axes in the main curvature directions) by a parabolic function in the form

$$g_{par}(\boldsymbol{x}) = g_n(x_n^0 - x_n) + \frac{1}{2}\sum_{i=1}^{n-1} g_{ii}^2(x_i - x_i^0)^2. \tag{5.65}$$

By numerical methods as Fourier transformations it is also possible to calculate the exact probability content of a failure domain which is exactly quadratic (see [48] and [52]). This result improves the asymptotic approximation only if the true limit surface is exactly quadratic. If it is not, we have to consider that the deviation of this surface from the exact quadratic surface gives an additional approximation error. Therefore, it cannot be said that the calculation of the exact distribution of a quadratic function, approximating the limit state function, always improves the quality of the approximation in comparison with the asymptotic approximation.

5.3.3 Approximations for Non-Normal Distributions

The results of the last part of section 5.3.2 are derived for the case of normal p.d.f., but they can be used for non-normal p.d.f.'s too. This should avoid a transformation from the original space of random variables into the space of standard normal random variables.

To find this generalization, we need a probabilistic interpretation of the meaning of the minimal distance point. At this point we have not only a minimal distance to the origin but, due to the special structure of the standard normal p.d.f. a maximal value of the p.d.f. in the failure domain. Therefore we make a Taylor expansion around the point of maximum likelihood (PML) in the failure domain. Shinozuka proposed in [46] to make expansions around this point for non-normal p.d.f.'s.

If we consider a failure domain $I\!F$ in a space of non-normal random variables and its transformed domain $\mathcal{T}(I\!F)$, where $\mathcal{T}(\boldsymbol{x})$ is defined in equation (5.55), in the standard normal space, it can be shown that asymptotically, i.e., for large distances of the domain $\mathcal{T}(I\!F)$ from the origin, the PML $\boldsymbol{x}^*$ in the original space and the minimal distance point $\boldsymbol{x}^0$ in the standard normal space correspond to each other, i.e., $\boldsymbol{x}^0 \approx \mathcal{T}(\boldsymbol{x}^*)$. The following derivations are based on the results in [9] and [10]. It should be noted that this is a heuristic approach, the quality of the derived approximations depend on the tail structure of the random vectors. This problem is partially discussed in chapter 4 of this book.

For large values of the standardized variable z_i and of the original one x_i the squared distance of the transformed point $\boldsymbol{z}$ in the standard normal space is approximately equal to $2l(\boldsymbol{x})$, i.e., the double of the log-likelihood function at the untransformed point $\boldsymbol{x}$ (see appendix 5.7.4). Therefore a point on the limit state surface with minimal distance to the origin in the standard normal space corresponds to a point on the limit state surface in the original space with maximal log-likelihood.

If this correspondence is clear, we can make asymptotic approximations also in the space of non-normal p.d.f.'s by finding the maximum likelihood points in the original failure domain and then making a second order Taylor expansion around these points.

To find a form similar to the normal p.d.f., we consider the log-likelihood function of a p.d.f. $f(\boldsymbol{x})$, which is defined by

$$l(\boldsymbol{x}) = \ln(f(\boldsymbol{x})). \tag{5.66}$$

We can write the p.d.f. then in the form

$$f(\boldsymbol{x}) = \exp(\beta_0^2 h(\boldsymbol{x})) \tag{5.67}$$

with $h(\boldsymbol{x}) = \beta_0^{-2} l(\mathbf{x})$. This gives for the failure probability integral the form

$$I\!P(I\!F) = \int\limits_{g(\boldsymbol{x})\le 0} e^{\beta_0^2 h(\boldsymbol{x})}\, d\boldsymbol{x}. \tag{5.68}$$

Now we can apply again the theory of Laplace type integrals to obtain an approximation for this integral.

If there is exactly one point $\boldsymbol{x}^* \in G$ where the p.d.f. in the failure domain is maximal, we obtain the following asymptotic approximation for the probability of failure

$$\mathbb{P}(\mathbb{F}) \approx (2\pi)^{(n-1)/2} \frac{f(\boldsymbol{x}^*)}{|\nabla l(\boldsymbol{x}^*)| |\det(\tilde{\boldsymbol{H}}(\boldsymbol{x}^*))|^{1/2}}. \tag{5.69}$$

Here the matrix $\tilde{\boldsymbol{H}}(\boldsymbol{x}^*)$ is defined by

$$\tilde{\boldsymbol{H}}(\boldsymbol{x}^*) = \boldsymbol{A}^T \boldsymbol{H}_1(\boldsymbol{x}^*) \boldsymbol{A} \tag{5.70}$$

with

$$\boldsymbol{H}_1(\boldsymbol{x}^*) = \left(l_{ij}(\boldsymbol{x}^*) - \frac{|\nabla l(\boldsymbol{x}^*)|}{|\nabla g(\boldsymbol{x}^*)|} g_{ij}(\boldsymbol{x}^*) \right)_{i,j=1,\ldots,n} \tag{5.71}$$

and $\boldsymbol{A} = (\boldsymbol{a}_1, \ldots, \boldsymbol{a}_{n-1})$ with the $\boldsymbol{a}_i$'s being an orthonormal basis of the tangential space of G at $\boldsymbol{x}^*$.

As outlined in appendix 5.7.3, this can be written in a simpler form. We get for the determinant of $\tilde{\boldsymbol{H}}(\boldsymbol{x}^*)$ that it is equal to the determinant of $(\boldsymbol{I}_n - \boldsymbol{P})\boldsymbol{H}_1(\boldsymbol{x}^*)(\boldsymbol{I}_n - \boldsymbol{P}) - \boldsymbol{P}$ with $\boldsymbol{P} = |\nabla g(\boldsymbol{x}^*)|^{-2} \nabla g(\boldsymbol{x}^*)(\nabla g(\boldsymbol{x}^*))^T$. Thence we can write the approximation in the following form

$$\begin{aligned} \mathbb{P}(\mathbb{F}) &\approx \frac{(2\pi)^{(n-1)/2} f(\boldsymbol{x}^*)}{|\nabla l(\boldsymbol{x}^*)| |\det((\boldsymbol{I}_n - \boldsymbol{P})\boldsymbol{H}_1(\boldsymbol{x}^*)(\boldsymbol{I}_n - \boldsymbol{P}) - \boldsymbol{P})|^{1/2}} \\ &= \frac{(2\pi)^{(n-1)/2} \exp(l(\boldsymbol{x}^*))}{|\nabla l(\boldsymbol{x}^*)| |\det((\boldsymbol{I}_n - \boldsymbol{P})\boldsymbol{H}_1(\boldsymbol{x}^*)(\boldsymbol{I}_n - \boldsymbol{P}) - \boldsymbol{P})|^{1/2}} \end{aligned} \tag{5.72}$$

If the boundary of the failure domain is linear at the point $\boldsymbol{x}^*$, the second derivatives of g vanish and we get for the matrix $\boldsymbol{H}_1(\boldsymbol{x}^*)$ the form

$$\boldsymbol{H}_1(\boldsymbol{x}^*) = \left(l_{ij}(\boldsymbol{x}^*) \right)_{i,j=1,\ldots,n}. \tag{5.73}$$

If the accuracy of asymptotic approximations is not sufficient, suitable importance sampling methods can be used to improve the estimation. Such methods were proposed, for example, by Hohenbichler and Rackwitz [37] and Bucher [14]. Details are provided in the appendix 5.7.5.

5.4 Parameter Optimization and Uncertainty

5.4.1 Derivatives with Respect to Parameters

In this section some results about the derivatives of the failure probability integral with respect to the components of the parameter vector are given.

The partial derivatives are often no good indicator for the sensitivities of failure probabilities with respect to parameter changes, since they depend on the scales which are used. A better tool is the partial elasticities, which are widely used in economics.

For a function $f(x_1, \ldots, x_n)$ the partial elasticity $\epsilon_i(f(\boldsymbol{x}))$ with respect to the component x_i at the point $\boldsymbol{x} = (x_1, \ldots, x_n)$ is given by

$$\epsilon_i(f(\boldsymbol{x})) = x_i \frac{f_i(\boldsymbol{x})}{f(\boldsymbol{x})} = x_i \frac{\partial \ln(f(\boldsymbol{x}))}{\partial x_i} \tag{5.74}$$

where f_i denotes the derivatives of f with respect to x_i.

The partial elasticity gives approximately the change in percents of the function $f(\boldsymbol{x})$ if the value of x_i is changed by 1%. We can express the change $C(h)$ in percents of $f(\boldsymbol{x})$ if the value of the variable x_i is changed $h\%$ by the formula

$$\begin{aligned} C(h) &= \frac{f(x_1, \ldots, (1 + h/100)x_i, \ldots, x_n) - f(x_1, \ldots, x_n)}{f(x_1, \ldots, x_n)/100} \\ &\approx \frac{f_i(x_1, \ldots, x_n)}{f(x_1, \ldots, x_n)/100}(h/100)x_i \\ &= hx_i \frac{f_i(x_1, \ldots, x_n)}{f(x_1, \ldots, x_n)}. \end{aligned} \tag{5.75}$$

This shows partial elasticity measures the change in percents. This measure for parameter sensitivity is dimensionless and does not depend on the scales. An application of this sensitivity measure is given in [17].

In many reliability problems, not only the failure probability itself, but its derivatives with respect to the components of the parameter vector are of interest. In [2] and [23] this problem is considered for the case of a parameter dependent limit state function. Here we consider the general case of an integral as in equation (5.1).

The simplest reliability optimization problem is given in the form

$$\min_{\boldsymbol{\tau}} I\!P(\boldsymbol{\tau}) \tag{5.76}$$

under the condition

$$r(\boldsymbol{\tau}) = 0. \tag{5.77}$$

Here the $r_i(\boldsymbol{\tau})$ represent restrictions on the possible parameter values.

The constrained gradient $\nabla_r I\!P(\boldsymbol{\tau})$ of the function $I\!P(\boldsymbol{\tau})$ under these restrictions is then

$$\nabla_r I\!P(\boldsymbol{\tau}) = \nabla I\!P(\boldsymbol{\tau}) - (|\nabla r(\boldsymbol{\tau})|^{-2}(\nabla r(\boldsymbol{\tau}))^T \nabla I\!P(\boldsymbol{\tau}))\nabla r(\boldsymbol{\tau}). \tag{5.78}$$

The change of the approximation depends now on the change of the PML as the parameter values change and of the derivatives.

We consider an integral as in equation (5.1), but with only a scalar parameter τ. To simplify the notation, we write instead of $f(\boldsymbol{x}, \tau) = f(x_1, \ldots, x_n, \tau)$ and $g(\boldsymbol{x}, \tau) = g(x_1, \ldots, x_n, \tau)$ in a shorthand way f and g.

Following from the results of the previous section the failure probability is approximated making a Taylor expansion of the log-likelihood function around the PML and using the Laplace method. We assume that for all feasible values of τ there is exactly one PML $\boldsymbol{x}^* = \boldsymbol{x}^*(\tau)$ on the limit state surface $G(\tau)$.

The gradients of these functions with respect to the first n variables $x_1, \ldots, x_n$ are denoted by $\nabla_{\boldsymbol{x}} f$ (resp. $\nabla_{\boldsymbol{x}} g$) and the partial derivative with respect to τ by f_τ (resp. g_τ). The Hessian of a function f with respect to the first n variables is written as $\boldsymbol{H}_f$.

We assume that for a fixed value of τ there is a unique PML $\boldsymbol{x}^*(\tau)$. Again we write $\boldsymbol{x}^*$. The vector of the first derivatives of $\boldsymbol{x}^*$ with respect to τ is written as $\boldsymbol{x}^*_\tau$. This point is a stationary point of the Lagrangian function $L(\boldsymbol{x}, \lambda, \tau)$ defined by

$$L = f + \lambda g. \tag{5.79}$$

Therefore for the point $\boldsymbol{x}^*(\tau)$ the following equation system

$$\nabla_{\boldsymbol{x}} f + \lambda \nabla_{\boldsymbol{x}} g = \mathbf{o}_n$$

$$g = 0. \tag{5.80}$$

must be fulfilled.

To find the derivatives of the coordinates of the PML with respect to parameter changes, we differentiate this system with respect to τ and set all derivatives equal to zero. This gives then

$$\boldsymbol{H}_f \boldsymbol{x}_\tau^* + \nabla_{\boldsymbol{x}} f_\tau + \lambda_\tau \nabla_{\boldsymbol{x}} g + \lambda(\boldsymbol{H}_g \boldsymbol{x}_\tau^* + \nabla_{\boldsymbol{x}} g_\tau) = \mathbf{o}_n$$

$$(\nabla_{\boldsymbol{x}} g)^T \boldsymbol{x}_\tau^* + g_\tau = 0. \tag{5.81}$$

Rearranging the terms we get

$$(\boldsymbol{H}_f + \lambda \boldsymbol{H}_g)\boldsymbol{x}_\tau^* + \lambda_\tau \nabla_{\boldsymbol{x}} g = -\nabla_{\boldsymbol{x}}(f_\tau + \lambda g_\tau) \tag{5.82}$$

$$(\nabla_{\boldsymbol{x}} g)^T \boldsymbol{x}_\tau^* = -g_\tau. \tag{5.83}$$

We define now the matrix $\boldsymbol{L}$ by

$$\boldsymbol{L} = \begin{pmatrix} \boldsymbol{H}_f + \lambda \boldsymbol{H}_g & \nabla_{\boldsymbol{x}} g \\ \nabla_{\boldsymbol{x}} g^T & 0 \end{pmatrix}. \tag{5.84}$$

If the constrained maximum of g at $\boldsymbol{x}^*$ is regular, we have that

$$\det(\boldsymbol{L}) \neq 0, \tag{5.85}$$

i.e., the matrix $\boldsymbol{L}$ is non-singular (see [11], p. 20). This gives then for the vector $((\boldsymbol{x}_\tau^*)^T, \lambda_\tau)$ the form

$$\begin{pmatrix} \boldsymbol{x}_\tau^* \\ \lambda_\tau \end{pmatrix} = -\boldsymbol{L}^{-1} \begin{pmatrix} \nabla_{\boldsymbol{x}} f_\tau + \lambda \nabla_{\boldsymbol{x}} g_\tau \\ g_\tau \end{pmatrix}. \tag{5.86}$$

The change of the value $f^* = f(\boldsymbol{x}^*(\tau), \tau)$ is given by

$$f_\tau^* = (\nabla_{\boldsymbol{x}} f)^T \boldsymbol{x}_\tau^* + f_\tau. \tag{5.87}$$

Due to the Lagrange multiplier theorem, we can replace the gradient of f by the gradient of g, giving

$$f_\tau^* = \lambda (\nabla_{\boldsymbol{x}} g)^T \boldsymbol{x}_\tau^* + f_\tau. \tag{5.88}$$

From equation (5.83) we get then

$$f_\tau^* = -\lambda g_\tau + f_\tau. \tag{5.89}$$

If $f_\tau = 0$, the Lagrange multiplier λ gives the change of the value of f^* relative to the negative change of g with respect to τ. If the constraint is given in the form

$g(\boldsymbol{x}) - \tau = 0$ and the p.d.f. does not depend on τ, we obtain the simple form

$$f_\tau^* = \lambda. \tag{5.90}$$

The value of λ can be computed easily from the relation

$$\nabla_{\boldsymbol{x}} f = -\lambda \nabla_{\boldsymbol{x}} g. \tag{5.91}$$

If we consider the approximation for the failure probability given in equation (5.69) and we neglect the change of the quantities in the denominator we get approximately

$$\frac{\partial \mathbb{P}(F)}{\partial \tau} \approx \frac{(2\pi)^{(n-1)/2}}{|\nabla l(\boldsymbol{x}^*)| |\det(\tilde{\boldsymbol{H}}(\boldsymbol{x}^*))|^{1/2}} (-\lambda g_\tau + f_\tau). \tag{5.92}$$

5.4.2 Approximate Bayesian Analysis for Parameter Uncertainties

We consider now the case that the "true" value $\boldsymbol{\tau}$ of the parameter vector is unknown, but there is a p.d.f. $p(\boldsymbol{\tau})$ over the parameter space. The means and covariances are given by

$$\mathbb{E}[\boldsymbol{\tau}] = \boldsymbol{\tau}_m, \tag{5.93}$$

$$\mathbb{E}[(\boldsymbol{\tau} - \boldsymbol{\tau}_m)(\boldsymbol{\tau} - \boldsymbol{\tau}_m)^T] = \boldsymbol{\Sigma}. \tag{5.94}$$

To make an approximate Bayesian analysis we need the first derivatives of $\mathbb{P}(\boldsymbol{\tau})$ with respect to the components of the parameter vector at the point $\boldsymbol{\tau}_m$. We find then as first order approximations

$$\mathbb{E}[\mathbb{P}(\boldsymbol{\tau})] \approx \mathbb{P}(\boldsymbol{\tau}_m), \tag{5.95}$$

$$\mathbb{E}[(\mathbb{P}(\boldsymbol{\tau}) - \mathbb{P}(\boldsymbol{\tau}_m))^2] \approx (\nabla \mathbb{P}(\boldsymbol{\tau}_m))^T \boldsymbol{\Sigma} \mathbb{P}(\boldsymbol{\tau}_m). \tag{5.96}$$

Further details can be found in [20].

5.5 Numerical Examples

We show in this section some examples which clarify the procedure and the potentialities of the approach to structural reliability analysis discussed in this chapter. We discuss separately the two alternatives listed in section 5.2.1.

5.5.1 Function Approximation on Subspaces

Reference is made here to the problem of modeling the structure of the limit state function in a more accurate way. For this purpose we use simple cases for which either the analytical solution is already known or a simulation can be easily conducted.

EXAMPLE: **The product of lognormal variables.** It is well known that the product of independent lognormal variables is a lognormal random variable with median the product of the medians and variance of its logarithm the sum of the variances of the logarithms [15]. We introduce two lognormal random variables with parameters 1.0 and 2.0 for the means, and 0.1 and 0.15 for the standard deviations. A third variable, still lognormal of parameters 1.0 and 0.01, is regarded as secondary random variable. The product of the previous three variables is investigated through a response surface approach. The main results, for which the SORM correction is not used except where explicitly specified, are summarized in the following figures:

1. Figure 5.2 provides the comparison between the response surface approximation (dashed line) and the true result (solid line; if the third variable is neglected in computing it, the resulting curve cannot be distinguished). This first result was obtained without any functional transformation of the dependent variable (the product y; it is only divided by 0.01 to enlarge the numerical range), but with four replications of the experiments. The value of λ in equation (5.37) is 1.058. When just one replication is conducted, this value increases to 1.245.

2. For single replication, the optimization of the functional transformation is pursued among the following ones:

$$y^{0.05};\ y^{0.25};\ y^{0.5};\ y;\ y^{1.25};$$

$$\ln(y)^{0.05};\ \ln(y)^{0.25};\ \ln(y)^{0.5};\ \ln(y);\ \ln(y)^{1.25}.$$

The optimal value of λ was 1.216 and corresponds to the transformation $y^{0.5}$; the comparison of the resulting approximation of the probability distribution of y with the true one is shown in Figure 5.3 Although the scale

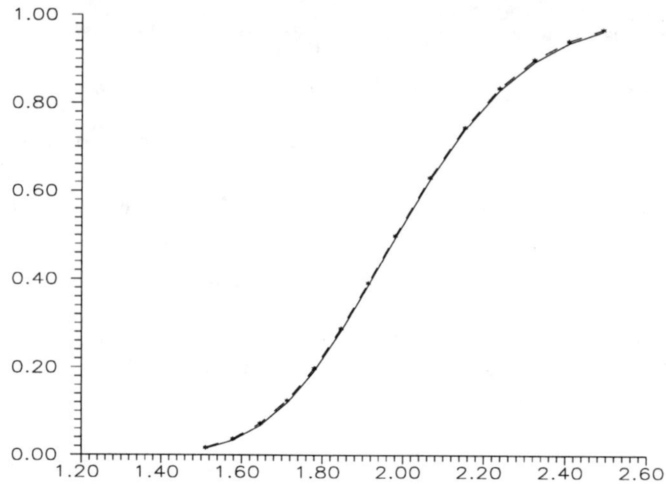

FIGURE 5.2
Product of three lognormal variables: 0.1 is the standard deviation of the third one. Comparison of the true probability distribution function (solid line) with the response surface approximation obtained with no transformation and four replications.

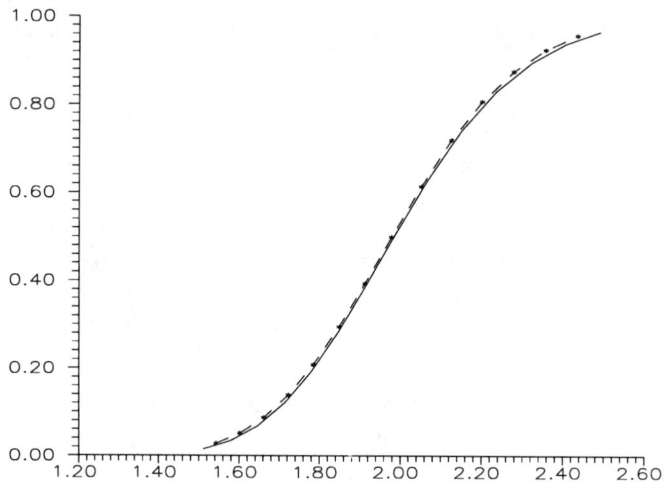

FIGURE 5.3
Product of three lognormal variables: 0.1 is the standard deviation of the third one. Comparison of the true probability distribution function (solid line) with the response surface approximation obtained with optimal transformation and no replication.

of the picture does not emphasize it, a larger discrepancy between the two curves characterizes the absence of replications of this new approximation. All the values of λ for the different transformations are greater than 1 except the value associated with the transformation $\ln(y)$:

$$1.38; 1.34; 1.22; 1.24; 1.25$$

$$8.00; 2.58; 1.82; 0.65; 1.36$$

and the probability in equation (5.33) is always lower than the confidence value $1 - \alpha = .975$ except for $\ln(y)^{0.05}$:

$$0.73; 0.71; 0.65; 0.67; 0.67$$

$$0.999; 0.95; 0.86; 0.26; 0.72$$

3. Again for a single replication, the optimization of the functional transformation is pursued among the ones listed above but after dividing y by 0.001 instead of 0.01. The optimal value of λ becomes 1.06 and corresponds to the transformation $\ln(y)^{1.25}$; nevertheless, the comparison of the resulting approximation and the previous one, shown in Figure 5.4, emphasizes the uselessness of a search for the best approximation, the discrepancy with the true curve being due to the use of a single replication, i.e., to the bias in simulating the secondary variable.

4. The problem is re-investigated by correcting the bias in the response which arises from the bias on the mean of the simulated values of the secondary variable; Figure 5.5 provides the comparison between this approximation (for which the optimal transformation is again $\ln(y)^{0.05}$) and the true result; it is evident in this case the correction of the undesired shift of the curve estimate which characterizes the plots without replication. The λ value in this case becomes 0.975 and should be compared with 1.058 obtained with four replications and no transformation and 1.019 with four replications and optimal transformation ($\ln(y)^{0.25}$).

The calculations are repeated for a similar problem with standard deviation of the secondary variable four times the original one. Figure 5.6 provides the comparison between the true curve (solid line; no mark) and the curve (long-dashed line) computed by neglecting the variability of the third variable. These two curves are then compared with the response surface approximations with four replications but without transformation (solid line with marks; $\lambda = 1.046$) and with optimal transformation ($\ln(y)^{0.25}$) but no replication (short-dashed line; $\lambda = 0.981$; $\lambda = 1.229$ without transformation). In this case the replications do not avoid an error in the lower and upper tails and, hence, a SORM estimate of the ordinates was

Table 5.1 SORM vs. FORM approximation in the upper tail of the distribution; no replication.

y	β-value	FORM	SORM	True value
2.262	-1.22	0.889	0.839	0.844
2.337	-1.53	0.938	0.916	0.896
2.412	-1.85	0.968	0.959	0.933

Table 5.2 SORM vs. FORM approximation in the upper tail of the distribution; four replications.

y	β-value	FORM	SORM	True value
2.312	-1.30	0.902	0.857	0.881
2.425	-1.72	0.957	0.941	0.939
2.539	-2.13	0.983	0.978	0.971
2.652	-2.53	0.994	0.992	0.987

made. The results are provided in Table 5.1 for the case with a single replication and Table 5.2 for the case with four replications.

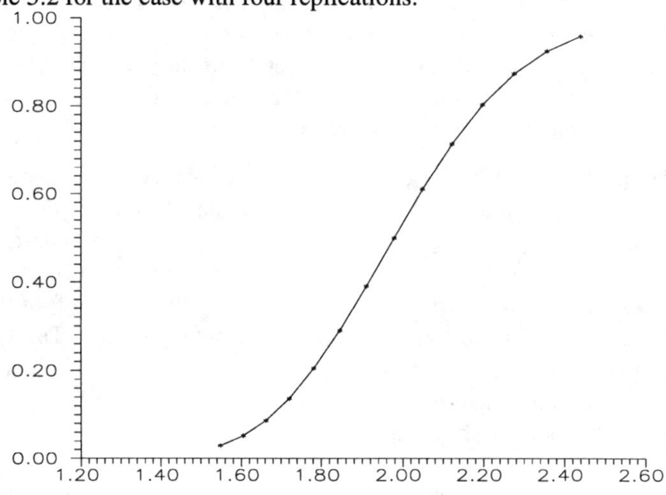

FIGURE 5.4
Product of three lognormal variables: 0.1 is the standard deviation of the third one. Comparison of the probability distribution function obtained with the response surface approximation corresponding to two different amplifications of y (100 and 100 times); no replication.

□

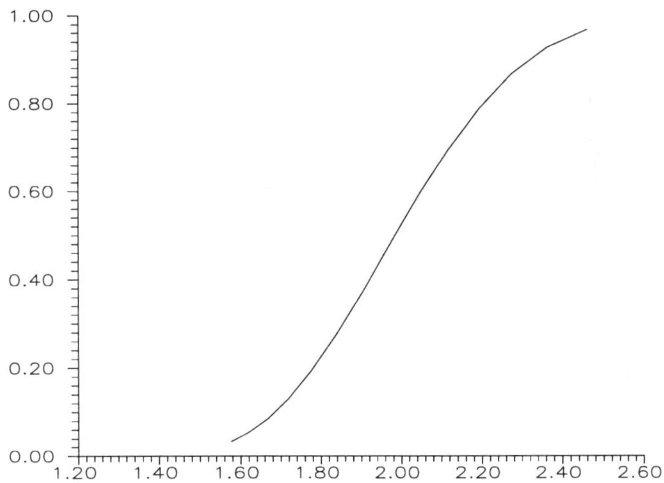

FIGURE 5.5
Product of three lognormal variables: 0.1 is the standard deviation of the third one. Comparison of the true probability distribution function (solid line) with the response surface approximation obtained with no transformation, no replication, and bias correction.

5.5.2 Reliability Assessment

We search now for models of the limit state function restricted to the part (or parts) of the state space important for reliability analysis.

EXAMPLE: **Elliptical limit state.** We consider as limit state function the ellipse of equation:

$$g(x_1, x_2) = 16 - ax_1^2 - x_2^2; \qquad 1. \leq a \leq 1.5$$

Let X_1 and X_2 be two independent standardized Gaussian random variables and let the region outside the ellipse be the failure domain. Thus, the probability of failure coincides with the probability that any realization (x_1, x_2) falls outside the ellipse as shown in Figure 5.7 (for $a = 1.5$). Numerical integration of the joint probability density function gives the results of Table 5.3. An approximation of the probabilities of failure as a increases from 1.1 to 1.5 can be achieved using the correction factor in equation (5.62) computed at the end of the shorter axis; the resulting estimate is multiplied by 2 to allow for the symmetry of the function. Note that the correction term goes to ∞ as a approaches 1 and therefore a circular limit state function cannot be studied by the SORM approximation (of course a perturbation (say an increase of 10%) into one of the axis makes the problem again accessible and the error is still reasonable).

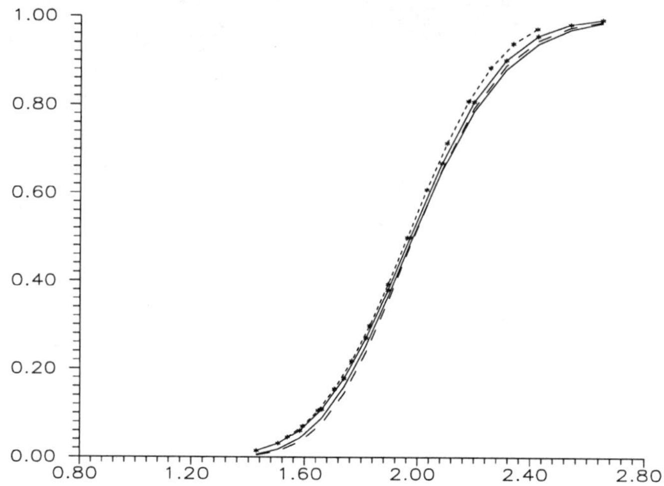

FIGURE 5.6
Product of three lognormal variables: 0.4 is the standard deviation of the third one. Comparison of the true probability distribution function (solid line; no mark) with the one which neglects the contribution of the third variable (long-dashed line). The two approximations were obtained with four replications (marked solid line) and with no replication (short-dashed line).

Table 5.3 Probability of failure for the elliptical domain in Figure 5.7 the values of the first column were obtained by numerical integration; the ones of the second column by a SORM approximation

a	Num. integr.	SORM
1.0	3.4e-4	–
1.1	5.0e-4	4.5e-4
1.2	7.5e-4	6.4e-4
1.3	1.1e-3	9.4e-4
1.4	1.5e-3	1.4e-3
1.5	2.1e-3	1.9e-3

Assume now that we ignore the actual shape of the limit state function: we can only conduct experiments which associate a value of g to each couple (x_1, x_2). Let a be equal to 1.5. The procedure of Figure 5.1 suggests to start with a linear response surface approximation of the g-function and this is pursued through a factorial design plus a few experiments in the origin. This first step interpolates the results of the experiments with a plane parallel to the plane of x_1 and x_2 and hence the design point (which coincides for this example with the maximum like-

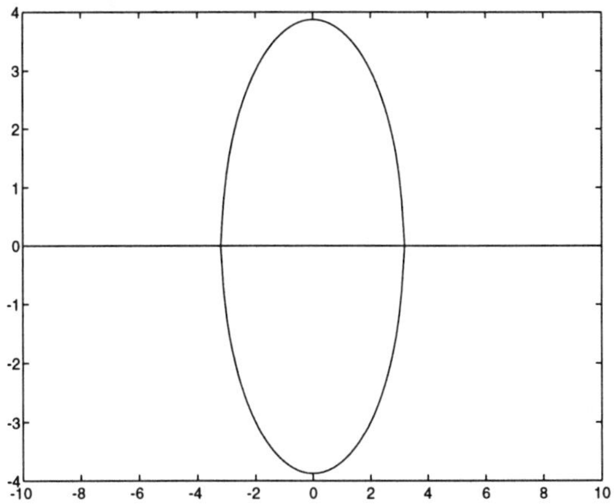

FIGURE 5.7
Elliptical limit state function.

lihood point) goes to ∞. We introduce therefore a perturbation to the experiment results giving a slope to the interpolating plane and, hence, a finite fictitious design point is found. It will be the support of the second step, adopting a quadratic response surface, even if the center of the experiments will be maintained close to the origin. As a secondary variable we introduce a third independent, Gaussian, random variable with mean 0.06 and standard deviation 0.0006. This computer run immediately identifies the true coefficients of the ellipse and the following FORM and SORM approximation for the probability of failure: $1.194\ 10^{-3}$ and $2.180\ 10^{-3}$, respectively. □

5.6 Conclusions

This chapter emphasizes a way for incorporating large random vectors, stochastic processes, and random fields in standard reliability assessment procedures. For this purpose special model validation tests are introduced and theoretically justified. The main steps of the reliability assessment procedure are the parametric identification of the limit state function within the class of second order polynomial functions and the search for the maximum likelihood point under the quadratic limit state constraint.

References

[1] Berger, J.O., *Statistical Decision Theory and Bayesian Analysis.* Springer, New York, second edition, 1985.

[2] Bjerager, P. and Krenk, S., Parametric Sensitivity in First Order Reliability Theory. *Journal of the Engineering Mechanics Division ASCE,* 115, 1989, pp. 1577–1582.

[3] Bleistein, N. and Handelsman, R.A., *Asymptotic Expansions of Integrals.* Dover Publications Inc., New York, 1986. Reprint of the edition by Holt, Rinehart and Winston, New York, 1975.

[4] Böhm, F. and Brückner-Foit, A., On Criteria for Accepting a Response Surface Model. *Probabilistic Engineering Mechanics,* 7, 1992, pp. 183–190.

[5] Box, G.E.P. and Tiao, C., *Bayesian Inference in Statistical Analysis.* Addison-Wesley, Reading, Massachusetts, 1973.

[6] Box, G.E.P. Hunter, W.G., and Hunter, J.S., *Statistics for Experimenters.* Wiley, New York, 1978.

[7] Box, G.E.P. and Draper, N.R., *Empirical Model Building and Response Surface.* Wiley, New York, 1987.

[8] Breitung, K., Asymptotic Approximations for Multinormal Integrals. *Journal of the Engineering Mechanics Division ASCE,* 110(3), 1984, pp. 357–366.

[9] Breitung, K., Asymptotic Approximations for Probability Integrals. *Journal of Probabilistic Engineering Mechanics,* 4(4), 1989, pp. 187–190.

[10] Breitung, K., Probability Approximations by Log-likelihood Maximization. *Journal of the Engineering Mechanics Division ASCE,* 117(3), 1991, pp. 457–477.

[11] Breitung, K., *Asymptotic Approximations for Probability Integrals.* Springer, New York, 1994. Springer Lecture Note in Mathematics No. 1592.

[12] Breitung, K. and Faravelli, L., Log-likelihood Maximization and Response Surface Reliability Assessment. *Nonlinear Dynamics,* 5(3), 1994, pp. 273–286.

[13] Breitung, K. and Hohenbichler, M., Asymptotic Approximations for Multivariate Integrals with an Application to Multinormal Probabilities. *Journal of Multivariate Analysis*, 30, 1989, pp. 80–97.

[14] Bucher, C.G., Adaptive Sampling - An Iterative Fast Monte Carlo Method. *Structural Safety*, 5, 1988, pp. 119–126.

[15] Casciati, F., and Faravelli, L., *Fragility Analysis of Complex Structural Systems*. Research Studies Press Ltd., Taunton, U.K., 1991.

[16] Casciati, F. and Faravelli, L., Randomness Effects in Crashworthiness Analysis. *Journal of Nonlinear Mechanics*, 1991, pp. 827–834.

[17] Cherng, R.-H. and Wen, Y.K., Reliability of Uncertain Nonlinear Trusses under Random Excitation. II. *Journal of the Engineering Mechanics Division ASCE*, 120(4), 1994, pp. 748–757.

[18] Copson, E.T., *Asymptotic Expansions*. Cambridge University Press, London, 1965.

[19] Debreu, G., Definite and Semidefinite Quadratic Forms. *Econometrica*, 20, 1952, pp. 295–300.

[20] Der Kiureghian, A., Measures of Structural Safety under Imperfect States of Knowledge. *Journal of the Engineering Mechanics Division ASCE*, 115(5), 1989, pp. 1119–1140.

[21] Ditlevsen, O., Structural Reliability and the Invariance Problem. Technical Report Research Report No. 22, Solid Mechanics Division, University of Waterloo, Waterloo, Canada, 1973.

[22] Draper, N. and Smith, H., *Applied Regression Analysis*, Second Edition. Wiley, New York, 1981.

[23] Enevoldsen, I., Sensitivity Analysis of Reliability-based Optimal Solution. *Journal of the Engineering Mechanics Division ASCE*, 120(1), 1994, pp. 198–205.

[24] Erdélyi, A., *Asymptotic Expansions*. Dover, New York, 1956.

[25] Faravelli, L., A Response Surface Approach for Reliability Analysis. *Journal of the Engineering Mechanics Division ASCE*, 115(12), 1989, pp. 2763–2781.

[26] Faravelli, L., Finite Element Analysis of Stochastic Nonlinear Continua. *Computational Mechanics of Probabilistic and Reliability Analysis*, Eds. W.K. Liu, T. Belitschko, ElmePress Int., Lausanne, 1989, pp. 263–280.

[27] Faravelli, L., Structural Reliability via Response Surface. *Nonlinear Stochastic Mechanics*, Eds. N. Bellomo & F. Casciati, Springer Verlag, Berlin, 1992, pp. 213–224.

[28] Faravelli, L., Reliability Assessment via Log-likelihood Maximization. *Probabilistic Structural Mechanics: Advances in Structural Reliability Methods*, Eds. P.D. Spanos & Y-T. Wu, Springer Verlag, Berlin, 1994, pp. 175–188.

[29] Faravelli, L., Blocking Problems in the Analysis of Random Fields. *Probabilities and Materials*, Ed. D. Breysse, NATO ASI Series 269, Kluwer, Dordrecht, 1994, pp. 177–196.

[30] Fiessler, B., Neumann, H.-J., and Rackwitz, R., Quadratic Limit States in Structural Reliability. *Journal of the Engineering Mechanics Division ASCE*, 105(4), 1979, pp. 661–676.

[31] Freudenthal, A.M., Safety and the Probability of Structural Failure. *Transactions of the ASCE*, 121, 1956, pp. 1337–1397.

[32] Graybill, F.A., *Matrices with Applications in Statistics*, Wadsworth, Belmont, CA, 1969.

[33] Guttman, I., *Linear Models: An Introduction.* Wiley, New York, 1982.

[34] Hasofer, A.M. and Lind, N.C., An Exact and Invariant First-order Reliability Format. *Journal of the Engineering Mechanics Division ASCE*, 100(1), 1974, pp. 111–121.

[35] Hohenbichler, M., Gollwitzer, S., Kruse, W., and Rackwitz, R., New Light on First- and Second-order Reliability Methods. *Structural Safety*, 4, 1987, pp. 267–284.

[36] Hohenbichler, M. and Rackwitz, R., Non-normal Dependent Vectors in Structural Safety. *Journal of the Engineering Mechanics Division ASCE*, 107(6), 1981, pp. 1227–1241.

[37] Hohenbichler, M. and Rackwitz, R., Improvement of Second-order Reliability Estimates by Importance Sampling. *Journal of the Engineering Mechanics Division ASCE*, 114(12), 1988, pp. 2195–2198.

[38] Johnson, N.L. and Kotz, S., *Continuous Univariate Distributions-2*. Wiley, New York, 1970.

[39] Madsen, H., Krenk, S., and Lind, N.C., *Methods of Structural Safety.* Prentice-Hall Inc., Englewood Cliffs, N.J., 1986.

[40] Maes, M.A., Breitung, K., and Dupuis, D.J., Asymptotic Importance Sampling. *Structural Safety*, 12, 1993, pp. 167–183.

[41] Magnus, J.R. and Neudecker, H., *Matrix Differential Calculus with Applications in Statistics and Econometrics.* Wiley, New York, 1988.

[42] Mallows, C.L., Some Comments on C_p. *Technometrics*, 15, 1973, pp. 661–675.

[43] Myers, R.H., *Response Surface Methodology.* Allyn and Bacon, Boston, 1971.

[44] Papoulis, A., *Probability, Random Variables and Stochastic Processes.* McGraw-Hill, New York, third edition, 1991.

[45] Petersen, R.G., *Design and Analysis of Experiments.* M. Decker Inc., New York, 1985.

[46] Shinozuka, M., Basic Analysis of Structural Safety. *Journal of the Structural Division ASCE*, 109(3), 1983, pp. 721–740.

[47] Thorpe, J.A., *Elementary Topics in Differential Geometry.* Springer, New York, 1979.

[48] Tvedt, L., Distribution of Quadratic Forms in Normal Space - Application to Structural Reliability. *Journal of the Engineering Mechanics Division ASCE*, 116(6), 1990, pp. 1183–1198.

[49] Walpole, R.E. and Myers, R.H., *Probability and Statistics for Engineers and Scientists.* Macmillan, New York, 1990.

[50] Westlake, J.R., *A Handbook of Numerical Matrix Inversion and Solution of Linear Equations.* Wiley, New York, 1968.

[51] Wong, R., *Asymptotic Approximations of Integrals.* Academic Press, San Diego, 1989.

[52] Wu, Y.-T. and Wirsching, P.H., New Algorithm for Structural Reliability Estimation. *Journal of the Engineering Mechanics Division ASCE*, 113(9), 1987, pp. 1319–1336.

5.7 Appendix A - Analytical Details

This appendix groups some analytical details removed from the main text in order to increase its global readability.

5.7.1 Projections and Projection Matrices

Let $U \subset \mathbb{R}^n$ be a linear subspace of the n-dimensional Euclidean space $\mathbb{R}^n$. Its orthogonal complement, i.e., the set of all vectors $\boldsymbol{w}$ such that $\boldsymbol{w}^T\boldsymbol{u} = 0$ for all $\boldsymbol{u} \in U$ will be denoted by $U^\perp$.

For each vector $\boldsymbol{x} \in \mathbb{R}^n$ there is a unique orthogonal projection onto this subspace. This means there exists a vector $\boldsymbol{u} \in U$ and a vector $\boldsymbol{w} \in U^\perp$ such that $\boldsymbol{x} = \boldsymbol{u} + \boldsymbol{w}$. The vector $\boldsymbol{u}$ is called the projection of $\boldsymbol{x}$ on U and is obtained by multiplying $\boldsymbol{x}$ by the $n \times n$ projection matrix $\boldsymbol{P}_U$ of the subspace U (see for example [32], theorem 4.4.1 p. 73, or [11], lemma 2, pp. 9 and 10).

Let be given a subspace U spanned by k linearly independent vectors $\boldsymbol{a}_1, \ldots, \boldsymbol{a}_k$. Then we have for the projection

$$\boldsymbol{u} = \sum_{i=1}^{k} \gamma_i \boldsymbol{a}_i = \boldsymbol{A}\boldsymbol{\gamma}^T. \tag{5.97}$$

with $\boldsymbol{\gamma} = (\gamma_1, \ldots, \gamma_k)^T$ and $\boldsymbol{A} = (\boldsymbol{a}_1, \ldots, \boldsymbol{a}_k)$. The vector $\boldsymbol{x} - \boldsymbol{u}$ must be orthogonal to all vectors $\boldsymbol{a}_1, \ldots, \boldsymbol{a}_k$. This gives

$$\boldsymbol{a}_i^T(\boldsymbol{x} - \boldsymbol{u}) = \boldsymbol{a}_i^T(\boldsymbol{x} - \boldsymbol{A}\boldsymbol{\gamma}^T) = 0. \tag{5.98}$$

Written together in a matrix form, it gives (with $\mathbf{o}_k$ a k-dimensional zero vector)

$$\boldsymbol{A}^T(\boldsymbol{x} - \boldsymbol{A}\boldsymbol{\gamma}^T) = \mathbf{o}_k,$$

$$\boldsymbol{A}^T\boldsymbol{x} = \boldsymbol{A}^T\boldsymbol{A}\boldsymbol{\gamma}^T,$$

$$\boldsymbol{\gamma}^T = (\boldsymbol{A}^T\boldsymbol{A})^{-1}\boldsymbol{A}^T\boldsymbol{x}. \tag{5.99}$$

Therefore the projection matrix for this subspace is given by

$$\boldsymbol{P}_U = \boldsymbol{A}(\boldsymbol{A}^T\boldsymbol{A})^{-1}\boldsymbol{A}^T. \tag{5.100}$$

For a projection matrix $\boldsymbol{P}_U$ the following relations hold

1. $P_U = P_U^2$, i.e., P_U is idempotent,
2. $P_{U^\perp} = I_n - P_U$,
3. $P_U^T = P_U$,
4. If U and V are orthogonal subspaces, then for their projection matrices $P_U P_V = P_V P_U = o_{n,n}$, with $o_{n,n}$ an $n \times n$ matrix of zeros.

They follow immediately from the definition of projection matrices.

5.7.2 Definiteness under Constraints

An $n \times n$ matrix A is called positive definite under the linear constraint $Bx = o_k$ with B being a $k \times n$ matrix if for all x with $Bx = o_k$ we have $x^T A x > 0$.

Let A be a symmetric $n \times n$-matrix and B a $k \times n$-matrix with $k < n$. The subspace $U \subset I\!R^n$ is defined by $U = \{x; Bx = o_k\}$.

With the notation adopted in the previous subsection, we give a proof for the equivalence of the following two statements

a) The $n \times n$ matrix A is positive definite under the constraint $B \cdot x = o_k$.
b) The $n \times n$-matrix $A_+ = P_U^T A P_U + P_{U^\perp}^T P_{U^\perp} = P_U A P_U + P_{U^\perp}$ is positive definite.

If the matrix A_+ is positive definite, we have for all x with $x \neq o_n$ and $B \cdot x = o_k$ that

$$x^T A_+ x = x^T (P_U^T A P_U + P_{U^\perp}^T P_{U^\perp}) x = x^T P_U^T A P_U x + x^T P_{U^\perp}^T P_{U^\perp} x.$$

Since $x \in U$, we have $P_U x = x$ and $P_{U^\perp} x = o_n$. This gives

$$x^T A_+ x = x^T A x + 0 = x^T A x > 0. \tag{5.101}$$

The last inequality is just the definition of the positive definiteness of A under the constraint $B \cdot x = o_k$. This proves that b) $\Rightarrow$ a).

To show a) $\Rightarrow$ b), we consider an arbitrary vector $z \in I\!R^n$:

$$z^T A_+ z = z^T (P_U^T A P_U + P_{U^\perp}^T P_{U^\perp}) z \tag{5.102}$$

$$= z^T P_U^T A P_U z + z^T P_{U^\perp}^T P_{U^\perp} z.$$

Then we can write this vector as the sum of its projections onto U and $U^\perp$, i.e., $z = u + w$ with $u \in U$ and $w \in U^\perp$. We have $P_U z = u$ and $P_{U^\perp} z = w$.

From this it follows then

$$z^T A_+ z = u^T A u + |w|^2. \tag{5.103}$$

Since $u \in U$, it follows from a) that $u^T A u > 0$ for $u \neq o_n$. Since $|w|^2 > 0$ iff $w \neq o_n$, we get

$$z^T A_+ z > 0$$

for all $z \neq o_n$. But this means that the matrix A_+ is positive definite, i.e., the proposition b). Therefore we have that a) $\Rightarrow$ b).

5.7.3 Quadratic Forms on Subspaces

In this appendix we derive a simple method for approaching the following problems:

1. to identify a local extremum under the constraint $g(x) = 0$;
2. to calculate the curvature factor in equation (5.62).

If there are given two twice differentiable functions f and g defined on an open set $D \subset \mathbb{R}^n$, the Lagrange multiplier theorem gives a necessary condition for local extrema of f under the constraint $g(x) = 0$. If at a point x^0 there is a local extremum of f under the constraint $g(x) = 0$, then there exists a number λ (the Lagrange multiplier) with $\nabla f(x^0) = -\lambda \nabla g(x^0)$. This means that the gradients of f and g are parallel at x^0 and that the point is a stationary point of the Lagrangian function $L(x, \lambda) = f(x) + \lambda g(x)$.

Similar as in the case of an extremum without restrictions, it is possible to give a sufficient criterion for a local extremum using the second derivatives of f and g.

Let x^0 be a point with $g(x^0) = 0$. Define the matrix $H(x^0)$ by

$$H(x^0) = \left(f_{ij}(x^0) + \lambda g_{ij}(x^0) \right)_{i,j=1,\dots,n}. \tag{5.104}$$

Then the function f has a local minimum under the constraint $g(x) = 0$ at x^0, if

$$\nabla f(x^0) = -\lambda \nabla g(x^0) \tag{5.105}$$

and $H(x^0)$ is positive definite under the constraint

$$(\nabla g(x^0))^T x = 0. \tag{5.106}$$

A proof of this result can be found in [47](p. 98-99). We consider in the following only the regular case that $\nabla g(x^0) \neq o_n$ and that the Hessian is definite under the linear constraint (5.106).

To check the definiteness of a matrix under linear constraints usually a criterion of Debreu published in [19] is used. A proof for this result is given for example in [41](p. 136). Since it requires the calculation of an $(n+1)\times(n+1)$ determinant for one linear constraint and for k constraints the calculation of k determinants of matrices with dimensions $(n+1)\times(n+1), \ldots, (n+k)\times(n+k)$, in the following a simpler criterion is given for definiteness under linear constraints.

For this we need some results about projections of vectors onto subspaces. They are summarized in appendix 5.7.1.

Then as shown in the last paragraph the following two statements are equivalent:

1. The matrix $\boldsymbol{A}$ is positive definite under the constraint $\boldsymbol{B}\cdot\boldsymbol{x} = o_k$.
2. The $n \times n$-matrix $\boldsymbol{A}_+ = \boldsymbol{P}_U^T \boldsymbol{A} \boldsymbol{P}_U + \boldsymbol{P}_{U^\perp}^T \boldsymbol{P}_{U^\perp}$ is positive definite.

To find if a symmetric matrix is positive definite under linear constraints, it is therefore only necessary to examine if another suitably defined matrix is positive definite. This result reduces the proof of definiteness under an arbitrary number of constraints to the problem of showing that *one* $n \times n$ matrix is definite. In the reliability problem with one linear constraint $(\nabla g(\boldsymbol{x}))^T\boldsymbol{x} = 0$, the projection matrix $\boldsymbol{P}$ for the one-dimensional subspace spanned by the gradient $\nabla g(\boldsymbol{x})$ is given by

$$\boldsymbol{P} = |\nabla g(\boldsymbol{x})|^{-2}\nabla g(\boldsymbol{x})(\nabla g(\boldsymbol{x}))^T \tag{5.107}$$

(see appendix 5.7.1).

Define the matrix $\tilde{\boldsymbol{H}}(\boldsymbol{x}^0)$ by

$$\tilde{\boldsymbol{H}}(\boldsymbol{x}^0) = \boldsymbol{P}\boldsymbol{H}(\boldsymbol{x}^0)\boldsymbol{P} + |\nabla g(\boldsymbol{x}^0)|^{-2}\nabla g(\boldsymbol{x}^0)(\nabla g(\boldsymbol{x}^0))^T \tag{5.108}$$

with

$$\boldsymbol{P} = \boldsymbol{I}_n - |\nabla g(\boldsymbol{x}^0)|^{-2}\nabla g(\boldsymbol{x}^0)(\nabla g(\boldsymbol{x}^0))^T. \tag{5.109}$$

The function f has a local minimum under the constraint $g(\boldsymbol{x}) = 0$ at $\boldsymbol{x}^0$, if

$$\nabla f(\boldsymbol{x}^0) = \lambda \nabla g(\boldsymbol{x}^0) \text{ and} \tag{5.110}$$

$$\tilde{\boldsymbol{H}}(\boldsymbol{x}^0) \text{ is positive definite.} \tag{5.111}$$

In the same way we obtain a sufficient criterion for extrema under several constraints. If at the point $\boldsymbol{x}^0$ there are k functions $g^i(\boldsymbol{x})$ with $g^1(\boldsymbol{x}^0) = \ldots = g^k(\boldsymbol{x}^0) = 0$ and the gradients $\nabla g^1(\boldsymbol{x}^0), \ldots, \nabla g^k(\boldsymbol{x}^0)$ (in matrix form

$\boldsymbol{A} = (\nabla g^1(\boldsymbol{x}^0), \ldots, \nabla g^k(\boldsymbol{x}^0))$ are linearly independent, we have the following result.

The function f has under the constraints $g^1(\boldsymbol{x}) = \ldots = g^k(\boldsymbol{x}) = 0$ a local minimum at $\boldsymbol{x}^0$ if

1. The gradients $\nabla g^1(\boldsymbol{x}^0), \ldots, \nabla g^k(\boldsymbol{x}^0)$ are linearly independent and there are numbers $\lambda_1, \ldots, \lambda_k$ with

$$\nabla f(\boldsymbol{x}^0) = -\sum_{m=1}^{k} \lambda_m \nabla g^m(\boldsymbol{x}^0), \tag{5.112}$$

2. The matrix defined by

$$(\boldsymbol{I}_n - \boldsymbol{P})\boldsymbol{H}^*(\boldsymbol{I}_n - \boldsymbol{P}) + \boldsymbol{P} \tag{5.113}$$

with $\boldsymbol{P} = \boldsymbol{A}(\boldsymbol{A}^T\boldsymbol{A})\boldsymbol{A}^T$ and

$$\boldsymbol{H}^* = \left(f_{ij}(\boldsymbol{x}^0) + \sum_{m=1}^{k} \lambda_m g_{ij}^m(\boldsymbol{x}^0) \right)_{i,j=1,\ldots,n} \tag{5.114}$$

is positive definite.

An $n \times n$-matrix $\boldsymbol{C}$ is positive definite if and only if there exists a non-singular lower triangular $n \times n$ matrix $\boldsymbol{R}$ such that $\boldsymbol{C} = \boldsymbol{R} \cdot \boldsymbol{R}^T$, (see [50] (theorem 22, p. 129) and the references given there). Whether a symmetric $n \times n$ matrix $\boldsymbol{C}$ is positive definite can be decided by applying the numerical algorithm for the Cholesky decomposition described in [50] (p. 13). In the case that the matrix $\boldsymbol{C}$ is positive definite, the algorithm gives the Cholesky decomposition, i.e., a lower triangular matrix $\boldsymbol{R}$ such that $\boldsymbol{C} = \boldsymbol{R} \cdot \boldsymbol{R}^T$. If $\boldsymbol{C}$ is not positive definite, the algorithm will terminate before the end, since then at one step square roots of negative numbers will appear (see [50](p. 13)). This shows then that the matrix is not positive definite. The Cholesky algorithm for a symmetric $n \times n$ matrix $\boldsymbol{C} = (c_{ij})_{i,j=1,\ldots,n}$ is for the first column

$$r_{11} = \sqrt{c_{11}} \tag{5.115}$$

$$r_{j1} = \frac{c_{j1}}{r_{11}}, \; j = 2, \ldots, n \tag{5.116}$$

and for the following columns

$$r_{ii} = \sqrt{c_{ii} - \sum_{k=1}^{i-1} r_{ik}^2}, \; i = 2, \ldots, n \tag{5.117}$$

$$r_{ji} = \frac{1}{r_{ii}} \left(c_{ij} - \sum_{k=1}^{i-1} r_{ik} r_{jk} \right) \text{ for } i \leq j. \tag{5.118}$$

The determinant of $\boldsymbol{C}$ is then

$$\det(\boldsymbol{C}) = (\det(\boldsymbol{R}))^2 = \prod_{i=1}^{n} r_{ii}^2. \tag{5.119}$$

This result can be used also for calculating the value of such a determinant. The SORM-approximation for the failure probability $I\!P(I\!F)$ is given by

$$I\!P(I\!F) \sim \Phi(-\beta(I\!F)) \prod_{i=1}^{n-1} (1 - \beta(I\!F)\kappa_i)^{-1/2}. \tag{5.120}$$

In this form an eigenvalue analysis of the matrix $|\nabla g|^{-1}\tilde{G}$ with $\tilde{G} = (g_{ij}(\boldsymbol{x}^0))_{i,j=1,\ldots,n-1}$ is necessary to find the main curvatures $\kappa_1, \ldots, \kappa_{n-1}$.

But it is possible to derive another form for the curvature correction factor $\prod_{i=1}^{n-1}(1-\kappa_i)^{-1/2}$ by using again projection matrices.

By a rotation of the coordinates it can always be achieved that the x_n axis is in the direction of the normal vector of the surface G at $\boldsymbol{x}^0$ and the tangential space is spanned by the vectors in the directions of the $x_1, \ldots, x_{n-1}$ axes. In the following we will assume for simplicity that $\beta = 1$; elsewhere the κ_i's and the second derivatives have to be multiplied by β.

The curvature factor is [8] (equation (25)):

$$\prod_{i=1}^{n-1}(1 - \kappa_i)$$

$$= \det \begin{pmatrix} 1 + h_{11}(\boldsymbol{x}^0) & h_{12}(\boldsymbol{x}^0) & \ldots & h_{1,n-1}(\boldsymbol{x}^0) \\ h_{21}(\boldsymbol{x}^0) & 1 + h_{22}(\boldsymbol{x}^0) & \ldots & h_{2,n-1}(\boldsymbol{x}^0) \\ \vdots & \ldots & \ddots & \vdots \\ h_{n-1,1}(\boldsymbol{x}^0) & \ldots & \ldots & 1 + h_{n-1,n-1}(\boldsymbol{x}^0) \end{pmatrix}. \tag{5.121}$$

Here $h_{ij}(\boldsymbol{x}^0)$ is equal to $|\nabla g(\boldsymbol{x}^0)|^{-1} g_{ij}(\boldsymbol{x}^0)$.

By adding an n-th row and an n-th column with zeros everywhere and only 1 in the main diagonal, the value of the determinant is not changed and we get

$$\prod_{i=1}^{n-1}(1-\kappa_i)$$

$$= \underbrace{\det\begin{pmatrix} 1+h_{11}(\boldsymbol{x}^0) & h_{12}(\boldsymbol{x}^0) & \dots & h_{1,n-1}(\boldsymbol{x}^0) & 0 \\ h_{21}(\boldsymbol{x}^0) & 1+h_{22}(\boldsymbol{x}^0) & \dots & h_{2,n-1}(\boldsymbol{x}^0) & 0 \\ \vdots & \dots & \ddots & \vdots & \vdots \\ h_{n-1,1}(\boldsymbol{x}^0) & \dots & \dots & 1+h_{n-1,n-1}(\boldsymbol{x}^0) & 0 \\ 0 & \dots & \dots & 0 & 1 \end{pmatrix}}_{=\boldsymbol{D}}. \tag{5.122}$$

We define

$$\boldsymbol{H}(\boldsymbol{x}^0) = (\boldsymbol{I}_n + |\nabla g(\boldsymbol{x}^0)|^{-1}\boldsymbol{G})_{i,j=1,\dots,n}, \tag{5.123}$$

with $\boldsymbol{G} = (g_{ij}(\boldsymbol{x}^0))_{i,j=1,\dots,n}$. Due to the special choice of the coordinate system, $\boldsymbol{e}_n$ is the normal vector of the surface G at $\boldsymbol{x}^0$ and $\boldsymbol{P}$ is the projection matrix onto the tangential space of G at $\boldsymbol{x}^0$.

But this formulation is invariant under linear coordinate changes; we just have to replace $\boldsymbol{e}_n$ by the normal vector $\boldsymbol{n}$ in the new coordinates $\boldsymbol{n} = |\nabla g(\boldsymbol{x}^0)|^{-1}\nabla g(\boldsymbol{x}^0)$. For an arbitrary coordinate system the following expression for the curvature factor is valid

$$\prod_{i=1}^{n-1}(1-\kappa_i) = \det(\boldsymbol{P}\boldsymbol{H}(\boldsymbol{x}^0)\boldsymbol{P} +$$

$$|\nabla g(\boldsymbol{x}^0)|^{-2}\nabla g(\boldsymbol{x}^0)(\nabla g(\boldsymbol{x}^0))^T) \tag{5.124}$$

with

$$\boldsymbol{P} = \boldsymbol{I}_n - (|\nabla g(\boldsymbol{x}^0)|^{-1}\nabla g(\boldsymbol{x}^0)(|\nabla g(\boldsymbol{x}^0)|^{-1}\nabla g(\boldsymbol{x}^0))^T$$

$$= \boldsymbol{I}_n - |\nabla g(\boldsymbol{x}^0)|^{-2}\nabla g(\boldsymbol{x}^0)(\nabla g(\boldsymbol{x}^0))^T. \tag{5.125}$$

Since the function $|\boldsymbol{x}|^2$ has a minimum at the point $\boldsymbol{x}^0$ under the constraint $g(\boldsymbol{x}) = 0$, the matrix $\boldsymbol{H}(\boldsymbol{x}^0)$ is positive definite under the linear constraint

$n^T x = 0$, if the extremum is regular. Therefore in this case the determinant of D can be found from the Cholesky decomposition of $PH(x^0)P + |\nabla g(x^0)|^{-2}\nabla g(x^0)(\nabla g(x^0))^T$ using equation (5.119).

The same result is valid in the non-normal case. Here we have the maximum of the log-likelihood function $l(x)$ under the constraint $g(x) = 0$. The asymptotic approximation is given in equation (5.69). The determinant of the matrix appearing in this equation can be written then as $\det((I_n - P)H_1(x^*)(I_n - P) + P)$ (see equation 5.72).

5.7.4 Maximum Likelihood for Non-Gaussian Distribution

We consider the transformation to standard normal random variables given in the form

$$z_i = \Phi^{-1}[F_i(x_i)] \tag{5.126}$$

The symmetry of the standard normal distribution permits one to write

$$-z_i = \Phi^{-1}[1 - F_i(x_i)]. \tag{5.127}$$

We write now the complementary c.d.f. $(1 - F_i(x_i))$ as an integral

$$1 - F_i(x_i) = \int_{x_i}^{\infty} f_i(y_i)dy_i = \int_{x_i}^{\infty} \exp(l_i(y_i))dy_i. \tag{5.128}$$

with $l_i(y) = \ln(f_i(y))$. This last integral can be approximated again by asymptotic methods. We write, making a Taylor expansion of l_i at x_i:

$$\int_{x_i}^{\infty} f_i(y_i)dy_i = \int_{x_i}^{\infty} \exp\left(l_i(x_i) + (y_i - x_i)l_i'(x_i + (y_i - x_i)\vartheta(y_i))\right)dy_i \tag{5.129}$$

where $0 \le \vartheta(y_i) \le 1$.

Replacing the derivative at the points between x_i and y_i by the derivative at x_i

gives then approximately

$$\int_{x_i}^{\infty} f_i(y_i)dy_i \approx \int_{x_i}^{\infty} \exp\left(l_i(x_i) + l_i'(x_i)(y_i - x_i)\right) dy_i$$

$$= \frac{\exp(l_i(x_i))}{|l_i'(x_i)|}. \tag{5.130}$$

Then we have for the transformed value of x_i

$$-z_i \approx \Phi^{-1}\left(\frac{\exp(l_i(x_i))}{|l_i'(x_i)|}\right) \tag{5.131}$$

or

$$\Phi(-z_i) \approx \frac{\exp(l_i(x_i))}{|l_i'(x_i)|}. \tag{5.132}$$

If z_i is large, we can use Mill's ratio, giving

$$(2\pi)^{-1/2}\exp(-z_i^2/2)/z_i \approx \frac{\exp(l_i(x_i))}{|l_i'(x_i)|}. \tag{5.133}$$

Neglecting terms of lower order, we find then

$$-z_i^2/2 \approx l_i(x_i). \tag{5.134}$$

Collecting these terms for $i = 1, \dots, n$ gives then

$$-|\boldsymbol{z}|^2 \approx 2l(\boldsymbol{x}) \tag{5.135}$$

with $l(\boldsymbol{x}) = \sum_{i=1}^{n} l_i(x_i)$.

5.7.5 Improvement by Importance Sampling Methods

In this appendix we will derive the asymptotic distribution of the random vector $\boldsymbol{X}$ in the failure domain.

We consider first the case of a linear limit state function g_L. By a rotation we can always achieve that $g_L(\boldsymbol{x}) = 1 - x_n$. Then the minimal distance point is the point $\boldsymbol{x}^0 = (0, \dots, 0, 1)$. For the sequence $\beta\mathbb{F}$ of failure domains defined by

$\beta \mathbb{F} = \{\boldsymbol{x}; x_n > \beta\}$ we study the conditional distribution of the random vector $\boldsymbol{X}$ in the failure domain $\beta \mathbb{F}$. Its p.d.f. is given by

$$f_\beta(\boldsymbol{x}) = \frac{f(\boldsymbol{x})}{\mathbb{P}(\beta \mathbb{F})} \quad \text{for } \boldsymbol{x} \in \beta \mathbb{F} \tag{5.136}$$

and zero elsewhere (see for example [44], p. 162, equation (7-39)). We obtain

$$f_\beta(\boldsymbol{x}) = (2\pi)^{-n/2} \frac{\exp(-|\boldsymbol{x}|^2/2)}{\Phi(-\beta)}. \tag{5.137}$$

and this can be written using Mill's ratio as $\beta \to \infty$

$$f_\beta(\boldsymbol{x}) \sim (2\pi)^{-(n-1)/2} \beta \exp\left(-\frac{1}{2}(x_n^2 + \sum_{i=1}^{n-1} x_i^2 - \beta^2)\right). \tag{5.138}$$

Now we have

$$x_n^2 = (x_n^0 + (x_n - x_n^0))^2 = \beta^2 + 2\beta(x_n - x_n^0) + (x_n - x_n^0)^2. \tag{5.139}$$

This gives

$$f_\beta(\boldsymbol{x}) \sim (2\pi)^{-(n-1)/2} \beta \times \exp(-\tfrac{1}{2}(2\beta(x_n - x_n^0)$$

$$+(x_n - x_n^0)^2 + \textstyle\sum_{i=1}^{n-1} x_i^2)). \tag{5.140}$$

If we now make a transformation $(x_1, \dots, x_n) \to (x_1, \dots, x_{n-1}, \beta(x_n - x_n^0)) = \boldsymbol{u}$, we have for the joint density $f_\beta^u(\boldsymbol{u})$ of this random vector under the condition $U_n > 0$ the form

$$f_\beta^u(\boldsymbol{u}) \sim (2\pi)^{-(n-1)/2} \exp(-\frac{1}{2} \sum_{i=1}^{n-1} u_i^2) \cdot \exp(-u_n). \tag{5.141}$$

This shows that the random vector $\boldsymbol{U}$ has, under the condition $u_n > 0$, a distribution, which is the product of an $(n-1)$-dimensional standard normal distribution for the components $U_1, \dots, U_{n-1}$ and a standard exponential distribution for the component U_n.

Now we consider the case of a non-linear function g; but again we assume that the minimal distance point is $\boldsymbol{x}^0$. To derive a similar result as above in the case of a linear limit state function

$$f_\beta(\boldsymbol{x}) = \frac{f(\boldsymbol{x})}{I\!P(\beta I\!F)}. \tag{5.142}$$

The asymptotic approximation for $I\!P(\beta I\!F)$ is given by $\Phi(-\beta)\prod_{i=1}^{n-1}(1-\kappa_i)^{-1/2}$. Inserting this in the last equation and using again Mill's ratio we obtain

$$\begin{aligned} f_\beta(\boldsymbol{x}) \sim \quad & (2\pi)^{-(n-1)/2}\prod_{i=1}^{n-1}(1-\kappa_i)^{-1/2} \\ & \times \exp\left(-\tfrac{1}{2}(x_n^2 + \textstyle\sum_{i=1}^{n-1} x_i^2 - \beta^2)\right). \end{aligned} \tag{5.143}$$

We now make a transformation $\boldsymbol{x} \to \boldsymbol{u} = (u_1, \ldots, u_{n-1}, u_n)$, with $u_n = \beta(x_n - x_n^0)$. Here u_i, $(i = 1, \ldots, n-1)$, is the arc length, on the surface G, of a curve on G from the point $\boldsymbol{x}^0$ to the point on the surface G lying above $(x_1^0, \ldots, x_i^0 + (x_i - x_i^0), \ldots, x_{n-1}^0)$.

Using asymptotic methods (see [11]) we obtain for the p.d.f. of the random vector $\boldsymbol{U}$ the asymptotic form

$$\begin{aligned} f_\beta^u(\boldsymbol{u}) \sim \quad & (2\pi)^{-(n-1)/2}\prod_{i=1}^{n-1}(1-\kappa_i)^{-1/2} \\ & \exp\left(-\tfrac{1}{2}(\textstyle\sum_{i=1}^{n-1}(1-\kappa_i)u_i^2)\right)\exp(-u_n). \end{aligned} \tag{5.144}$$

In the case we see that we have again a product of an $(n-1)$-dimensional normal distribution with independent zero mean components and variances $(1-\kappa_1)^{-1}, \ldots, (1-\kappa_{n-1})^{-1}$ for the arc lengths in the directions of the axes and a standard exponential distribution for u_n. The normal distribution is for the arc lengths on the surface in the directions of the main curvatures, which are parallel to the first $n-1$ axes (see appendix 5.7.6).

This result can be used to improve the asymptotic approximations by importance sampling methods using this density as approximate importance sampling density.

Similar results hold for the case of non-normal densities. We consider a surface with a unique PML $\boldsymbol{x}^*$. If we assume again that the $u_1, \ldots, u_{n-1}$ are local coordinates on the surface G, defined by the arc lengths in the direction of the main curvatures and u_n is distance from the PML in the direction of the x_n-axis, we obtain, after a suitable scaling, the form

$$f(u_1, \ldots, u_{n-1}, u_n) \propto \exp(-|\nabla l(\boldsymbol{x}^*)|u_n) \times \exp\left(-\frac{1}{2}\boldsymbol{u}^*\boldsymbol{L}\boldsymbol{u}^*\right) \tag{5.145}$$

for the joint density of the local coordinates $u_1, \ldots, u_n$. Here we have that $\boldsymbol{u}^* = (u_1, \ldots, u_{n-1})$. The matrix $\boldsymbol{L}$ is composed of the second derivatives of the log-likelihood with respect to the local coordinates

$$\boldsymbol{L} = \left(l_{ij}(\boldsymbol{x}^*) - \frac{|\nabla l(\boldsymbol{x}^*)|}{|\nabla g(\boldsymbol{x}^*)|} g_{ij}(\boldsymbol{x}^*) \right)_{i,j=1,n-1} . \tag{5.146}$$

An algorithm to use this result for improved reliability estimates was developed in [40]. The idea is to produce random points in the failure domain, which have the p.d.f. given in equation (5.145), and then to calculate at these points the quotient of the actual p.d.f. and this approximating p.d.f.

5.7.6 Differential Geometry of a Surface

Let G be a surface defined by $G = \{\boldsymbol{x}; g(\boldsymbol{x}) = 0\}$ and oriented by $\boldsymbol{n}(\boldsymbol{x}) = |\nabla g(\boldsymbol{x})|^{-1} \nabla g(\boldsymbol{x})$. Consider a point $\boldsymbol{x}_0 \in G$ such that $\boldsymbol{n}(\boldsymbol{x}_0) = \boldsymbol{e}_n$ ($\boldsymbol{e}_i$ denotes the unit vector in the direction of the x_i-axis). Then the tangential space $T = \{\boldsymbol{y}; \boldsymbol{n}(\boldsymbol{x})^T \boldsymbol{y} = 0\}$ is spanned by the vectors $\boldsymbol{e}_1, \ldots, \boldsymbol{e}_{n-1}$. The Weingarten map $L_p(\,)$, which describes the change of the normal $\boldsymbol{n}(\boldsymbol{x})$ in the tangential direction of G at $\boldsymbol{x}_0$, is a linear mapping from T into T. It has the following matrix

$$L_p(\boldsymbol{e}_i) = -|\nabla g(\boldsymbol{x}_0)|^{-1} \boldsymbol{G} \boldsymbol{e}_i , \tag{5.147}$$

where $\boldsymbol{G}$ denotes the Hessian matrix of g at $\boldsymbol{x}_0$. The eigenvalues $\kappa_1, \ldots, \kappa_{n-1}$ of the Weingarten map are called the main or principal curvatures of the surface G at $\boldsymbol{x}$. (It can be shown that the Weingarten map has $n-1$ eigenvalues (see [47], chapter 12).) The eigenvectors $\boldsymbol{v}_1, \ldots, \boldsymbol{v}_{n-1}$ for these eigenvalues are called the main or principal curvature directions. In such a direction, the directional derivative $\nabla_{\boldsymbol{v}} \boldsymbol{n}$ of the normal is parallel to the direction itself.

5.7.7 Asymptotic Approximations

To estimate the error in the approximation methods we need some basic concepts of asymptotic analysis. The main topic of asymptotic analysis is the approximation of functions near limit points of their domains of definition. Usually the important and interesting cases are those in which the function is not defined in the limit points.

Often the function $f(x)$ has a complicated form and the intention is to find a simpler function $\varphi(x)$ with the property that it has near the limit point "approximately" the same behavior as $f(x)$. Certainly, if a function is replaced by a simpler function, it is not the same. The trick of asymptotic approximations is to find a simpler function, which retains the essential structure of the original function.

Let a function $f(x)$ and a function $\varphi(x)$ be given. We say that $\varphi(x) = o(f(x))$ as $x \to \infty$ if for all $k > 0$ there is a $C_k > 0$ such that $|\varphi(x)| \leq k|f(x)|$ for all $x > C_k$. This means that for large x the magnitude of $\varphi(x)$ is negligible in comparison with the magnitude of $f(x)$.

We say that $\varphi(x) = O(f(x))$ as $x \to \infty$ if for one $K > 0$ there is a $C > 0$ such that always $|\varphi(x)| \leq K|f(x)|$ for all $x > C$. That means that for large x the magnitude of $\varphi(x)$ is not larger than the magnitude of $f(x)$ multiplied by a constant.

A further relation between two functions is the following. We write

$$f(x) \sim \varphi(x), \; x \to \infty, \tag{5.148}$$

if we have that

$$\lim_{x\to\infty} \frac{f(x)}{\varphi(x)} = 1. \tag{5.149}$$

The last relation can also be written in the form

$$\lim_{x\to\infty} \frac{f(x) - \varphi(x)}{\varphi(x)} = 0. \tag{5.150}$$

This means that the relative error of approximating $f(x)$ by $\varphi(x)$ as $x \to \infty$ approaches zero, i.e., for large x we can approximate $f(x)$ by $\varphi(x)$.

For example, we have for the functions $f(x) = x + \sin(x)$ and $\varphi(x) = x$ that $f(x) \sim \varphi(x)$, since

$$\lim_{x\to\infty} f(x)/\varphi(x) = \lim_{x\to\infty} \frac{x + \sin(x)}{x} = 1 + \lim_{x\to\infty} \frac{\sin(x)}{x} = 1. \tag{5.151}$$

Why do we need asymptotic approximations? In reliability problems failure occurs in general if there are extremely large loads or extremely small resistances. Let us consider a very simple example: the resistance is deterministic and the load is random with a standard normal distribution. If the element will break when the load L is larger than r_0, the probability of failure is

$$\mathbb{P}(F) = \Phi(-r_0). \tag{5.152}$$

If r_0 is large, we can use Mill's ratio, an asymptotic approximation for the standard normal integral. Mill's ratio states that

$$\Phi(-x) \sim \phi(x)/x, \; x \to \infty. \tag{5.153}$$

Here the standard normal integral, which can be calculated only numerically, is approximated by a simple analytic function as $x \to \infty$. In our example we have then for large r_0

$$\mathbb{P}(F) \sim \phi(r_0)/r_0. \tag{5.154}$$

The basic philosophy for asymptotic approximations in reliability is to find for the integrals over the failure domains simple approximations, which are "good" for "large" values of the random influences.

Elementary textbooks about asymptotic methods are the ones by Erdélyi [24] and Copson [18]. More advanced texts, which treat the multidimensional case, are the books by Bleistein and Handelsman [3] and Wong [51].

5.7.8 Asymptotic Approximations for Multidimensional Integrals

We consider an integral in the form

$$I(\beta) = \int\limits_{g(\boldsymbol{x})\leq 0} h(\boldsymbol{x}) \exp\left(\beta^2 f(\boldsymbol{x})\right)\, d\boldsymbol{x} \tag{5.155}$$

such that the integrand f has only one maximum equal to zero in the integration domain, which lies on the boundary $g(\boldsymbol{x}) = 0$ at the point $\boldsymbol{x}_0$. We assume that the maximum is regular, the maximum point is the origin, and that the gradient of the limit state function $g(\boldsymbol{x})$ points into the direction of the x_n-axis. By making a Taylor expansion around this point we get

$$f(\boldsymbol{x}) = f_n x_n + \frac{1}{2}\sum_{i,j=1}^{n-1} f_{ij} x_i x_j$$

$$+\frac{1}{2}\sum_{i=1}^{n-1} f_{in} x_i x_n + \frac{1}{2} f_{nn} x_n^2 + o(|\boldsymbol{x}|^2). \tag{5.156}$$

Here f_n and f_{ij} are the first and second derivatives of f with respect to the components x_n, and x_i and x_j, respectively, at the origin. By introducing the transformation

$$(x_1, \ldots, x_{n-1}, x_n) \to (u_1, \ldots, u_{n-1}, u_n) \tag{5.157}$$

with $u_i = \beta x_i$ for $i = 1, \ldots, n-1$ and $u_n = \beta^2 x_n$ we obtain for the integral the form

$$I(\beta) \approx \beta^{-(n+1)} \int_{g(\boldsymbol{x}) \leq 0} h(\beta^{-1}u_1, \ldots, \beta^{-1}u_{n-1}, \beta^{-2}u_n)$$

$$\times \exp\left(u_n f_n + \frac{1}{2}\sum_{i,j=1}^{n-1} f_{ij}u_i u_j + \frac{\beta^{-1}}{2}\sum_{i=1}^{n-1} f_{in}u_i u_n + \frac{\beta^{-2}}{2} f_{nn}u_n^2\right) d\boldsymbol{u}. \quad (5.158)$$

For a fixed point $(u_1, \ldots, u_n)$, we have, as $\beta \to \infty$, that

$$h(\beta^{-1}u_1, \ldots, \beta^{-1}u_{n-1}, \beta^{-2}u_n) \to h(0, \ldots, 0), \quad (5.159)$$

$$\frac{\beta^{-1}}{2}\sum_{i=1}^{n-1} f_{in}u_i u_n \to 0, \quad (5.160)$$

$$\frac{\beta^{-2}}{2} f_{nn}u_n^2 \to 0. \quad (5.161)$$

Therefore, asymptotically, we have for the integral the form

$$I(\beta) \sim \beta^{-(n+1)} h(0, \ldots, 0)$$

$$\int_{g(\boldsymbol{x}) \leq 0} \exp(f_n u_n + \frac{1}{2}\sum_{i,j=1}^{n-1} f_{ij}u_i u_j)\, d\boldsymbol{u}. \quad (5.162)$$

This can be split into two integrals, giving

$$I(\beta) \sim \beta^{-(n+1)} h(0, \ldots, 0) \int_0^\infty \exp(f_n u_n) du_n$$

$$\times \int_{\mathbb{R}^{n-1}} \exp\left(\frac{1}{2}\sum_{i,j=1}^{n-1} f_{ij}u_i u_j\right) du_1 \ldots u_{n-1}. \quad (5.163)$$

We obtain for the first integral

$$\int_0^\infty \exp(f_n u_n)\, du_n = |f_n|^{-1}. \tag{5.164}$$

and for the second integral with $\boldsymbol{H} = (f_{ij})_{i,j=1,\dots,n-1}$ that

$$\int_{\mathbb{R}^{n-1}} \exp\left(\frac{1}{2}\sum_{i,j=1}^{n-1} f_{ij}u_iu_j\right) du_1\dots u_{n-1} =$$

$$(2\pi)^{(n-1)/2}|\det \boldsymbol{H}|^{-1/2}. \tag{5.165}$$

(see for example [5], equation (A2.1.10), p. 145).

This gives the final result

$$I(\beta) \sim (2\pi)^{(n-1)/2}\beta^{-(n+1)}h(0,\dots,0)|f_n|^{-1}|\det \boldsymbol{H}|^{-1/2}. \tag{5.166}$$

This result can be generalized to the case of surfaces, whose curvatures do not vanish at the maximum point. We assume again that the maximum is regular, the maximum point is the origin, and that the gradient of the limit state function points into the direction of the x_n-axis. Then we make a transformation into local coordinates on the surface in the form

$$(x_1,\dots,x_{n-1},x_n) \to (u_1,\dots,u_{n-1},u_n). \tag{5.167}$$

Here the $u_1,\dots,u_{n-1}$ are a local coordinate system on the surface G near the origin and they are given by $u_i = h_i(x_1,\dots,x_n)$. The new n-th variable is simply the negative of the value of the limit state function at $\boldsymbol{x}$, i.e., $u_n = -g(\boldsymbol{x})$. Since the gradient of g at the origin does not vanish and the gradients of the h_i's are lying in the tangential space T of G, the Jacobian of the so defined transformation is not equal to zero at the origin. Therefore this defines locally a coordinate transformation. Now since the function g is constant equal to zero on the surface G, we get by differentiating the equation

$$g(T(u_1,\dots,u_{n-1},0)) = 0 \tag{5.168}$$

that

$$\frac{\partial g(T(u_1,\dots,u_{n-1},0))}{\partial u_i} = \sum_{l=1}^{n}\frac{\partial g}{\partial x_l}\frac{\partial x_l}{\partial u_i} = 0. \tag{5.169}$$

By differentiating further we find then

$$\frac{\partial^2 g(T(u_1,\ldots,u_{n-1},0))}{\partial u_i \partial u_j} = \sum_{l,k=1}^{n} \frac{\partial^2 g}{\partial x_l \partial x_k} \frac{\partial x_l}{\partial u_i} \frac{\partial x_k}{\partial u_j}$$

$$+\sum_{l=1}^{n} \frac{\partial g}{\partial x_l} \frac{\partial^2 x_l}{\partial u_i \partial u_j} = 0. \tag{5.170}$$

This gives

$$\sum_{l,k=1}^{n} \frac{\partial^2 g}{\partial x_l \partial x_k} \frac{\partial x_l}{\partial u_i} \frac{\partial x_k}{\partial u_j} = -\sum_{l=1}^{n} \frac{\partial g}{\partial x_l} \frac{\partial^2 x_l}{\partial u_i \partial u_j}. \tag{5.171}$$

We have for the second derivatives of f with respect to the local coordinates

$$\frac{\partial^2 f(T(u_1,\ldots,u_{n-1},0))}{\partial u_i \partial u_j} = \sum_{l,k=1}^{n} \frac{\partial^2 f}{\partial x_l \partial x_k} \frac{\partial x_l}{\partial u_i} \frac{\partial x_k}{\partial u_j}$$

$$+\sum_{l=1}^{n} \frac{\partial f}{\partial x_l} \frac{\partial^2 x_l}{\partial u_i \partial u_j}. \tag{5.172}$$

Since due to the Lagrange multiplier theorem we have that at $\boldsymbol{x}_0$ for the gradients $\nabla f = \lambda \nabla g$, we can write this as

$$\frac{\partial^2 f(T(u_1,\ldots,u_{n-1},0))}{\partial u_i \partial u_j} = \sum_{l,k=1}^{n} \frac{\partial^2 f}{\partial x_l \partial x_k} \frac{\partial x_l}{\partial u_i} \frac{\partial x_k}{\partial u_j}$$

$$+\frac{|\nabla f|}{|\nabla g|} \sum_{l=1}^{n} \frac{\partial g}{\partial x_l} \frac{\partial^2 x_l}{\partial u_i \partial u_j}. \tag{5.173}$$

Using equation (5.171), we obtain

$$\frac{\partial^2 f(T(u_1,\ldots,u_{n-1},0))}{\partial u_i \partial u_j} = \sum_{l,k=1}^{n} \frac{\partial^2 f}{\partial x_l \partial x_k} \frac{\partial x_l}{\partial u_i} \frac{\partial x_k}{\partial u_j}$$

$$-\frac{|\nabla f|}{|\nabla g|} \sum_{l,k=1}^{n} \frac{\partial^2 g}{\partial x_l \partial x_k} \frac{\partial x_l}{\partial u_i} \frac{\partial x_k}{\partial u_j} = \sum_{l,k=1}^{n} \Big(\frac{\partial^2 f}{\partial x_l \partial x_k}$$

$$-\frac{|\nabla f|}{|\nabla g|}\frac{\partial^2 g}{\partial x_l \partial x_k})\frac{\partial x_l}{\partial u_i}\frac{\partial x_k}{\partial u_j}. \tag{5.174}$$

Now we have assumed that the main curvature directions at the minimal distance point are in the directions of the $x_1, \ldots, x_{n-1}$-axes. If we take as local coordinates on the surface the arc lengths in the direction of these axes, we get

$$\frac{\partial x_k}{\partial u_j} = \delta_{kj} \tag{5.175}$$

for $k, j = 1, \ldots, n-1$.
This gives for the Hessian $\tilde{\boldsymbol{H}}_{f(T(\boldsymbol{u}))} = (\partial^2 f(T(\boldsymbol{u}))/\partial u_i u_j)_{i,j=1,\ldots,n-1}$ the form

$$\boldsymbol{H}_{f(T(\boldsymbol{u}))} = \tilde{\boldsymbol{F}} - |\nabla f|/|\nabla g|\tilde{\boldsymbol{G}} \tag{5.176}$$

with $\tilde{\boldsymbol{F}}$ and $\tilde{\boldsymbol{G}}$ the $(n-1) \times (n-1)$ matrices of the second derivatives of f and g with respect to the first $n-1$ coordinates.

Therefore we get for the asymptotic approximation in this case the form

$$I(\beta) \sim \beta^{-(n+1)}(2\pi)^{(n-1)/2}h(0, \ldots, 0)|f_n|^{-1}|\det(\tilde{\boldsymbol{H}}_{f(T(\boldsymbol{u}))})|^{-\frac{1}{2}}. \tag{5.177}$$

We consider now the case of the standard normal density, i.e., $f(\boldsymbol{x}) = -|\boldsymbol{x}|^2/2$ and $h(\boldsymbol{x}) = (2\pi)^{-n/2}$. Further we assume now that $\boldsymbol{x}^0 = (0, \ldots, 0, 1)$ and that $\nabla g(\boldsymbol{x}^0) = (0, \ldots, 0, -1)$. Then we find for the modified Hessian, since $|\nabla f(\boldsymbol{x}^0)| = 1$

$$\boldsymbol{H}_{f(T(\boldsymbol{u}))} = -\boldsymbol{I}_{n-1} - |\nabla g|^{-1}\tilde{\boldsymbol{H}}_g = -(\boldsymbol{I}_{n-1} + |\nabla g|^{-1}\boldsymbol{G}). \tag{5.178}$$

But the eigenvalues of the matrix $-|\nabla g|^{-1}\boldsymbol{G}$ are the main curvatures of the surface G at $\boldsymbol{x}_0$ (see appendix 5.7.6) and so we get for the determinant of this matrix the result

$$\det(\tilde{\boldsymbol{H}}_{f(T(\boldsymbol{u}))}) = -\prod_{i=1}^{n-1}(1-\kappa_i)^{-1/2}. \tag{5.179}$$

This gives the approximation result

$$I(\beta) \sim \beta^{-(n+1)}(2\pi)^{-1/2}\prod_{i=1}^{n-1}(1-\kappa_i)^{-1/2}\exp(-\frac{\beta^2}{2}). \tag{5.180}$$

5.8 Appendix B - Notation

We use the following notation:

$\mathbb{E}[\,]$ = expected value
$\mathbb{F}$ = failure set
$\mathbb{P}$ = probability of a set
$\mathbb{R}^k$ = Euclidean space of dimension k
$\mathcal{T}$ = standardization transformation

C_p = C_p statistic defined in equation 5.39
$f(\,)$ = probability density function
F = F – distribution variable
$g(\,)$ = limit state function
G = limit state surface
I = integral result
$\boldsymbol{I}_n$ = n-dimensional unit matrix
m = number of midpoint experiments
n = number of regressor variables
n_e = number of experiments
n_c = number of coefficients in the model
$\boldsymbol{n}(\boldsymbol{x})$ = orientation vector

$\mathbf{o}_n$ = n-dimensional zero vector
$\mathbf{o}_{n,m}$ = $n \times m$ matrix consisting of zeros
$\boldsymbol{P}$ = projection matrix
SS_R = sum of residual squares
SS_E = sum of pure error squares
SS_L = $SS_R - SS_E$
T = tangential space
Var[ ] = variance

$$\boldsymbol{x}^{(i)} = \begin{pmatrix} x_1^{(i)} \\ x_2^{(i)} \\ \vdots \\ x_n^{(i)} \end{pmatrix} = \text{i-th vector of regressor variables}$$

$\boldsymbol{X}$ = vector of random design variables

$$\boldsymbol{y} = \begin{pmatrix} y_1 \\ y_2 \\ \vdots \\ y_{n_e} \end{pmatrix} = \text{observed vector of response variables}$$

$\boldsymbol{Z}$ = standardized random variables

Δ = non-centrality parameter for lack of fit
ϵ_L = lack of fit terms
ϵ = pure error terms
Φ = standard Gaussian distribution function
$\kappa_1, \ldots, \kappa_{n-1}$ = main curvatures of limit state surface
λ = defined in equation 5.37
ν_E $= m - 1$
ν_L $= n_e - n_c - (m - 1)$
$\boldsymbol{\vartheta}$ = regression coefficients vector
ϑ_0 = constant regression coefficient
$\boldsymbol{\vartheta}_1$ = regression coefficients of the linear terms
Θ = regression coefficients of the quadratic terms
$\boldsymbol{\tau}$ = vector of parameters
σ^2 = model error variance

6

Stochastic Methods for Offshore Structures

R.S. Langley
University of Southampton, U.K.

S. McWilliam
University of Nottingham, U.K.

6.1 Introduction

An offshore structure is by definition surrounded by the ocean environment, and it must therefore be designed to withstand severe random loading arising from wind, waves, and current. For many years it has been recognized that a realistic description of the environment can only be achieved by stochastic modeling, and the basic principles of random vibration theory are now used routinely throughout the offshore industry. In more recent years it has been recognized that structural uncertainties can also have a significant effect on the system reliability, and there have been moves toward the development of a design procedure which incorporates both environmental and structural uncertainties. The need for cost-effective systems in ever-increasing water depths has lead to a requirement for improved design and analysis techniques which can produce structures having the minimum amount of redundancy subject to a minimum acceptable reliability. The aim of the present chapter is to provide an overview of the analysis techniques which are currently available, as well as to highlight a number of "specialist" stochastic methods which are applicable to certain types of structures.

A brief review of the more common types of offshore structure is given in Section 6.2, while the nature of the environmental loading is reviewed in Section 6.3. Methods for the stochastic response analysis of offshore structures are then presented in Section 6.4; this section provides an overview of the currently available methods. Details of specialist statistical techniques for (i) fixed offshore structures and (ii) large floating structures are given in Sections 6.5 and 6.6, respectively. Fatigue analysis is then described in Section 6.7, while in Section 6.8 a brief discussion of reliability theory (which incorporates structural uncertainties) is presented.

6.2 Types of Offshore Structures

Offshore structures are often classified as being either fixed or compliant; this classification is shown schematically in Figure 6.1. The description given here is necessarily brief; for more details the reader is referred to [1]. The aim of the present section is to give an overview of the range of offshore systems for which a stochastic dynamic analysis must be performed at the design stage.

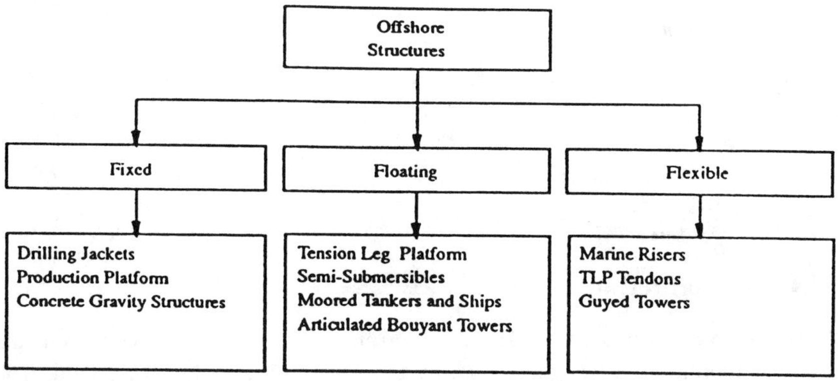

FIGURE 6.1
Schematic for the classification of offshore structures.

Fixed offshore structures are usually referred to as being piled or gravity type structures depending upon the way in which they are held in position. Of the examples given in Figure 6.1, the drilling jacket is an example of a piled structure (as shown in Figure 6.2) in which the steel framed tubular structure is connected to the sea-bed by piles. By the nature of its construction this type of structure has a large self-weight and will encounter large steady environmental forces arising from the wind and current and large cyclic loading arising from the waves.

An alternative to the piled type structure is the gravity type structure which uses concrete to create a large heavy construction which can rest on the sea-bed and remain stable under its own weight. A typical design for a gravity structure is shown in Figure 6.3 and consists of a concrete base with three columns supporting the upper structure. Although gravity type structures are expensive compared to pile type structures, they provide an alternative to jackets for use in extreme conditions,

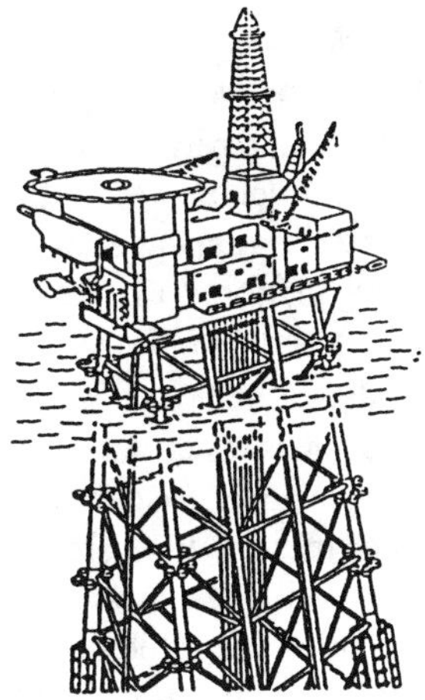

FIGURE 6.2
A typical drilling jacket (reproduced from [1] with permission from Patel).

where isolation of the structure makes it a primary requirement to have oil storage facilities on the platform.

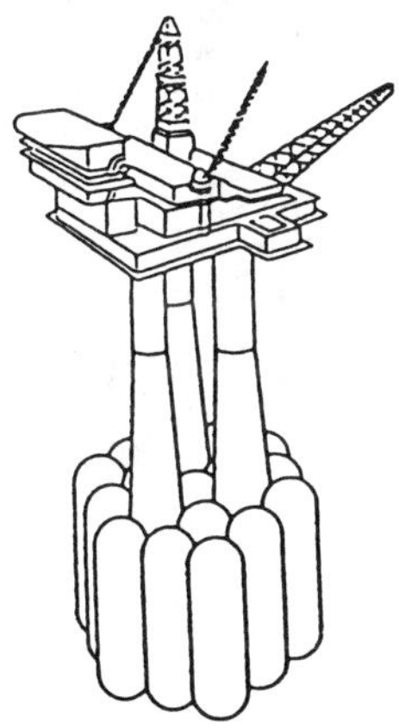

FIGURE 6.3
A typical gravity type structure (reproduced from [1] with permission from Patel).

The material and manufacturing costs of fixed structures increase dramatically with water depth. It is therefore not surprising that fixed installations are not well suited to the development of oil and gas fields in water depths greater than 500m. For this reason a new breed of offshore structure has been developed over the past decade which allows the operator to drill and produce at much greater depths at low cost. These new structures are designed to comply with the environmental forces and are usually referred to as compliant offshore structures.

Compliant offshore structures can be divided into two different categories, these being towers and freely-floating structures. Since the majority of existing compliant platforms are freely-floating, the remainder of this section considers this category, which includes tension leg platforms (TLP's), semisubmersible, and moored tankers (e.g., single anchor leg storage (SALS) systems).

The tension leg platform shown in Figure 6.4 consists of a surface platform tethered to the sea bed by vertical legs which are kept in tension by excess buoyancy in the platform. The tethers are tensioned to such an extent that the heave, pitch, and roll motions of the platform are virtually eliminated, and the platform is constrained to remain on station above the wells. An array of marine risers connects the oil wells to the surface, where the oil is processed before being exported along the sea bed. Due to the success of the first operational TLP, situated in the Hutton field (by Conoco Ltd. 1984) at a water depth of 148m, TLP's are being considered for many future developments, with the possibility of producing in water depths of up to 1500m.

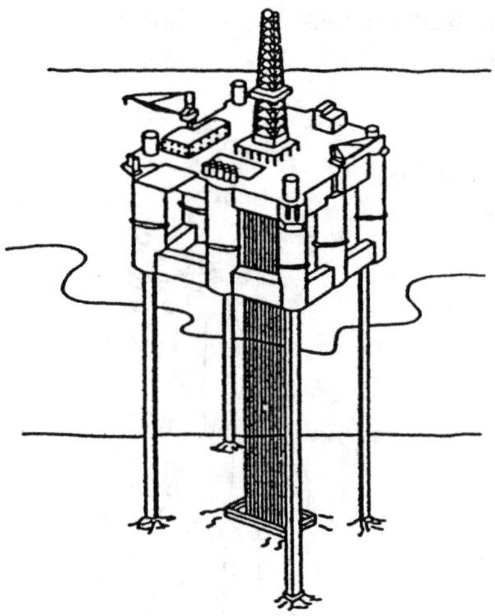

FIGURE 6.4
A typical tension leg platform (TLP) (reproduced from [1] with permission from Patel).

The semi-submersible shown in Figure 6.5 consists of a twin pontoon vessel which is catenary moored to the sea bed. The oil is transported from the wells to the surface either by flexible or tensioned multi-tube risers. Due to the relatively small payload, this type of structure cannot provide oil storage facilities and has limited oil processing equipment. Although the compliancy of the structure and the catenary mooring system mean that the vessel is liable to large mean offsets, running the risk of riser damage, these structures provide an economical means of exploiting small reservoirs of oil for water depths in the range 70-250m.

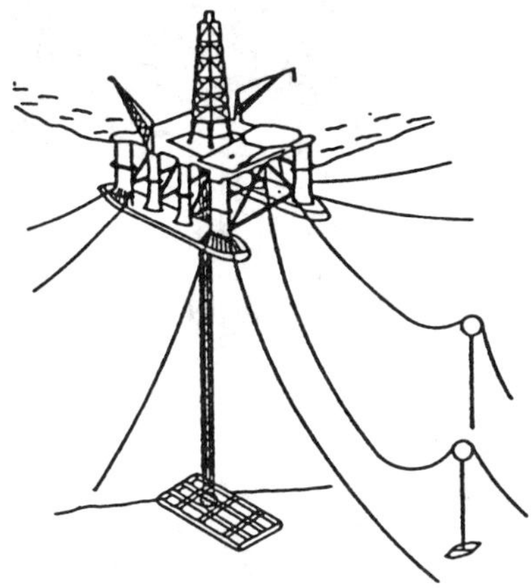

FIGURE 6.5
A typical semi-submersible vessel (reproduced from [1] with permission from Patel).

The single anchor leg storage system shown in Figure 6.6 uses a yoke structure, a buoyancy tank, and a tensioned riser connection to moor the tanker. The crude oil processing and storage facilities are accommodated within the vessel, with the oil being exported via a shuttle tanker or pipeline. Such systems are ideally suited to the development of small isolated fields in water depths of up to 170m.

The main difference between fixed and compliant offshore structures is that fixed structures respond quasi-statically to environmental forces, while compliant structures respond dynamically. As will be seen later, this difference means that distinct approaches are needed to analyze and design these structures. However, before considering these approaches in detail, it is necessary to consider the environmental loading experienced by offshore structures.

6.3 Environmental Loading

As mentioned in the introduction, offshore structures must be designed to withstand severe environmental forces arising from wind, waves, and current. The starting point of any structural reliability analysis must be an assessment of the nature of the applied environmental loading, and this loading is reviewed briefly in the present section. The specification of the environmental conditions is addressed in Section 6.3.1 below, while the forces arising from the interaction between the environment and a structure are described in Section 6.3.2.

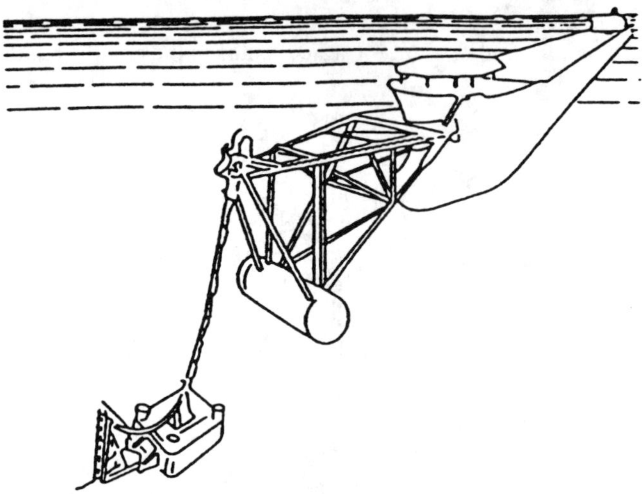

FIGURE 6.6
A typical single anchor leg storage (SALS) system (reproduced from [1] with permission from Patel).

6.3.1 The Offshore Environment

6.3.1.1 Waves

In reality, the offshore weather conditions are continually changing, and this is reflected in the changing severity of the wave environment: a full stochastic description of the sea surface elevation would therefore constitute a non-stationary random process. Such a description would lead to severe analytical complications; however, for design calculations it has been found that a "quasi-stationary" description of the wave environment is adequate. With this approach, it is assumed that the wave conditions over some duration (typically 3 hrs) can be regarded as a stationary random process, so that the true non-stationary environment is modeled as a succession of stationary intervals. This approach can be expected to yield accurate response predictions if the rate of change of the environmental conditions is slow when compared to (i) the natural periods of the structure, and (ii) the mean excitation period. Langley [2] has shown that these conditions are generally satisfied for offshore structures. The "stationary" wave conditions over some finite interval are normally described in terms of the spectrum $S_{\eta\eta}(\omega)$ of the surface elevation $\eta(t)$ at some particular location in the vicinity of the offshore structure. Many empirical forms for $S_{\eta\eta}(\omega)$ have been developed based on environmental measurements, and these have been summarized by Sarpkaya and Isaacson [3]. As an example, the International Ship and Offshore Structures Congress (ISSC) have recommended the following empirical formula for open sea conditions:

$$S_{\eta\eta}(\omega) = A\omega^{-5}\exp(-B\omega^{-4}), \tag{6.1}$$

where

$$A = 173H_s^2/T_1^4, \tag{6.2}$$

$$B = 691/T_1^4. \tag{6.3}$$

Here H_s is the significant wave height, which is defined as the mean height of the highest one-third waves, and T_1 is a measure of the average wave period given by $T_1 = 2\pi m_0/m_1$, where m_n is the nth spectral moment. Most other forms of empirical wave spectra are also dependent on the two parameters H_s and T_1, and together these parameters give a measure of the "severity" of the seastate. The likely occurrence of different combinations of H_s and T_1 over the design life of a structure is normally represented on a "wave scatter diagram", which essentially gives a numerical approximation to the joint probability density function (jpdf) $p(H_s, T_1)$ based on environmental measurements [3]. In a sense, this type of data completes the "non-stationary" description of the wave environment, in that it gives the long term distribution of the limited duration "stationary" conditions.

Although much information regarding the surface elevation is contained within the spectrum, this description is not statistically complete unless it is assumed that the surface elevation is Gaussian. If it is then assumed that linear wave theory is valid and that the seastate is composed of unidirectional long crested waves, then the statistical properties of the fluid motion at any point can be fully determined by making use of the transfer functions arising from linear wave theory. The assumption regarding unidirectional waves can be relaxed by the introduction of a "spreading function" $S(\vartheta)$ whereby the energy contained in $S_{\eta\eta}(\omega)$ is distributed amongst the wave headings ϑ. The "directional" wave spectrum $S_{\eta\eta}(\omega, \vartheta)$, representing the energy for waves of heading ϑ, is then written in the form

$$S_{\eta\eta}(\omega, \vartheta) = S_{\eta\eta}(\omega)S(\vartheta). \tag{6.4}$$

A number of semi-empirical forms have been proposed for $S(\vartheta)$ (for example, a cosine squared function) and full details of these have been given by Sarpkaya and Isaacson [3]. If equation (6.4) is accepted, and linear wave theory is taken to apply, then the statistical properties of the fluid motion at any point may be determined from $S_{\eta\eta}(\omega, \vartheta)$ by making use of the known properties of a linear wave of frequency ω and heading ϑ. In particular, the surface elevation (η) and the horizontal (u) and vertical (v) fluid velocities associated with a wave of amplitude a and frequency ω can be written in the form

$$\eta = a\cos(kx - \omega t), \tag{6.5}$$

$$u = a\omega \frac{\cosh k(d-z)}{\sinh(kd)} \cos(kx - \omega t), \tag{6.6}$$

$$v = \frac{a\omega \sinh k(d-z)}{\sinh(kd)} \sin(kx - \omega t), \tag{6.7}$$

where x is a horizontal coordinate measured in the direction of wave propagation, z is a vertical coordinate measured positive down from the mean sea level, and d is the water depth. Furthermore, k is the wavenumber, which is given by the solution to the dispersion relation $\omega^2(gk) = \tanh(kd)$. The requisite transfer functions between the fluid motions and the surface elevation may readily be deduced from equations (6.5-6.7).

As mentioned above, the completeness of the description afforded by the wave spectrum is dependent upon the assumption that (i) the surface elevation is Gaussian, and (ii) linear wave theory is valid. The validity of assumption (i) is often argued on the basis of the central limit theorem, and environmental measurements tend to confirm this assumption, at least for deep water. In shallow water the wave height may be an appreciable fraction of the water depth, and in this case neither assumption (i) or (ii) is likely to be satisfied. In this case a perturbation solution to the hydrodynamic equations may be sought, and the statistics of the surface elevation derived on the *a priori* assumption that the "first order" surface elevation is Gaussian — a second order analysis of this type has been presented by Longuet-Higgins [4]. For very shallow water the wave motion may be highly non-linear, and in this case the only option may be to perform a deterministic analysis based on a non-linear wave theory; this aspect is considered further in Section 6.4.5.

6.3.1.2 Current

Currents may arise from wind action, tides, and a number of secondary causes including non-linear wave effects, local density variations, and ocean circulation [5]. Normally the wind induced current velocity is assumed to have a linear profile ranging from U_w on the surface to zero at a depth of 50m (and below). Typically $U_w$is around 2% of the hourly mean wind velocity at 10m above sea level (U_{10}). The tidal current is normally taken to have a uniform velocity U_t down to 10m above the sea bed; below this point the velocity is taken to be $U_t \log[1+9(y-d)/(10-d)]$, where d is the water depth and y is the distance above the seabed. Clearly, U_w and U_t will vary with time in both magnitude and direction, and environmental measurements of U=U_w+U_t at various depths are normally presented in the form of "frequency of occurrence" tables which effectively give a numerical estimate of the jpdf $p(U,\vartheta)$, where ϑ is the current direction.

6.3.1.3 Wind

The horizontal component of the wind velocity consists of a steady component and a fluctuating (gust) component. The steady component is normally specified in terms of hourly mean wind speed at a height of 10m above sea level, U_{10}; as in the case of current, environmental measurements of the hourly mean wind speed can be presented in the form of a frequency of occurrence table which yields a numerical estimate of $p(U_{10},\vartheta)$. The fluctuating component of the wind velocity can be represented by a velocity spectrum, and empirical models include the Harris Spectrum [6] and the Ochi-Shin Spectrum [7]. These spectra normally include U_{10} as a parameter, so that the properties of the fluctuating component are dependent on the magnitude of the steady component. The forces arising from the fluctuating wind component are normally small when compared to the wave forces acting on a structure; however, in some cases the wind forces may excite low frequency resonant motions, and this possibility should be borne in mind when considering the dynamic response of floating structures (see for example Kareem [8]).

6.3.2 Environmental Forces

6.3.2.1 Wave Forces

The nature of the wave forces which act on an offshore structure is strongly dependent on the size of the structure in relation to the wavelength of the incident waves. If D represents a typical dimension of the structure and λ is the incident wavelength, then for $D/\lambda>0.2$ the wave forces can generally be calculated on the basis of potential flow theory, as described in (a) below. For $D/\lambda<0.2$, viscous effects and flow separation can be important and a semi-empirical fluid loading model is generally required, as discussed in (b) below.

a. $D/\lambda>0.2$: Potential Flow Theory. If the structure is sufficiently large in relation to the fluid wavelength then the fluid motion can be predicted by using potential flow theory, in which it is assumed that the flow is inviscid and irrotational. The problem is then to find the velocity potential ϕ when an incident regular wave impinges on the structure, given that this potential satisfies Laplace's equation and also that the flow must satisfy the boundary conditions on the seabed, the free surface, the surface of the structure, and at infinity. Offshore structures (including ships) usually have a complex geometry, and it is generally not possible to develop a closed form solution for ϕ. Rather, numerical methods must be employed, and the most commonly used method is the boundary element (or "source-sink") method (see for example, van Oortmerssen [9], Faltinsen [5]).

This approach is based on Green's Theorem, which in the present context states that for harmonic motion the fluid potential may be expressed in the form

$$\phi(x, y, z) = (1/4\pi) \int_S f(\alpha, \beta, \gamma) G(x, y, z; \alpha, \beta, \gamma) ds. \tag{6.8}$$

Here (x,y,z) is a point within the fluid while (α,β,γ) is a point on the submerged surface of the offshore structure. Furthermore, G is a known "Green's function" which satisfies Laplace's equation together with the boundary conditions on the seabed, the free surface, and at infinity. The integral which appears in equation (6.8) is taken over the submerged surface S of the structure, and the function $f(\alpha, \beta, \gamma)$ represents an unknown "source" distribution which must be found in order to determine the velocity potential. An integral equation for f can be established by applying the body surface boundary conditions, which yields

$$u_n(x, y, z) = -f(x, y, z)/2 + (1/4\pi) \int_{S'} f(\alpha, \beta, \gamma) \frac{\partial}{\partial n_x} G(x, y, z; \alpha, \beta, \gamma) ds, \tag{6.9}$$

for $(x,y,z) \in S$. Here u_n $(=\partial\phi/\partial n)$ is the normal velocity on the body surface, and the derivative of the Green's function which appears inside the integral is taken along the normal with respect to the (x,y,z) coordinates. The first term which appears on the right of equation (6.9) arises from a singularity in the integrand which occurs when $(x,y,z) = (\alpha,\beta,\gamma)$; the integration surface is written as S' to indicate that this singularity has been removed from the integral. If the normal velocity on the body surface u_n is specified, then in principle equation (6.9) may be solved to yield the source distribution $f(\alpha,\beta,\gamma)$; in practice this is often achieved by dividing the body surface into a finite number of panels and assuming that the function f is constant over each panel — the application of equation (6.9) on each panel then yields a set of linear algebraic equations which can be solved to yield the source strengths. Further developments in this area include the use of higher order shape functions to describe the source distribution over each panel (see for example Hearn and Donati [10]).

In the analysis of a rigid floating structure, it is convenient to express the total fluid potential in the form

$$\phi = \sum_{j=0}^{7} \phi_j = \sum_{j=0}^{7} i\omega a_j \psi_j. \tag{6.10}$$

Here ϕ_0 is the potential of the incident wave, ϕ_7 is the potential of the diffracted wave (i.e., the disturbance potential which would arise were the structure held stationary), and ϕ_1 to ϕ_6 are the potentials arising from motion in the six rigid body degrees of freedom; furthermore, the terms a_1 to a_6 represent the complex amplitudes of the body motions, while a_0 and a_7 represent the amplitudes of the incident and diffracted waves (nominally, a_7 can be placed equal to a_0). The potentials ψ_j which appear in equation (6.10) are essentially the potentials ϕ_j scaled to a unit velocity amplitude (the time dependency $\exp(i\omega t)$ has been implicitly assumed). If the structure is flexible, then the sum in equation (6.10) can be extended to include a number of generalised coordinates.

The potential ψ_0 which appears in equation (6.10) is known from linear wave theory; the potentials ψ_1 to ψ_6 can be calculated from equations (6.8) and (6.9) by employing the appropriate normal velocity on the left side of equation (6.9) — for example, to calculate ψ_1 the normal velocity corresponding to a unit velocity surge motion ($a_1 \equiv$ surge) is used in equation (6.9), which can then be solved to yield the "surge" source distribution f_1. The use of f_1 in equation (6.8) then yields ψ_1. The diffraction potential ψ_7 can be calculated in a similar way — in this case the normal velocity u_n is chosen so that the incident and diffracted waves sum to give zero normal velocity on the body surface. It can be noted that only the diffraction potential is actually dependent on the incident wave; the potentials ψ_1 to ψ_6 relate to the fluid motions produced by body oscillations in otherwise still water. Once the fluid potential ϕ is known, the fluid pressure follows from the relation $p = -\rho\partial\phi/\partial t = -i\omega\rho\phi$ where ρ is the fluid density. The fluid excitation force acting in degree of freedom j can then be expressed in the form

$$F_j = -\int_S pn_j ds = \sum_{k=0}^{7} T_{jk}a_k, \tag{6.11}$$

$$T_{jk} = -\rho\omega^2 \int_S \psi_k n_j ds, \tag{6.12}$$

where n_j is the component of velocity along the local surface normal produced by a unit velocity in degree of freedom j. Given the excitation force in the form of equation (6.11), the equations of motion of the structure can be written as

$$\sum_{k=1}^{6} \left\{ -\omega^2(M_{jk} + A_{jk})a_k + i\omega(C_{jk} + B_{jk})a_k + K_{jk}a_k \right\} = T_{j0}a_0 + T_{j7}a_7, \tag{6.13}$$

where

$$A_{jk} = (1/\omega^2)Re(T_{jk}), \tag{6.14}$$

$$B_{jk} = (-1/\omega)Im(T_{jk}). \tag{6.15}$$

Here j = 1,..,6, and the terms M_{jk}, C_{jk}, and K_{jk} represent the mass, damping, and stiffness matrices of the structure in the absence of any dynamic fluid effects. The motion dependent fluid induced terms A_{jk} and B_{jk} which appear on the left of equation (6.13) are known, respectively, as the added mass and potential damping matrices. The first term on the right of equation (6.13) is the fluid force arising from the undisturbed incident wave potential, while the second term is the force due to the diffraction potential; a well known formula known as Haskind's Relation states that $T_{j7} = -T_{0j}$, which in practice means that it is unnecessary to calculate ψ_7, at least as far as equation (6.13) is concerned.

To summarize this section, equation (6.13) represents the equations of motion of an offshore structure which is subjected to an incident linear wave. The various terms which appear in this equation require knowledge of the tensor T_{jk}, and this, via equation (6.12), depends upon the potentials ψ_j. The potentials ψ_1 to ψ_6 (and ψ_7 if required) can be calculated from a numerical scheme based on equations (6.8) and (6.9). Ultimately, equation (6.13) yields the transfer functions between the structural response and the sea surface elevation, and as discussed in Section 6.4.5, these can be used in conjunction with the statistical properties of the seastate to yield the statistics of the response.

The analysis presented in this section has assumed linearity, both in terms of the fluid loading and the structural properties. In some cases nonlinear second-order fluid loading can have a very significant effect on the response, and this is discussed in some detail in Section 6.6. The effect of structural non-linearities is discussed in Section 6.4.5.

b. D/λ<0.2: Semi-Empirical Loading Formulae. If a structure is small in relation to the wavelength of the incident waves, then flow separation and viscosity can have an important effect on the fluid loading. For a fixed cylinder of diameter D which is subjected to flow of velocity u normal to the cylinder axis, the force (per unit cylinder length) in the flow direction is normally expressed in terms of the following three components:

$$F_1 = (\pi/4)\rho D^2 \dot{u}, \tag{6.16}$$

$$F_2 = (\pi/4)\rho D^2 C_a \dot{u}, \tag{6.17}$$

$$F_3 = (1/2)\rho C_d D u \mid u \mid . \tag{6.18}$$

The force F_1 (known as the Froude-Krylov force) is equal to the fluid acceleration multiplied by the mass of the fluid displaced, and is essentially the force which would be experienced by a cylinder of fluid having the same dimensions as the structure. The fact that the cylinder has rigid walls means that, unlike an equivalent cylinder of fluid, the structure disturbs the fluid flow, and F_2 is the resulting "diffraction force". The coefficient C_a which appears in F_2 is known as the added mass coefficient, for reasons which will become apparent. While F_1 and F_2 account mainly for potential flow effects, the final force component F_3 is the "drag force" arising from flow separation and other effects. If the cylinder itself moves with velocity x, then the total fluid force per unit length is generally written in the form

$$F = (\pi/4)\rho D^2 C_m \dot{u} + (1/2)\rho C_d D(u - \dot{x}) \mid u - \dot{x} \mid -(\pi/4)\rho D^2 C_a \ddot{x}, \tag{6.19}$$

where $C_m = 1 + C_a$ is known as the inertia coefficient. Equation (6.19) has been derived by using the relative velocity in F_3 and the relative acceleration in F_2; the use of the terminology "added mass coefficient" for C_a is now clear, since the final term in equation (6.19) has the effect of "adding" a mass to the structural mass of the cylinder.

Equation (6.19) is known as Morison's equation [11] and it is appropriate to note that Newman [12] states "the justification for Morison's formula is strictly pragmatic and must rest with experimental confirmation". A vast literature exists regarding the appropriate choice of the empirical coefficients C_m and C_d, and much of the available data have been reviewed by Sarpkaya and Isaacson [3]. The coefficients are generally found to depend upon: (i) the Reynolds number, $Re = u_{max}D/\nu$, where u_{max} is the maximum fluid velocity during the wave cycle and ν is the kinematic viscosity; (ii) the Keulegan-Carpenter number, $KC = u_{max}T/D$, where T is the wave period; (iii) the surface roughness of the structure. Typical values for C_m are around 2, while C_d typically ranges between 0.6 and 1.4. For $10 < KC < 20$ great uncertainty exists, and in recent years there have been numerous attempts to replace the Morison equation with a more robust alternative; as yet however, no such alternative has received general acceptance.

Vortex shedding around an offshore structure can also lead to lateral forces (i.e., forces perpendicular to the flow direction); these forces can have a significant effect on cables, risers, and mooring lines in the sense that strumming oscillations can be induced. A detailed discussion of such effects has been given by Sarpkaya and Isaacson [3] and Faltinsen [5]. The latter reference also contains a description of wave slam loads.

One important aspect of the Morison equation from a random vibration point of view is that the drag force is a non-linear function of the fluid velocity. This means that standard spectral analysis procedures based on response transfer functions cannot be applied, and some alternative approach is required. One such approach is to replace the drag force with an "equivalent" linear drag force in such a way as to minimise the resultant error, and this is discussed in the following sub-section. With this approach it is possible to perform an approximate spectral analysis based on linearized transfer functions, as discussed in Section 6.4.5.

c. *Drag Force Linearization.* Various schemes exist whereby the drag force F_3 which appears in equation (6.18) can be replaced by an equivalent linear drag force. For one dimensional random flow u in the presence of a current V_c, the following linearization has been suggested by Wu and Tung [13]:

$$(1/2)\rho C_d D(u + V_c) \mid u + V_c \mid \rightarrow (1/2)\rho C_d D(au + b), \tag{6.20}$$

where

$$a = \sqrt{8/\pi} \left\{ \sigma_u \exp[-(V_c/\sigma_u)^2/2] + \sqrt{2\pi} V_c erf(V_c/\sigma_u) \right\}, \tag{6.21}$$

$$b = \sqrt{2/\pi}\left\{V_c\sigma_u \exp[-(V_c/\sigma_u)^2/2] + \sqrt{2\pi}(\sigma_u^2 + V_c^2)erf(V_c/\sigma_u)\right\}. \quad (6.22)$$

Here $\mathrm{erf}(x) = (1/\sqrt{2\pi})\int_0^x \exp(-y^2/2)dy$, and σ_u is the standard deviation of u. Equations (6.20-6.22) are based on the "equivalent linearization method" in which the coefficients a and b are chosen to minimise the mean squared difference between the original drag force and the linearized version. In order to apply this method, it is necessary to assume certain statistical properties of u, and the above results are based on the assumption that u is Gaussian — if u is the *relative* motion between the structure and the fluid, then this assumption is clearly an approximation, as the system response will not be Gaussian in the presence of the non-linear drag force. It can be noted that for zero current equations (6.21) and (6.22) reduce to $a = \sigma_u\sqrt{(8/\pi)}$ and $b = 0$; this version of the linearized drag force is often referred to as Borgman's approximation.

If the environment is modeled by a regular wave, rather than a random seastate, then an alternative drag linearization scheme is available [14]. In this case the coefficients a and b are expressed in terms of the amplitude and phase of u, and equation (6.20) effectively contains the first two terms in a Fourier series expansion of the drag force.

If the flow direction is not normal to the axis of the cylindrical member, then strictly a three-dimensional model of the drag force is required. The linearization of three-dimensional drag forces has been considered by Hamilton [15] for regular waves and Langley [16] for random waves.

One disadvantage of drag linearization is that the higher frequency content of the drag force is neglected; for example, the non-linear drag force has a frequency component at 3ω (where ω is the frequency of u) which could potentially excite a structural resonance. Less approximate treatments of the drag force have been proposed which are capable of capturing this kind of effect (Eatock Taylor and Rajagopalan [17], Koliopoulis [18], Eatock Taylor [19]).

6.3.2.2 Current and Wind Forces

The current and wind forces which act on an offshore structure must generally be calculated by semi-empirical means. A significant data-base exists regarding current and wind forces on oil tankers [5], and "drag" and "lift" type formulae can be applied to other classes of structure. It can be noted that a steady current can produce a significant *dynamic* response due to vortex shedding and instabilities arising from fluid/structure interaction; the "fishtailing" motion of a Single Point Mooring (SPM) is a good example of this [20].

6.4 Stochastic Response Analysis

6.4.1 Overview

In order to perform a reliability analysis of an offshore structure it is necessary to select:

(1) the environmental model,

(2) the wave force model,

(3) the structural model, and

(4) the analytical solution technique.

The various options available to the designer are summarized in the following sub-sections.

6.4.2 The Environmental Model

The main issue to be decided regarding the environmental model concerns the representation of the wave conditions. The following options are available.

a. The Design Wave Approach. In this approach the environment is modeled by a single regular wave of "design" amplitude — this wave is intended to represent the worst case which can reasonably be expected over the design life of the structure. Such waves are normally specified in terms of the "return period": the 100-year wave, for example, is a wave whose amplitude will be exceeded on average only once in every 100 years — for the North Sea the 100-year wave has an amplitude of around 15m. The calculation of the amplitude of the design wave is based on long-term wave records. If $P(H)$ represents the long term probability distribution function of wave heights H, and it is estimated that there are around m waves per year, then the n-year wave height is given by the solution of the equation $P(H) = 1 - 1/nm$. Assuming that the wave heights are statistically independent, the probability that the design wave will be exceeded is approximately 1-exp($-r/n$) in any r-year period regardless of the detailed form of $P(H)$.

The design wave approach has the advantage of simplicity, and furthermore a highly non-linear wave theory may be used to describe the design wave if necessary. The disadvantage is that the approach does not fully account for the range of frequencies and wavelengths which occur in a random seastate; the design wave may "miss" a structural resonance, for example, or the wavelength may not yield the most damaging distribution of fluid velocities and accelerations. These effects can be accounted for to some extent by considering a range of design waves of equal height but varying steepness ratio H/λ, in an effort to determine the steepness ratio which produces the worst response. Generally a range of design wave headings ϑ will also need to be considered to ensure that allowance has been made for the least favorable case. Even when parameter studies are performed over the steepness ratio H/λ and wave heading ϑ, there is no guarantee that the design wave approach will accurately model the maximum response produced by a random seastate. For this reason a random wave model is often preferred, as explained in the following sub-sections.

b. The Design Storm Approach. The design storm approach replaces the "design wave" with a stationary random seastate of maximum expected severity. Accordingly, the n-year storm is defined as the storm whose significant wave height H_s will be exceeded on average once every n years — for the North sea the 100-year storm has a significant wave height of around 15m. In principle the severity of the n-year storm can be calculated from the condition $P(H_s) = 1 - 1/nm$, where m is the average number of storms in a year; however, there are certain difficulties associated with this approach as pointed out by Battjes [21]. Whereas it is possible to identify a single discrete wave, and thus estimate the number of waves m in a year, the identification of a single storm is much more problematical. As mentioned in Section 6.3, the offshore environment is in reality non-stationary, and the division of the environmental conditions into a series of stationary increments is an artificial process. Normally it is assumed that each set of conditions lasts for around 3 hrs, giving $m = 2920$ "storms" in each year. As highlighted by Battjes [21], any change in the assumed storm duration can arbitrarily change the severity of the n-year design storm, and thus some care is needed in the statistical interpretation of this approach.

Various mean headings ϑ of the design storm can be investigated to determine the least favorable case; as discussed in Section 6.3.1 the storm can be considered to be either uni-directional or "spreading" — the latter model has a particularly important effect for structures which are far from rotationally symmetric, such as a moored tanker. The design storm approach has the advantage of catering for the range of frequencies and wavelengths which can occur in a random seastate. The disadvantage is that in some cases it may not be the most severe seastate which causes the most severe dynamic response; the mean period of the seastate tends to increase with increasing H_s, and it is possible that a lower seastate may cause the largest response in a particular resonant mode. Second-order motions are particularly sensitive to the frequency content of the seastate, and in this case the maximum response is unlikely to occur in the design storm. In such cases the "long-term" approach outlined below is more appropriate.

c. Long Term Environmental Modeling. With this approach the wave conditions are modeled as a succession of stationary storms, with each storm being described by the significant wave height and the average wave period (H_s, T_1). The long term distribution of the seastates $p(H_s, T_1)$ is then estimated from the wave scatter diagram to build up a consistent picture of the environmental conditions over the design life of the structure. This is then used in conjunction with a response analysis (performed for each stationary storm) to yield the lifetime reliability of the structure, as explained in Section 6.4.5 below.

6.4.3 The Wave Force Model

As discussed in Section 6.3.2, the wave force model adopted may be based upon either potential flow theory or the semi-empirical Morison's equation (or, in fact, a combination of the two approaches). If potential flow theory is used, then significant computation time may be required to calculate the wave excitation forces; if Morison's equation is used, then non-linearity will be introduced into the analysis via the presence of the drag force. It is also important to decide whether second-order wave forces are likely to induce a significant response; if so, then these forces will need to be computed by using potential flow theory (see Section 6.6 below).

6.4.4 The Structural Model

The structural model employed will depend to a large extent upon the location of the structural resonances relative to the wave excitation frequencies. In this regard it can be recalled from Section 6.3 that the presence of the surrounding fluid can lead to a significant "added mass" effect; the fact that this can cause a significant reduction in the natural frequencies relative to "in vacuo" conditions should be taken into account when considering the choice of the structural model. The various types of structural model which can be adopted are summarized below.

a. Static Model. If the structural natural frequencies are all well above the wave excitation frequency range then it is unnecessary to consider structural dynamic effects. In this case a "static" model can be adopted in which structural inertia and damping effects are neglected. This situation generally arises for jacket structures, although in deep water the lower "cantilever" type modes of the structure may call for dynamic modeling. Jacket structures are almost invariably modeled by using the finite element method; generally beam elements will yield a sufficiently accurate estimate of the overall stress levels, while a complex shell element model may be required for the detailed stressing of joints.

b. Rigid Body Model for Floating Structures. For moored structures such as semi-submersibles, production tankers, or tension leg platforms, it is likely that the modes of vibration which involve elastic deformation will lie well above the six "rigid body" modes corresponding to surge, sway, heave, roll, pitch, and yaw. *If* this is the case then the dynamic response can be computed by employing a six degree of freedom rigid body model, and subsequently the detailed stresses can be obtained by using finite element techniques in a quasi-static mode.

c. Structural Dynamic Model. If the flexible modes of the structure lie sufficiently close to the excitation frequency range, then the structural model must allow for dynamic elastic deformation. Typically the structure would in this case be modeled by using the finite element method, and the dynamic response calculation would generally be performed by using a reduced basis of "modal" (or principal) coordinates. If the structure is large in relation to the incident wavelength, then special care is needed in the calculation of the added mass and potential damping terms associated with the flexible degrees of freedom — such terms can be computed by using the source-sink numerical method, as described in Section 6.3. For flexible ship-type structures there is an extensive established literature regarding hydroelasticity, which owes much to the work of Bishop and Price [22].

d. Structural Non-Linearities. It is not unusual for the equations of motion of an offshore structure to contain non-linearities arising from structural effects. For example, semisubmersibles and oil tankers are often moored by catenary anchor chains; such chains have a highly non-linear restoring force characteristic for motion in the horizontal plane. In this case it may be possible to perform a nonlinear static analysis to obtain the mean position of the vessel, following which a linear small motion dynamic response analysis could be applied. If, however, the vessel is susceptible to second order forcing then large amplitude dynamic motions can occur, and a simple linear restoring force model may not be appropriate. Another example of structural non-linearity arises in the dynamic analysis of a flexible riser (essentially a flexible hose which forms part of pipework carrying oil from the seabed to the sea surface). Wave forces can cause gross changes in the riser geometry, and thus a large displacement dynamic analysis is required, which is usually implemented via a non-linear finite element model. As discussed in Section 6.4.5 below, the presence of structural non-linearities can cause significant analytical complications when a stochastic environmental model is adopted.

e. Structural Uncertainties. Attention thus far has focused on the random nature of the offshore environment and the resulting environmental forces. In practice there may also be uncertainty regarding the precise properties of the offshore structure itself — examples of potentially uncertain quantities include: the all-up mass of an offshore rig; the geometry and pre-tension in a spread mooring system; the extent of marine growth on the submerged portion of a structure; the pre-stress in a complex structural joint; the yield stress of the structural material. If the system response is sensitive to any one of these parameters then some allowance for structural uncertainty should be incorporated within the analysis procedure.

6.4.5 The Analytical Solution Technique

The various options which are available regarding the environmental model, the wave force model, and the structural model are summarized in Table 6.1.

In this Table the symbols RL and RN represent linear and non-linear regular wave models, S represents a random (spectral) wave model, and L and N represent linear and non-linear wave force and structural models. With this notation the type of analysis adopted can be represented by three symbols, so that for example S-L-N is a random analysis involving a linear wave force model and a non-linear structural model. The solution technique adopted will depend upon the chosen combination of options, and the various possibilities are summarized in what follows.

a. Regular Wave Analysis (RL–, RN–). A regular wave model presents the least difficulty regarding the solution of the equations of motion. If the adopted model is fully linear, then a simple frequency domain analysis may be performed

Table 6.1 Summary of the environmental, wave force, and structural models available.

Environment	Forces	Structure
Regular linear wave (RL)	Potential flow (L)	Linear (L)
Regular non-linear wave (RN)	Morison's equation (N)	Non-linear (N)
Wave spectrum (S)		Random properties (U)
Long-term random (S)		

to yield the structural response. If non-linearities are present in the wave loading or the structure, then an approximate "linearized" analysis could be performed — typically the non-linear terms would be expressed as a Fourier series, and only the mean and first harmonic term would be retained. This form of solution is necessarily iterative, in the sense that the coefficients of the Fourier terms are dependent on the system response, which is in turn dependent on the Fourier coefficients; this technique has been applied extensively to marine riser analysis (see for example Kirk [23]). If a non-linear wave theory is adopted, or if the linearized analysis is thought to be too approximate, then the equations of motion may be solved by numerical integration in the time domain. This yields a time history of the response from which the maximum values can readily be identified.

b. Linear Stochastic Analysis (S-L-L). In this case standard spectral analysis techniques can be applied. Thus if $H_{x\eta}(i\omega,\vartheta)$ represents the (complex) amplitude of a response quantity x produced by a regular wave of unit amplitude, frequency ω, and heading ϑ, then the spectrum of x can be expressed in the form

$$S_{xx}(\omega,\vartheta) = \mid H_{x\eta}(i\omega,\vartheta) \mid^2 S_{\eta\eta}(\omega)S(\vartheta), \tag{6.23}$$

where $S_{\eta\eta}(\omega)S(\vartheta)$ is the wave spectrum for spreading seas, as defined in Section 6.3. The mean squared values of x and the time derivative $\dot{x}$ are then given by

$$\sigma_x^2 = \int_0^\infty \int_0^{2\pi} S_{xx}(\omega,\vartheta)d\omega d\vartheta, \tag{6.24}$$

$$\sigma_{\dot{x}}^2 = \int_0^\infty \int_0^{2\pi} \omega^2 S_{xx}(\omega,\vartheta)d\omega d\vartheta. \tag{6.25}$$

The probability that x will exceed some critical level b within the storm duration T can be estimated by assuming that successive crossings of the level b are independent and constitute a Poisson process. If the seastate is assumed to be Gaussian then the response quantity x will also be Gaussian, and in this case the Poisson assumption yields the following well known result for the probability of exceedance $P(b,T)$ [24]:

$$P(b, T) = 1 - \exp(-\nu_b^+ T), \tag{6.26}$$

$$\nu_b^+ = (\sigma_{\dot{x}}/2\pi\sigma_x)\exp(-b^2/2\sigma_x^2). \tag{6.27}$$

Here $\nu_b{}^+$ is the mean rate at which x crosses the level b with positive velocity; the value given by equation (6.27) corresponds to the specific seastate used in equation (6.23) — as discussed in Section 6.3, the wave spectrum is dependent on the significant wave height H_s and the mean period T_1, and thus the crossing rate given by equation (6.27) can be written in the form $\nu_b{}^+(H_s,T_1)$.

If the seastate used in equation (6.23) corresponds to the "design storm" then the probability of a structural "failure" given by equation (6.26) represents the final result of a "design storm" analysis. This result can be extended to the case of "long term" environmental modeling by computing $\nu_b{}^+$ for a range of values of H_s and T_1: the long term probability of failure is then given by (Naess [25], Langley [26], Inglis et al. [27]):

$$P(b, T_L) = 1 - \exp\left\{-T_L \int_0^\infty \int_0^\infty \nu_b^+(H_s, T_1)p(H_s, T_1)dH_s dT_1\right\}, \tag{6.28}$$

where T_L is the design life of the structure and $p(H_s,T_1)$ is the long term jpdf of H_s and T_1, as defined in Section 6.3.

The Poisson process assumption which lies behind equations (6.26) and (6.28) is known to yield a conservative estimate of the probability of failure, particularly for narrow band processes and relatively small values of b ($b/\sigma_x<3$ say). If required, a number of more refined prediction methods are available which relax the Poisson process assumption (Roberts [28], Langley [29], Naess [30]).

c. Non-Linear Stochastic Analysis (S-L-N, S-N-L, S-N-N). The presence of non-linearities in the wave loading or the structural model leads to severe analytical difficulties, and in the majority of cases there is no exact theoretical approach whereby the response statistics may be derived. The various approximate/numerical techniques which are commonly applied in this situation are as follows.

(i) Linearized Spectral Analysis. With this approach the non-linear terms are replaced by "equivalent" linear terms in such a way as to minimise the mean squared error involved in the replacement. This technique, which is known as the equivalent linearization method, has been outlined for drag forces in Section 6.3. The method requires an assumption to be made regarding the statistics of the system response — when the response is assumed to be Gaussian (which is usually the case) the method is also known as "Gaussian closure" (a comprehensive review of the method has been given by Spanos [31]). With the assumption of a Gaussian response, the coefficients of the linearized terms are dependent on the mean values and the covariances of the response, which implies that an iterative solution is required: the mean values and covariances are given initial assumed values, which enables the linearized transfer functions of the system to be found — a spectral analysis is then performed which allows the mean values and covariances to be recalculated, and the process is repeated until convergence is obtained. Equations (6.26-6.28) can then be used to *estimate* the probability of failure, although it should be noted that the presence of non-linearities can lead to a crossing rate $\nu_b{}^+$ which differs significantly from the Gaussian formula. Linearized spectral analysis is perhaps most useful at the parameter study stage of initial design, where it can provide a rapid assessment of the relative merits of competing design concepts. The method has been applied to marine risers by Kirk [23] amongst others, and to a non-linearly moored vessel subjected to second order forces by Langley [32].

(ii) Time Domain Simulation. With this approach the wave spectrum is used to produce a time history of the random fluid velocities and accelerations. One way of doing this is to write the item of interest, say the fluid velocity $u(t)$ as some location, in the form

$$u(t) = Re\sum_{j=0}^{N}(a_j + ib_j)\exp[it(j\Delta\omega)], \tag{6.29}$$

where $\Delta\omega$ is a small frequency increment and a_j and b_j are Gaussian random variables of zero mean with

$$E[a_j^2] = E[b_j^2] = S_{uu}(j\Delta\omega)\Delta\omega, \tag{6.30}$$

where $S_{uu}(\omega)$ is the spectrum of u. In practice the random variables a_j and b_j would be evaluated for the surface elevation at some particular location, and transfer functions would then be used to calculate the coefficients associated with all other items of interest. Equation (6.29) can be evaluated efficiently over a fixed time span by using the Fast Fourier Transform (FFT) algorithm (Newland [33]). Once the random fluid environment has been established then the equations of motion of the structure can be integrated numerically to yield the response time histories; signal processing techniques can then be applied to yield an estimate of the response statistics. To perform a "long term" statistical analysis, the response must be simulated for a range of different wave conditions, thus enabling the lifetime reliability statistics to be inferred from knowledge of $p(H_s,T_1)$. Time domain simulation can be costly in terms of the computer time required, and the method is perhaps best employed towards the final stages of design when the structural configuration is more or less fixed. It can be noted that digital filtering techniques (Spanos [34]) may be used as an alternative to the environmental simulation technique represented by equation (6.29). It can be noted that there have been significant recent developments in improving the efficiency of simulation techniques for non-linear systems subjected to stochastic loads (see for example Pradlwarter [35]).

(iii) Problem Specific Methods. A range of advanced theoretical methods have been developed for non-linear random vibration problems. These include: Markov methods and the Fokker-Planck-Kolmogorov equation (see for example Lin [24]), non-Gaussian closure (see for example Nigam [36]), Volterra and Wiener methods (Schetzen [37]), and Statistical Quadratization (Donely and Spanos [38]). Unfortunately, each of these methods tends to be limited in terms of the type of problem which can be analyzed; even those methods which are in principle quite general are limited for practical reasons to systems with relatively few degrees of freedom. However, there are specific offshore dynamics problems which fall within the current capabilities of the methods — for example, Roberts ([39], [40]) has successfully applied the Fokker-Plank-Kolmogorov equation to the slow drift surge motions of a catenary moored tanker and to the non-linear rolling motion of a ship.

d. Allowance for Structural Uncertainties (-U). The effect of structural uncertainties can be examined by performing parameter studies in an attempt to identify the worst case situation. Alternatively, a more systematic approach based on a mathematical reliability theory could be adopted; one difficulty here is that a stochastic response analysis of a fully deterministic structure can be an extremely complex task, and the additional detailed consideration of structural uncertainty could be untenable. Approximate reliability methods such as the Level II theory offer some form of compromise, and this is discussed further in Section 6.8.

e. Specific Examples. To highlight some of the detailed aspects of the stochastic response analysis of offshore structures, two "problem specific" techniques are presented in the following sections. In Section 6.5 the statistics of the stresses induced in a "static" structure by Morison loading is considered: this is an example of a "problem specific method" for a non-linear system. In Section 6.6 the first and second order motions of a floating structure are analyzed by using a technique based on the Kac-Siegert method: this is a further example of a "problem specific method" for a non-linear system, which in this case makes an allowance for dynamic response.

6.5 Fixed Offshore Structures

For fixed offshore structures the environmental forces, which include wave, wind, and current forces are primarily balanced by the stiffness of the structure. By their nature, these structures are composed of members which have high values of stiffness, leading to natural frequencies which are well above the frequency range of the incident wave and wind spectra. For this reason, the structural members will not be excited at their natural frequencies and will therefore not be resonant. However, the structure will exhibit a quasi-static response leading to dynamic structural displacements and stresses. Knowledge of these stresses and their cyclic variation is necessary in the design of an offshore structure to ensure that it has sufficient structural strength and to estimate its fatigue life.

In practice it is usual to analyze the quasi-static response of an offshore structure by assuming that the wind and current contribute static forces, while the waves provide dynamic loading. The dynamic analysis of this type of structure under wave loading forms the main consideration of the present section, where the statistical distribution of the wave loading and response are considered in detail. The remainder of this section is considered in two parts: the first part considers the statistical distribution of the forces arising from Morison-type wave loading, while the second part considers the statistical distribution of the induced response.

6.5.1 Morison-Type Wave Loading Statistics

For fixed offshore structures with small diameter structural members, it is clear from the discussion given in Section 6.3 that Morison's equation provides an appropriate means of determining the wave loading. Accordingly the structure is divided into a number of "elements", and the wave loading on the ith element is expressed as the sum of an inertia force and a drag force

$$F_i = X_{2i-1} + X_{2i} \mid X_{2i} \mid, i = 1, ..., N \tag{6.31}$$

where

$$X_{2i-1} = \dot{u}_i K_{Ii}, \tag{6.32}$$

$$X_{2i} = u_i \sqrt{K_{Di}}, \tag{6.33}$$

and u_i and $\dot{u}_i$ are the fluid particle velocity and acceleration at the ith location. In the notation of Section 6.3, this gives $K_{Ii} = (14)\rho\pi D^2 C_{mi} L_i$ and $K_{Di} = (1/2)\rho D C_{di} L_i$, where L_i is the length of the element. The first term of equation (6.31) represents the (linear) inertia force, while the second term represents the (non-linear) drag term.

Assuming that the fluid particle velocity is a Gaussian random process, the inertia loading will be Gaussian, while the fluid drag loading will be non-Gaussian. In the absence of drag loading, the loading statistics can be obtained directly from a knowledge of the loading spectrum. However, as soon as the drag force contribution is included, additional knowledge is required to define the loading statistics. The distribution for the combined inertia and drag loading for a single element has been considered by Pierson and Holmes [41]. However, to be able to determine the loading distribution for a complex structure consisting of many components, it is necessary to consider the multivariate pdf of the loads acting over all the elements of the structure. Tickell [42] has considered this situation, in which the analysis developed can be considered to be a generalisation of the work performed for a single element. A summary of this work is given below. Further developments in this area have been reported recently by Najafian and Burrows [43].

Assuming that the fluid particle velocity u_i is Gaussian, the n-dimensional force pdf for all elements of the structure is obtained from the $2n$-dimensional Gaussian density function of $\boldsymbol{X}$ which is given by

$$p(\boldsymbol{X}) = \frac{1}{(2\pi)^n \sqrt{Det\,\boldsymbol{\mu}}} \exp\left\{-\frac{1}{2}\boldsymbol{X}^T \boldsymbol{\mu}^{-1} \boldsymbol{X}\right\}, \tag{6.34}$$

where $\boldsymbol{X}^T = (X_1, X_2, \ldots, X_{2n})$ and the matrix $\boldsymbol{\mu}$ is given by

$$\boldsymbol{\mu} = \begin{pmatrix} R_{11} & R_{12} & \ldots & R_{1,2n} \\ R_{21} & R_{22} & \ldots & R_{2,2n} \\ . & . & \ldots & . \\ . & . & \ldots & . \\ R_{2n,1} & R_{2n,2} & \ldots & R_{2n,2n} \end{pmatrix}. \tag{6.35}$$

The elements of equation (6.35) are the cross-covariances of $\boldsymbol{X}$ given by $R_{ij} = \mathrm{E}[X_i X_j]$, and can be calculated in terms of the wave spectrum [44]. The $2n$-dimensional force probability density function (pdf) is obtained from equation (6.34) by using the transformation of variables

$$p(F_1, F_2, \ldots, F_{2n}) = \frac{p(X_1, X_2, \ldots, X_{2n})}{\mid Det\boldsymbol{J} \mid}, \tag{6.36}$$

where $F_{n+1}, F_{n+2}, \ldots F_{2n}$ are dummy variables such that $F_{n+j} = X_{2j-1}$ for $j = 1, \ldots, N$, and $Det\boldsymbol{J}$ is the determinant of the Jacobian of the transformation from $\boldsymbol{X}$ to $\boldsymbol{F}$, which for this case is unity. The n-dimensional force pdf is obtained from equation (6.36) by integrating over the dummy variables from $-\infty$ to ∞, i.e.,

$$p(F_1, F_2, \ldots, F_n) = \int_{-\infty}^{\infty} \cdots\cdots \int_{-\infty}^{\infty} p(F_1, F_2, \ldots, F_{2n})\, dF_{n+1} dF_{n+2} \ldots dF_{2n}. \tag{6.37}$$

For *a single element of the structure*, Pierson and Holmes [41] have shown that the force pdf may be expressed as

$$p(F) = \frac{1}{2\pi\sigma_{X_1}\sigma_{X_2}} \int_{-\infty}^{\infty} \exp\left\{-\frac{1}{2}\left(\frac{X_1^2}{\sigma_{X_1}^2} + \frac{X_2^2}{\sigma_{X_2}^2}\right)\right\} dX_1, \tag{6.38}$$

where $X_2 = F\text{-}X_1 \mid X_1 \mid$. Equation (6.38) is dependent upon a knowledge of the second and fourth moments of the force only. This may be demonstrated by noting that the second and fourth moments of the force are

$$E[F^2] = 3\sigma_{X_1}^4 + \sigma_{X_2}^2, \tag{6.39}$$

$$E[F^4] = 105\sigma_{X_1}^8 + 18\sigma_{X_1}^4\sigma_{X_2}^2 + 3\sigma_{X_2}^4. \tag{6.40}$$

By rearranging equations (6.39) and (6.40), it is easily shown that

$$\sigma_{X_1}^2 = \left\{\frac{E[F^4] - 3E[F^2]^2}{78}\right\}^{\frac{1}{4}}, \tag{6.41}$$

$$\sigma_{X_2}^2 = E[F^2] - 3\sigma_{X_1}^4. \tag{6.42}$$

Consequently, a knowledge of $E[F^2]$ and $E[F^4]$ enables the force pdf to be calculated from equation (6.38).

From a practical view point, it is the peak distribution of the force which is of importance to the engineer. For analytic purposes it is convenient to distinguish between the "narrow band" and "broad band" case — a full description of these terms is given in Section 6.7. For the case in which the force is assumed to be narrow band and Gaussian, the distribution of the peaks can be derived analytically [45]. However, if either the drag or inertia contributions can be neglected, an analytic expression for the peak distribution can be obtained, even if the process is broad band [46]. In what follows the more general broad band case including drag and inertia effects is described as considered by Tickell [42].

For a broad band force F the distribution of the peaks is expressed in terms of the joint probability density function (jpdf) of F, $\dot{F}$, and $\ddot{F}$, where a dot represents differentiation with respect to time. The conditions for a peak to occur above a specified level b (say) are that $F>b$, $\dot{F}=0$, and $\ddot{F}<0$. Using these conditions it can be shown that the probability of a given peak force (acting on the ith element) having a value less than b is

$$P_p(b) = 1 - \frac{\int_{-\infty}^{0}\int_{b}^{\infty} \ddot{F}p(F, \dot{F}=0, \ddot{F})dFd\ddot{F}}{\int_{-\infty}^{0}\int_{-\infty}^{\infty} \ddot{F}p(F, \dot{F}=0, \ddot{F})dFd\ddot{F}}. \tag{6.43}$$

Here the jpdf is given by

$$p(F, \dot{F} = 0, \ddot{F}) = \frac{1}{(2\pi)^2 \sqrt{Det \boldsymbol{\mu} K_{Ii}}^3} \int_{-\infty}^{\infty} \exp\left\{-\frac{1}{2} \boldsymbol{u}^T \boldsymbol{\mu} \boldsymbol{u}\right\} du, \tag{6.44}$$

where $\boldsymbol{\mu}$ and $\boldsymbol{u}$ are given by

$$\boldsymbol{\mu} = \begin{pmatrix} m_0 & 0 & -m_2 & 0 \\ 0 & m_2 & 0 & -m_4 \\ -m_2 & 0 & m_4 & 0 \\ 0 & -m_4 & 0 & m_6 \end{pmatrix}, \tag{6.45}$$

$$\boldsymbol{u}^T = \left(u \ \frac{\partial u}{\partial t} \ \frac{\partial^2 u}{\partial t^2} \ \frac{\partial^3 u}{\partial t^3} \right), \tag{6.46}$$

and m_i are the moments of the velocity spectrum such that

$$m_i = \int_0^{\infty} \omega^i S_{uu}(\omega) d\omega. \tag{6.47}$$

A closed form expression for the peak distribution as given by equation (6.44) does not seem possible. However, the probability density function (pdf) for the individual inertia and drag cases can be evaluated analytically in terms of the spectral width parameter ϵ.

For the inertia dominated case the force is Gaussian and it may be shown that the peak pdf is [47]

$$\begin{aligned} p_p(V) &= \tfrac{1}{\sqrt{2\pi}} \epsilon \exp\left\{-\tfrac{1}{2}\left(\tfrac{V}{\epsilon}\right)^2\right\} \\ &+ V(1-\epsilon^2)^{\frac{1}{2}} \exp\left\{-\tfrac{1}{2} V^2\right\} \Phi\left(-\tfrac{V\sqrt{(1-\epsilon^2)}}{\epsilon}\right), \end{aligned} \tag{6.48}$$

where V is a non-dimensional form of b, such that $V = b/\sigma_F$, and Φ is the cumulative normal distribution function. For a narrow band process ($\epsilon \approx 0$), equation (6.48) reduces to a Rayleigh distribution, while for a broad band process ($\epsilon \approx 1$) it is a Gaussian distribution.

For the drag dominated case the force is non-Gaussian and it may be shown that the peak pdf is [46]

$$\begin{aligned} p_p(V) &= (1-\epsilon^2)^{\frac{1}{2}} \tfrac{\sqrt{3}}{4} \tfrac{V}{|V|} \exp\left[-\tfrac{\sqrt{3}}{2} \mid V \mid\right] \Phi\left(\frac{-\frac{3^{\frac{1}{4}}}{\sqrt{2}} V(1-\epsilon^2)}{|V|\epsilon}\right) \\ &+ \sqrt{\tfrac{2}{\pi}} \tfrac{3^{\frac{1}{4}}}{4} \tfrac{\epsilon}{\sqrt{|V|}} \exp\left[\tfrac{-\sqrt{3}}{2\epsilon^2 |V|}\right]. \end{aligned} \tag{6.49}$$

For a narrow band process ($\epsilon \approx 0$) this equation reduces to an exponential distribution.

In addition to the distribution of the forces acting on a structural member, the distribution of the stress (or displacement) is of primary importance to the design engineer. This situation has been considered by Tickell [42] and Burrows [44] in which the work described above is extended.

6.5.2 Quasi-Static Response of Linear Structures

For fixed offshore structures, the vector of nodal forces $\boldsymbol{F}_e$ in a finite element model is linearly related to the values of the distributed force acting at particular locations along each member $\boldsymbol{F}$ as

$$\boldsymbol{F}_e = \boldsymbol{A}\boldsymbol{F}, \tag{6.50}$$

where $\boldsymbol{A}$ is a transformation matrix. Assuming that the response of the structure is quasi-static and linear, the nodal displacements $\boldsymbol{Y}$ are obtained from the stiffness equation

$$\boldsymbol{F}_e = \boldsymbol{K}\boldsymbol{Y}, \tag{6.51}$$

where $\boldsymbol{K}$ is the structural stiffness matrix. Combining equations (6.50) and (6.51) gives

$$\boldsymbol{Y} = \boldsymbol{K}^{-1}\boldsymbol{A}\boldsymbol{F} = \boldsymbol{S}\boldsymbol{F}. \tag{6.52}$$

Using equation (6.52), the multi-dimensional pdf of nodal responses $\boldsymbol{Y}$ may be obtained from the multi-dimensional pdf of forces by using the transformation of variables

$$p(\boldsymbol{Y}) = \frac{p(\boldsymbol{F})}{\mid Det\boldsymbol{S} \mid}. \tag{6.53}$$

However, as in the derivation of the n-dimensional force pdf described in Section 6.5.1, it is necessary to introduce some dummy variables into equation (6.53) of the form $Y_i = F_j$ (i and j being arbitrary) to ensure that the transformation of variables is between spaces of equal dimension. The dummy variables are then integrated out in an identical manner to that used to derive equation (6.37). This approach can be extremely lengthy since to derive the n-dimensional pdf of force or response requires an n-fold integration to be performed. In addition, to obtain a more practically applicable marginal distribution requires an additional (n-1) fold integral to remove the unwanted (n-1) response variables. For example, the marginal pdf for Y_1 is obtained by evaluating

$$p(Y_1) = \int_{-\infty}^{\infty} ... \int_{-\infty}^{\infty} p(Y_1, Y_2, ..., Y_n) dY_2 dY_3 ... dY_n. \tag{6.54}$$

Given that equation (6.52) represents a linear transformation, it has been suggested that the distribution of the response Y might have an identical form to that of the distributed load F. Under this assumption the marginal distribution of the response will have the same form as that derived in equation (6.38).

The above heuristic argument for calculating the response pdf has been applied successfully by Tickell [42] to the analysis of stress records measured on actual members of an offshore structure. Results for the pdf of the bending stress experienced by one of the members considered are shown in Figure 6.7. By comparison with an equivalent Gaussian distribution it may be seen that the distribution is highly non-Gaussian, especially at the tails of the distribution. The theoretical non-Gaussian curve shown was obtained by using the measured second and fourth order statistical moments. For this technique to provide a theoretical prediction of the response distribution, it is necessary to be able to calculate the second and fourth order moments theoretically, without recourse to measured data. This problem has been solved by Burrows [44] who derived analytic expressions for the second and fourth moments of the response.

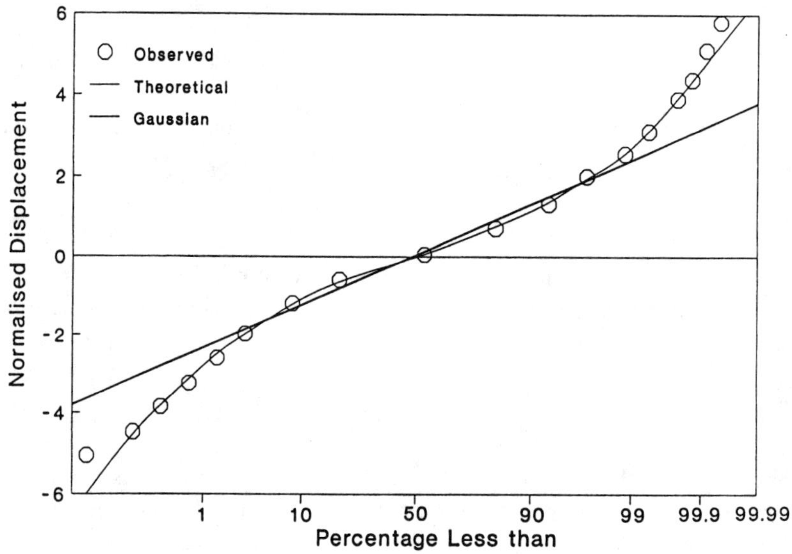

FIGURE 6.7
Cumulative distribution of the bending stress experienced by a member of an offshore structure.

Having discussed the means by which the distribution of the response can be determined, it is necessary to consider how to predict the distribution of the peak response. This is accomplished by using a similar argument to that used above, where it has been suggested that the peak response distribution will have the same form as the peak force distribution described in Section 6.5.1. However, given that the wide band peak distribution is dependent upon a knowledge of the higher order moments of the velocity spectrum given by equation (6.47), which can be difficult to evaluate accurately, it is convenient to simplify the analysis by introducing the assumption that the process, whether force or response, is narrow band. Using this assumption it can be shown that the cumulative peak force (and hence response) distribution is

$$P_p(b) = 1.0 - \frac{\int_0^\infty \dot{F} p(F = b, \dot{F}) d\dot{F}}{\int_0^\infty \dot{F} p(F = 0, \dot{F}) d\dot{F}} . \tag{6.55}$$

This equation has been used by Tung [48] to consider the fatigue damage due to wave loading. However, given that this distribution requires knowledge of the fourth moment of the velocity spectrum, which may also be difficult to evaluate accurately, it is convenient to assume that F and $\dot{F}$ are statistically independent [42]. Under this assumption it can be shown that the peak force (and hence response) distribution can be written as

$$P_p(b) = 1.0 - \frac{p(F = b)}{p(F = 0)} . \tag{6.56}$$

When it is assumed that the process is Gaussian, then equation (6.56) yields a Rayleigh distribution. The advantage of using this equation is that the peak distribution can be obtained from a knowledge of the initial distribution only, which is defined in terms of the second and fourth order moments.

Tickell [42] has applied equation (6.56) to the analysis of stress records measured on actual members of an offshore structure. Figure 6.8 shows results obtained for the pdf of the peak bending stress in a member. It is clear from these results that the measured process is not narrow band, since peaks occur well below the mean stress level. Consequently, the results obtained using equation (6.56) and the Rayleigh distribution are not in good agreement with the observed results at low stress levels. However, at high stress levels the results obtained using equation (6.56) approach the observed distribution, thus providing the design engineer with a reasonably accurate means of predicting the probability of failure.

6.6 Large Floating Structures

Most compliant systems are designed to have low stiffness in the horizontal plane so that the surge, sway, and yaw natural frequencies are well below the frequency range of the incident wave spectrum. Typically, the wave spectrum extends from 0.2 to 1.6 rad/sec, while the surge, sway, and yaw natural frequencies of a floating system are approximately 0.05 rad/sec. Such an arrangement is intended to ensure that the primary balancing forces will be the vessel inertia, meaning that the mooring system loads are kept to a minimum.

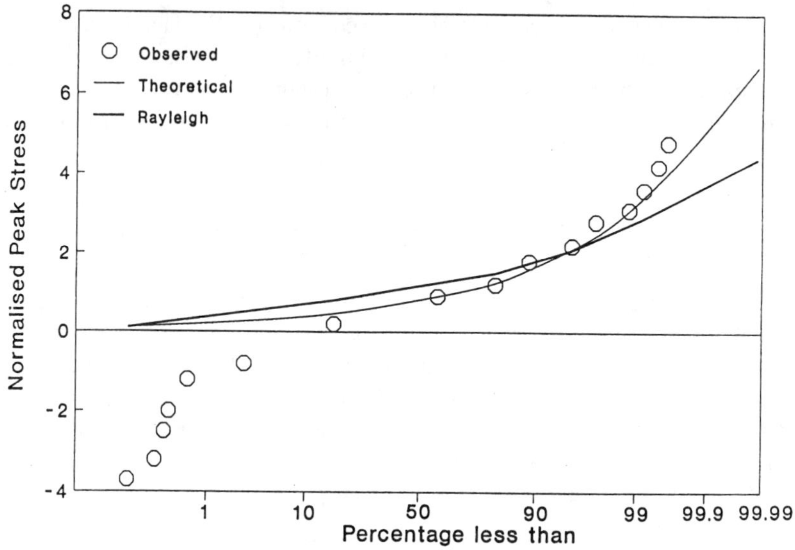

FIGURE 6.8
Cumulative distribution of the peak bending stress for a member of an offshore structure.

In principle, the above design philosophy should be valid in all cases. However, a major practical difficulty is that the wave forces produced by the sea-state contain a non-linear second order component. Although small, this component contains energy at all frequencies between zero and the bandwidth of the incident spectrum. It is therefore impossible to design a moored vessel which will not be resonated by these second order wave forces. In addition, first order wave forces, which are proportional to the wave height and act at the same frequencies as the incident wave spectrum, provide a force which acts at frequencies greater than the natural frequency of the mooring system. Consequently, a typical time history of the wave induced response of a compliant offshore structure in surge, sway, or yaw is characterised by a large amplitude low frequency oscillation occurring at the resonant frequency of the mooring system with a superimposed higher frequency oscillation.

In order to predict the extreme values of the response it is necessary to have knowledge of the statistical distribution of the response. The following two sub-sections consider the statistical distribution of the response and extreme motions of compliant offshore structures to these non-linear wave forces. The response of (approximately) linearly moored structures, such as tension leg platforms is considered in sub-section 6.6.1, while the response of non-linearly moored vessels, such as semi-submersibles, is considered in sub-section 6.6.2.

6.6.1 Response of Linearly Moored Structures to Non-Linear Wave Forces

For large volume compliant offshore structures it is necessary to take into account the reflection and radiation of waves when determining the forces acting upon the structure. For potential flow it can be shown that to second order the wave forces acting on such structures can be written as [49]

$$\begin{aligned}F(t) = {} & \sum_{j=1}^{N} c_j \left[R_j^c \cos(\omega_j t + \epsilon_j) + R_j^s \sin(\omega_j t + \epsilon_j) \right] + \sum_{j=1}^{N} \sum_{k=1}^{N} c_j c_k \\ & \Big[T_{jk}^c \cos((\omega_j - \omega_k)t + \epsilon_j - \epsilon_k) + T_{jk}^s \sin((\omega_j - \omega_k)t + \epsilon_j - \epsilon_k) + \\ & S_{jk}^c \cos((\omega_j + \omega_k)t + \epsilon_j + \epsilon_k) + S_{jk}^s \sin((\omega_j + \omega_k)t + \epsilon_j + \epsilon_k) \Big],\end{aligned} \tag{6.57}$$

where the random seastate is modeled as a linear combination of regular wave components, so that c_j and ϵ_j are the amplitude and random phase of the jth frequency component ω_j; ${R_j}^c$ and ${R_j}^s$ are first order force transfer functions; and ${T^c}_{jk}$, ${T^s}_{jk}$, ${S^c}_{jk}$, and ${S^s}_{jk}$ are second order force transfer functions. Further, ${T^c}_{jk}$ and ${T^s}_{jk}$ are associated with forces acting at difference-frequencies, while ${S^c}_{jk}$ and ${S^s}_{jk}$ are associated with forces acting at sum-frequencies. The first term in equation (6.57) represents the first order wave force, while the second term represents the second order wave force. Given that the first order force is proportional to the wave amplitude, it may be deduced that if the wave elevation is Gaussian then the first order force will also be Gaussian. Further examination of equation (6.57) indicates that the second order force is related to the square of the wave amplitude, meaning that the second order force will be non-Gaussian, if the wave elevation is Gaussian.

For a wave spectrum which extends from ω_{min} to ω_{max} (say) it may be deduced from equation (6.57) that the first order, difference- and sum-frequency second order wave forces provide excitations over the frequency ranges ω_{min} to ω_{max}, 0 to ω_{max}-ω_{min}, and $2\omega_{min}$ to $2\omega_{max}$, respectively. For a compliant offshore structure, where the surge, sway, and yaw natural frequencies tend to be much lower than ω_{min}, it may be deduced that the difference-frequency second order forces induce a low-frequency resonant response, while the first order and sum-frequency second order forces produce motions well above the natural frequency. Since the second order forces are much smaller than the first order forces, the sum-frequency second order wave forces are neglected in this case. Although the concern here is with the low frequency surge, sway, and yaw motions, it may be noted that the heave, pitch, and roll motions, which tend to have natural frequencies greater than ω_{max}, can be resonated by the sum-frequency second order wave forces, while the first order and difference-frequency second order wave forces induce motions well below the natural frequency; in this case, the second order difference-frequency forces are neglected. Although the combined first and sum-frequency second order response is not considered here, details of how the response pdf can be calculated are given by Naess and Ness [50].

The problem of predicting the pure difference-frequency second order force distribution has received much interest. By using various simplified analyses, Vinje [51] and Langley [52] were able to derive closed-form expressions for the pdf. However, by making use of the Kac-Siegert technique [53], Naess [54] [55], and Langley [56] have shown that it is possible to derive an exact analytic expression for the second order force pdf and second order response pdf of a linearly moored structure. The Kac-Siegert technique has also led to the development of a means of predicting the combined first and second order response statistics of a linearly moored structure. The application of the Kac-Siegert technique to this situation is described in what follows.

Assuming that the wave elevation is a zero-mean, stationary, Gaussian random process with wave spectrum $S_{\eta\eta}(\omega)$, it can be shown that the combined first order and difference-frequency second order wave force can be written as

$$F(t) = Re\sum_{j=1}^{N} A_j \hat{u}_j e^{i\omega_j t} + Re\sum_{j=1}^{N}\sum_{k=1}^{N} B_{jk}\hat{u}_j\hat{u}_k^* e^{i(\omega_j-\omega_k)t}, \tag{6.58}$$

where

$$A_j = (R_j^c - iR_j^s)\sqrt{2S_{\eta\eta}(\omega_j)d\omega}, \tag{6.59}$$

$$B_{jk} = (T_{jk}^c - iT_{jk}^s)\sqrt{2S_{\eta\eta}(\omega_j)d\omega}\sqrt{2S_{\eta\eta}(\omega_k)d\omega}, \tag{6.60}$$

and $\hat{\mathrm{u}}_j$ satisfy the statistical properties

$$E[|\ \hat{u}_j\ |^2] = 1, \tag{6.61}$$

$$E[\hat{u}_j\hat{u}_k] = 0,\ j \neq k, \tag{6.62}$$

$$E[\hat{u}_j\hat{u}_k^*] = 0,\ j \neq k. \tag{6.63}$$

The single degree of freedom equation of motion for a linearly moored structure is

$$\ddot{x} + 2\beta\omega_n\dot{x} + \omega_n^2 x = F(t), \tag{6.64}$$

where β is the non-dimensional damping ratio, ω_n is the undamped natural frequency, and $F(t)$ is the force (per unit mass). Solving this equation in the frequency domain, the response is

$$x(t) = Re\sum_{j=1}^{N} G_j \hat{u}_j e^{i\omega_j t} + Re\sum_{j=1}^{N}\sum_{k=1}^{N} H_{jk}\hat{u}_j\hat{u}_k^* e^{i(\omega_j-\omega_k)t}, \tag{6.65}$$

where

$$G_j = \frac{A_j}{(\omega_n^2 - \omega_j^2) + 2i\beta\omega_j\omega_n}, \tag{6.66}$$

$$H_{jk} = \frac{B_{jk}}{\omega_n^2 - (\omega_j - \omega_k)^2 + 2i\beta(\omega_j - \omega_k)\omega_n}. \tag{6.67}$$

For convenience, equation (6.65) is re-written in matrix form as

$$x(t) = Re\boldsymbol{G}^T\boldsymbol{u} + Re\boldsymbol{u}^T\boldsymbol{H}\boldsymbol{u}^*, \tag{6.68}$$

where $\boldsymbol{G}$ is a vector whose jth entry is G_j, $\boldsymbol{H}$ is a matrix whose jkth entry is H_{jk}, and $\mathbf{u}$ is the vector whose jth entry is $\hat{u}_j\exp(i\omega_j t)$. It may be shown that the matrix $\boldsymbol{H}$ is Hermitian, meaning that the following factorisation can be made

$$\boldsymbol{H} = \boldsymbol{R}^T\Lambda\boldsymbol{R}^*, \tag{6.69}$$

where Λ is a diagonal matrix containing the eigenvalues of $\boldsymbol{H}$, and $\boldsymbol{R}^*$ is a matrix whose rows are the eigenvectors of $\boldsymbol{H}$. The fact that $\boldsymbol{H}$ is Hermitian implies that the eigenvalues are real and that $\boldsymbol{R}$ is a unitary matrix, such that $\boldsymbol{R}^T\ \boldsymbol{R}^* = \boldsymbol{I}$. Substituting equation (6.69) into equation (6.68) and defining $\boldsymbol{X} = \boldsymbol{R}\boldsymbol{u}$, it may be shown that the response can be re-written as

$$x(t) = Re\sum_{j=1}^{N} \beta_j X_j + \lambda_j \mid X_j \mid^2, \tag{6.70}$$

where β_j is the jth entry of the vector $\boldsymbol{R}^*\boldsymbol{G}$, λ_j is the jth eigenvalue of the matrix $\boldsymbol{H}$, X_j is the jth entry of the vector $\boldsymbol{X}$, and N is the dimension of the matrix $\boldsymbol{H}$. In addition, the entries of the complex Gaussian random vector $\boldsymbol{X}$ can be shown to have the following statistical properties at any fixed time

$$E[\mid X_j \mid^2] = 1, \tag{6.71}$$

$$E[X_i X_j] = 0,\ i \neq j, \tag{6.72}$$

$$E[X_i X_j^*] = 0,\ i \neq j. \tag{6.73}$$

Equations (6.70-6.73) can be used to derive an analytic expression for the characteristic function *M(ϑ)*, such that

$$M(\vartheta) = E[e^{i\vartheta x}] = \prod_{j=1}^{N} (1 - i\vartheta\lambda_j)^{-1} \exp\left[\frac{-\mid \beta_j \mid^2 \vartheta^2}{4(1 - i\vartheta\lambda_j)}\right]. \tag{6.74}$$

Although it is equation (6.74) which is used later to determine the response pdf, it is worthwhile noting that it can be used directly to derive an analytic expression for the nth response cumulant K_n, such that

$$K_n = \frac{1}{i^n} \frac{d^n}{d\vartheta^n} [\ln M(\vartheta)] \mid_{\vartheta=0} = \sum_{j=1}^{N} \frac{1}{4} n! \lambda_j^{n-2} \mid \beta_j \mid^2 \delta_{n1} + (n-1)! \lambda_j^n, \tag{6.75}$$

where δ_{n1} is zero for $n = 1$ and unity otherwise. More specifically, the first four response cumulants are

$$K_1 = \sum_{j=1}^{N} \lambda_j, \tag{6.76}$$

$$K_2 = \sum_{j=1}^{N} \frac{1}{2} \mid \beta_j \mid^2 + \lambda_j^2, \tag{6.77}$$

$$K_3 = \sum_{j=1}^{N} \frac{3}{2} \lambda_j \mid \beta_j \mid^2 + 2\lambda_j^3, \tag{6.78}$$

$$K_4 = \sum_{j=1}^{N} 6\lambda_j^2 \mid \beta_j \mid^2 + 6\lambda_j^4. \tag{6.79}$$

In the above, the Kac-Siegert technique has been used to obtain analytic expressions for the characteristic function and the cumulants. Given that the characteristic function and probability density function (pdf) form a Fourier transform pair, the pdf can be expressed as

$$p(x) = \frac{1}{2\pi} \int_{-\infty}^{\infty} e^{-i\vartheta x} M(\vartheta) d\vartheta. \tag{6.80}$$

For a pure second order response it may be shown that equation (6.80) can be evaluated analytically and the pdf can be written as ([55], [56])

$$\begin{aligned} p(x_2) &= \sum_{j=1}^{M} \frac{\mu_j}{\lambda_j} e^{-\frac{x_2}{\lambda_j}} \qquad x_2 \geq 0 \\ &= \sum_{j=M+1}^{N} \frac{\mu_j}{|\lambda_j|} e^{-\frac{x_2}{\lambda_j}} \; x_2 < 0, \end{aligned} \tag{6.81}$$

where the eigenvalues have been ordered such that $\lambda_j, j = 1,2,..,M$ are positive and $\lambda_j, j = M+1,..,N$ are negative, and

$$\mu_j = \prod_{\substack{k=1 \\ k \neq j}}^{N} \left(\frac{1-\lambda_k}{\lambda_j} \right)^{-1} \tag{6.82}$$

Figure 6.9 shows results obtained for the pure second order response pdf of a semi-submerged horizontal cylinder of radius 10m and draft 10m, where the pdf has been plotted on a log-axis and the response has been normalized. The mooring system has a damping ratio $\beta = 0.1$ and natural frequency $\omega_n = 0.05$. The second order force is defined in terms of the force transfer function data given by Faltinsen and Løken [57] and the ISSC wave spectrum with significant wave height H_s = 2.0m and average wave period T_1 = 5.5s. Equation (6.81) has been used to calculate the "Exact" curve, while a Gaussian distribution is shown for comparison. It is immediately apparent from this figure that the second order response is highly non-Gaussian, especially at the tails of the distribution.

For the combined first and second order response there appears to be no exact analytic means of evaluating equation (6.80). In principle, it may be evaluated numerically using the Fast Fourier Transform (FFT) algorithm. However, the computational effort required to produce a reliable estimate of the tails of the distribution can be excessive. For this reason, much effort has been directed towards finding a means of calculating the combined first and second order response pdf. The approaches developed by Naess and Johnsen [58] and Langley and McWilliam [59] are briefly described below.

An efficient numerical means of determining the response pdf from equation (6.80) has been proposed by Naess and Johnsen [58], in which the so-called saddle-point integration technique has been adopted. Using this technique it may be shown that the pdf can be written as

$$p(x) = Re \left\{ \frac{hb}{\pi} \left[\frac{1}{2} e^{\rho(u_s)} + \sum_{j=1}^{\infty} e^{\rho(u_s + jhbi)} \right] \right\}, \tag{6.83}$$

where $\rho(w)$ and b are

$$\rho(w) = -wx + \sum_{j=1}^{N} \left[\frac{-\ln(1-\lambda_j w) + |\, \beta_j \,|^2 \, w^2}{4(1-\lambda_j w)} \right], \tag{6.84}$$

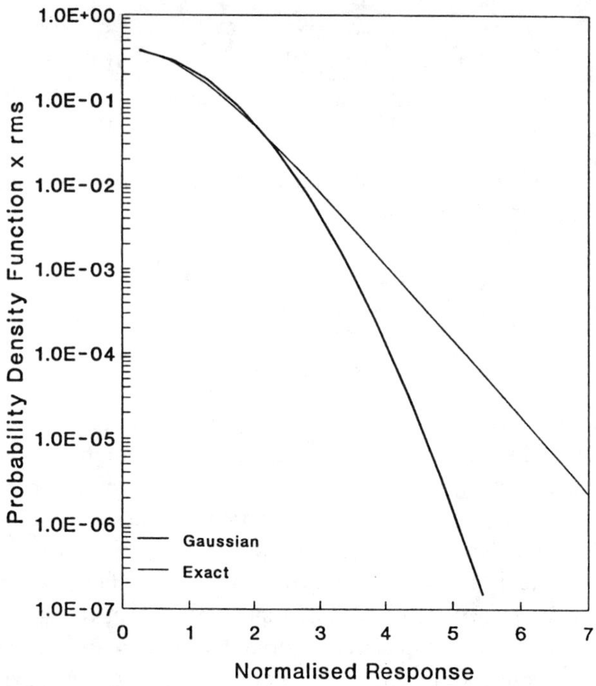

FIGURE 6.9
Pure second order response pdf for a moored horizontal cylinder.

$$b = \sqrt{2}\left\{\sum_{j=1}^{N} \frac{|\beta_j|^2}{2(1-u_s\lambda_j)^3} + \frac{\lambda_j^2}{(1-\lambda_j u_s)^2}\right\}^{-\frac{1}{2}}, \tag{6.85}$$

while u_s is given by the solution to

$$-x + \sum_{j=1}^{N}\left[\frac{\lambda_j}{(1-\lambda_j u_s)} + \frac{|\beta_j|^2}{4}\frac{u_s(2-u_s\lambda_j)}{(1-\lambda_j u_s)^2}\right] = 0 \tag{6.86}$$

subject to the condition that

$$\max_{\lambda_j < 0}\left(\frac{1}{\lambda_j}\right) < u_s < \min_{\lambda_j > 0}\left(\frac{1}{\lambda_j}\right). \tag{6.87}$$

The coefficient h is initially assigned a value of unity and is progressively halved until convergence of equation (6.83) is established. This method has the advantage that no approximations are introduced except those inherent in the numerical integration. Since this technique is based on a variable-step integration scheme the accuracy of this approach is "guaranteed" provided that a small enough value of h

is used. Consequently this method provides an accurate means of determining the response distribution, where the efficiency of the method is dependent upon the convergence of the numerical integration scheme.

An alternative approach has been developed by Langley and McWilliam [59] which yields an analytic series expansion for the response pdf. The first term of the series corresponds to the assumption that the first and second order responses are statistically independent, while subsequent terms are dependent upon the joint first and second order response cumulants. If only the first term in the series is used then the pdf of the combined response can be written as

$$p(x) = \sum_{j=1}^{M} \frac{\mu_j}{\lambda_j} e^{-\frac{x}{\lambda_j}+\frac{\sigma_1^2}{2\lambda_j^2}} \left[1 - \Phi\left[\frac{\sigma_1}{\lambda_j} - \frac{x}{\sigma_1}\right]\right] + \sum_{j=M+1}^{N} \frac{\mu_j}{|\lambda_j|} e^{-\frac{x}{\lambda_j}+\frac{\sigma_1^2}{2\lambda_j^2}} \Phi\left[\frac{\sigma_1}{\lambda_j} - \frac{x}{\sigma_1}\right], \tag{6.88}$$

where Φ is the cumulative normal distribution function.

Results obtained for the combined first and second order response pdf of the cylinder described earlier are shown in Figure 6.10. Here the first order force is defined in terms of the transfer function data given by Kato et al. [60] and the force transfer function data has been linearly scaled so that the ratio of the rms second order response to the rms first order response (σ_2/σ_1) is unity. For comparison results obtained from numerical simulation [61] and an equivalent Gaussian distribution are shown. For this case it can be seen that the results obtained using the assumption that the first and second order responses are statistically independent (i.e., equation (6.88)) slightly underestimate the simulated results. However, by using further terms in the series it can be shown that an accurate representation of the tails can be obtained [59]. The results obtained using Naess' numerical method (equation (6.83)) are seen to be in excellent agreement with the simulated results.

From a practical point of view the design engineer is primarily interested in predicting the extreme offset which will arise for a given mooring system. The extreme response is usually predicted using the assumption that successive up-crossings of a specified level are independent and constitute a Poisson process. Under this assumption, the probability $P(b,t)$ that the response $x(t)$ will cross a level $x = b$ at least once during a period t is given by equation (6.26). The mean up-crossing rate $\nu_b{}^+$ which appears in this equation may be written as

$$\nu_b^+ = \int_0^\infty \dot{x} p(b, \dot{x}) d\dot{x}, \tag{6.89}$$

where $p(x,\dot{x})$ is the joint displacement-velocity probability density function. An approximate method of calculating the mean up-crossing rate has been proposed by Langley [56] and Kato et al. [60], in which the displacement and velocity are assumed to be statistically independent, such that

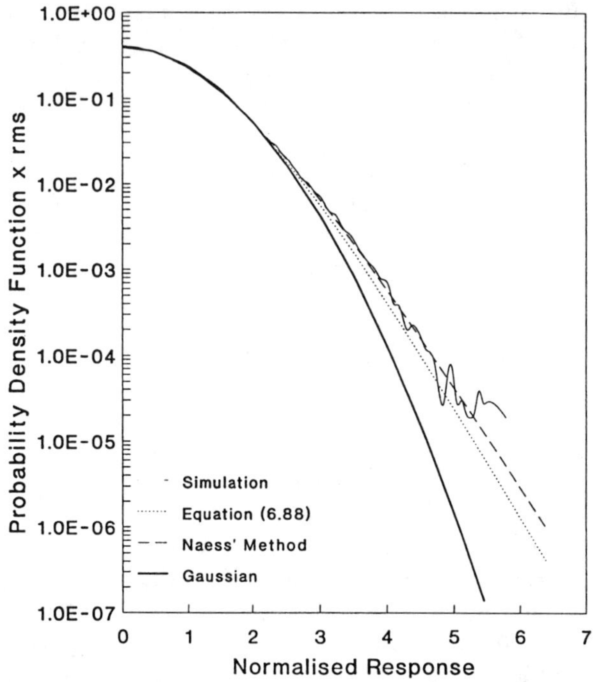

FIGURE 6.10
Combined first and second order response pdf of a moored horizontal cylinder.

$$\nu_b^+ = p(b) \int_0^\infty \dot{x} p(\dot{x}) d\dot{x}, \tag{6.90}$$

where $p(x)$ and $p(\dot{x})$ are the pdf's of the displacement and velocity, respectively. In fact, this approximate method forms the first term of an infinite series expansion for the mean up-crossing rate proposed by McWilliam and Langley [62]. This latter technique is based upon a general expansion for the jpdf of two variables, which can be used to express the joint displacement-velocity pdf as

$$p(x, \dot{x}) = p(x)p(\dot{x}) + \sum_{n=1}^{\infty} \sum_{m=1}^{\infty} (-1)^{n+m} A_{nm} \frac{\partial^n p(x)}{\partial x^n} \frac{\partial^m p(\dot{x})}{\partial \dot{x}^m}, \tag{6.91}$$

where the coefficients A_{nm} are related to the joint displacement-velocity cumulants. The first few A_{nm} terms are

$$A_{11} = K_{1+1}[x\dot{x}], \tag{6.92}$$

$$A_{12} = \frac{1}{2}K_{1+2}[x\dot{x}^2], \tag{6.93}$$

$$A_{21} = \frac{1}{2}K_{2+1}[x^2\dot{x}], \tag{6.94}$$

$$A_{22} = \frac{1}{4}K_{2+2}[x^2\dot{x}^2] + \frac{1}{2}K_{1+1}[x\dot{x}]^2, \tag{6.95}$$

where $K_{i+j}[x^i\dot{x}^j]$ denotes the ijth joint displacement-velocity cumulant.

Substituting equation (6.91) into equation (6.90) gives

$$\nu_b^+ = p(b)\int_0^\infty \dot{x}\,p(\dot{x})d\dot{x} + \sum_{n=1}^{\infty}\sum_{m=1}^{\infty}(-1)^{n+m}A_{nm}\,p^{(n)}(b)\int_0^\infty \dot{x}\,p^{(m)}(\dot{x})d\dot{x}, \tag{6.96}$$

where $p^{(n)}(b)$ is the nth derivative of $p(x)$ evaluated at level b and $p^{(m)}(\dot{x})$ is the mth derivative of $p\,(\dot{x})$. Using this equation, the mean up-crossing rate may be obtained from a knowledge of the individual displacement and velocity pdf's, and the joint displacement-velocity cumulants. Both the displacement and velocity pdf's and their derivatives can be calculated using the methods described earlier to calculate the response pdf, while the joint displacement-velocity cumulants can be calculated analytically using a technique first developed by Longuet-Higgins [4].

Figure 6.11 shows results for the mean up-crossing rate obtained using the method described above, where the response has been normalized and the mean up-crossing rate has been standardized with respect to the mean up-crossing rate of the mean level. The example considered is identical to that considered earlier in the section for the combined first and second order response pdf. For comparison, results obtained from numerical simulation and results corresponding to an equivalent Gaussian distribution are shown. For this example, it can be seen that the assumption that the displacement and velocity responses are statistically independent (denoted by "Indep.") provides a reasonably good estimate of the mean up-crossing rate, and that an improved estimate (denoted by "Refined") can be obtained by including further terms in equation (6.91); the "Refined" curve has been obtained by using two additional terms.

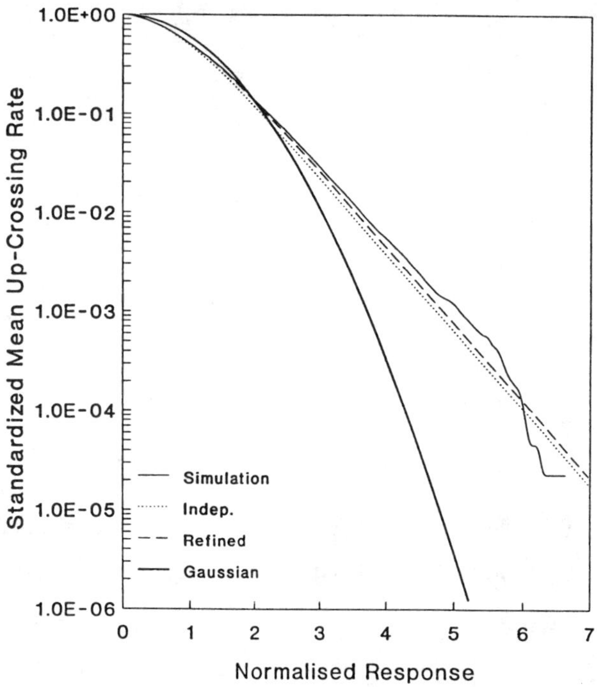

FIGURE 6.11
Mean up-crossing rate for a moored horizontal cylinder.

6.6.2 Response of Non-Linearly Moored Vessels to Non-Linear Wave Forces

The response of a non-linearly moored vessel to pure second order wave forces has been investigated by Roberts [39] using Markov methods, in which the response pdf was derived in terms of a multi-dimensional Fokker-Planck equation. The difficulties associated with solving this equation lead the author to adopt a Gaussian white-noise approximation for the excitation, enabling an analytic solution for the response pdf to be obtained. Although for the lightly damped system considered, this analytic solution gave reasonably good agreement with numerical simulation; the Gaussian white-noise assumption is unlikely to be true in practice. Marthinsen [63] has also used Markov methods to investigate the response of a non-linearly moored vessel to second order wave forces. A Galerkin method was used to solve the resulting Fokker-Planck equation. However, as the magnitude of the system non-linearity was increased, the solution became unstable. More recently Johnsen and Naess [64] have shown that the path integral solution technique provides a more suitable means of solving the Fokker-Planck equation for this case. However, the computer time required to obtain a solution was found to be excessive (i.e., 5 – 8 hrs on a DEC 3100 Workstation) and is therefore not well suited to the purposes of initial design. From a practical point of view, all of the above methods neglect the contribution of the first order wave forces. For a catenary moored vessel, such as a semisubmersible, these effects can contribute significantly to the response and should therefore be included within the analysis.

Techniques such as the Equivalent Linearisation method and the method of Statistical Quadratization are well suited to considering the response of a non-linearly moored vessel to combined first and second order wave forces. These methods are based on replacing the original non-linear system with an equivalent system which can be solved in the frequency domain. The method of Statistical Quadratization developed by Donely and Spanos [38] has been proposed as an improvement to the well-documented method of Statistical Linearisation [65] and is described below.

Consider the following non-linear, single degree of freedom system subject to stationary random excitation

$$\ddot{x} + 2\beta\omega_n\dot{x} + \omega_n^2 x + g(x, \dot{x}) = F(t), \tag{6.97}$$

where β is the non-dimensional damping ratio, ω_n is the undamped natural frequency, $g(x,\dot{x})$ is a general non-linear function, and $F(t)$ is the force (per unit mass).

In order to apply the method of Statistical Quadratization it is convenient to re-write equation (6.97) in a form such that the solution has zero mean. By expressing the force and the response as the sum of a mean component and a zero mean dynamic component, it may be shown that equation (6.97) can be written as

$$\ddot{x}_0 + 2\beta\omega_n\dot{x}_0 + \omega_n^2 x_0 + g_0(\mu_x + x_0, \dot{x}_0) = F_0(t), \tag{6.98}$$

where F_0 and x_0 are the zero mean force and response, respectively,

$$g_0(\mu_x + x_0, \dot{x}_0) \equiv g(\mu_x + x_0, \dot{x}_0) - E[g(\mu_x + x_0, \dot{x}_0)], \tag{6.99}$$

and

$$\mu_x = \frac{\mu_F - E[g(\mu_x + x_0, \dot{x}_0)]}{\omega_n^2}. \tag{6.100}$$

Here μ_x is the mean response and μ_F is the mean force. Equations (6.98) and (6.100) are coupled such that μ_x cannot be calculated without knowing the response statistics of x_0.

In the method of Statistical Quadratization the non-linearity is replaced by "equivalent" polynomials up to quadratic order. The form of the "equivalent" non-linear system is constructed as

$$\begin{aligned} &\ddot{x}_0 + 2\beta\omega_n\dot{x}_0 + \omega_n^2 x_0 + \alpha_1 x_0 + \alpha_2\dot{x}_0 + \alpha_3(x_0^2 - E[x_0^2]) \\ &+\alpha_4(\dot{x}_0^2 - E[\dot{x}_0^2]) + \alpha_5 x_0\dot{x}_0 = F_0(t), \end{aligned} \tag{6.101}$$

where the α_i's are the quadratization coefficients. These coefficients are obtained by minimising the error ϵ inherent from replacing equation (6.98) with equation (6.101) in the mean squared sense, where

$$\begin{aligned}\epsilon \equiv \omega_n^2\mu_x + g(\mu_x + x_0, \dot{x}_0) - \mu_F - \alpha_1 x_0 - \alpha_2\dot{x}_0 \\ -\alpha_3(x_0^2 - E[x_0^2]) - \alpha_4(\dot{x}_0^2 - E[\dot{x}_0^2]) - \alpha_5 x_0\dot{x}_0.\end{aligned} \tag{6.102}$$

The necessary conditions for the minimisation of ϵ can be shown to be given by

$$\begin{pmatrix} \mu_{x_0{}^2} & 0 & \mu_{x_0{}^3} & \mu_{x_0\dot{x}_0^2} & 0 \\ & \mu_{\dot{x}_0^2} & 0 & \mu_{\dot{x}_0^3} & \mu_{x_0\dot{x}_0^2} \\ & & \mu_{x_0^4} - (\mu_{x_0^2})^2 & \mu_{x_0^2\dot{x}_0^2} - \mu_{x_0^2}\mu_{\dot{x}_0^2} & 0 \\ & Symm. & & \mu_{\dot{x}_0^4} - (\mu_{\dot{x}_0^2})^2 & \mu_{x_0\dot{x}_0^3} \\ & & & & \mu_{x_0^2\dot{x}_0^2} \end{pmatrix} \begin{pmatrix} \alpha_1 \\ \alpha_2 \\ \alpha_3 \\ \alpha_4 \\ \alpha_5 \end{pmatrix} \tag{6.103}$$

$$= \begin{pmatrix} E[x_0 g] \\ E[\dot{x}_0 g] \\ E[x_0^2 g] - E[g]\mu_{x_0^2} \\ E[\dot{x}_0^2 g] - E[g]\mu_{\dot{x}_0^2} \\ E[x_0\dot{x}_0 g] \end{pmatrix}.$$

The matrix appearing in equation (6.103) is symmetric, the elements of which are

$$\mu_{x_0^m\dot{x}_0^n} = E[x_0^m\dot{x}_0^n]. \tag{6.104}$$

The statistical expectations occurring in equation (6.104) are evaluated by relying on the statistics of the "equivalent" non-linear system, where equation (6.101) is solved using the Volterra series method.

The method of Statistical Quadratization has been applied by Donely and Spanos [38] to the stochastic response of a compliant offshore structure under non-linear drag loading. This technique has also been used by McWilliam and Langley [66] to analyse the effect of a cubic stiffness nonlinearity (Duffing type system) on the response statistics, where the response statistics were calculated using the techniques described in Section 6.6.1. In this latter case the mooring system was modeled as a single degree of freedom, such that

$$\ddot{x} + 2\beta\omega_n\dot{x} + \omega_n^2 x(1 + \gamma x^2) = F(t), \tag{6.105}$$

where the coefficient γ governs the magnitude of the system non-linearity.

An example of the results obtained for the response pdf in this case is shown in Figure 6.12. This case corresponds to the example considered throughout Section 6.6.1, where the combined first and second order response of the linear system ($\gamma = 0.0$) is scaled to be unity and γ takes a value of 0.2. The results shown indicate that the method of Statistical Quadratization yields a reasonably accurate estimate of the response pdf in this case. However, it has also been found by McWilliam and Langley [66] that this is not always the case, and that for dominant second order motions and large system non-linearities, the accuracy of this method to predict the response distribution deteriorates. For example, if the contribution of the second order forces to the response is increased so that $\sigma_2/\sigma_1 = 5.0$ (in the linear case), then the Statistical Quadratization method yields the results shown in Figure 6.13.

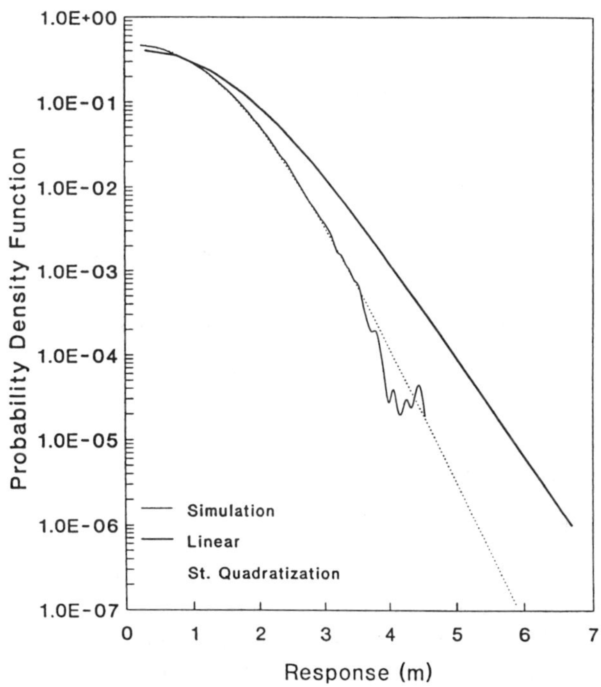

FIGURE 6.12
Response pdf for a moored horizontal cylinder with a cubic non-linearity.

The present section has served to illustrate the analytical difficulties which are introduced by non-linearities in either the fluid loading or the structural properties. Whilst analytical progress has been possible for the present problem, in many other cases the designer may need to rely on numerical simulation, as discussed in Section 6.4.5.

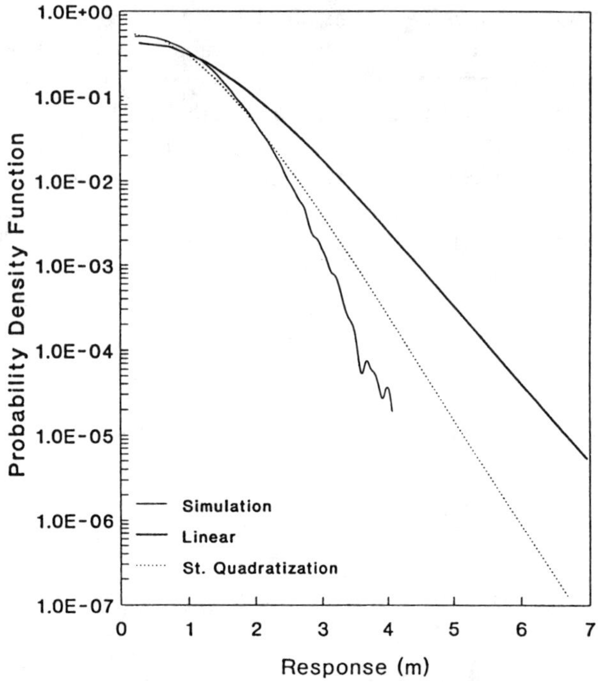

FIGURE 6.13
Response pdf for a moored horizontal cylinder with a cubic non-linearity.

6.7 Fatigue Analysis

6.7.1 Overview

The prediction of the fatigue life of an engineering structure can be approached either through a "Miner's Law" calculation or via a crack growth analysis. The former approach is the more common within the offshore industry, and this method alone is reviewed in the present work; details regarding stochastic crack growth analysis can be found in the text by Madsen et al. [67]. The Miner's Law approach is based on the fatigue endurance curve of a particular structural component — this "*S-N*" curve relates the number of cycles to failure N under constant amplitude loading to the applied stress amplitude S. For a welded offshore structure the *S-N* curve normally has the form

$$N = KS^{-r}, \tag{6.106}$$

where the constants K and r are determined by experiment — recommended values are available in the various design codes (for example DnV [68]). Under Miner's

Law, the fatigue "damage" D caused by variable amplitude loading is written in the form [69]

$$D = \sum_i n_i/N(S_i) = (1/K)\sum_i n_i S_i^r, \tag{6.107}$$

where n_i is the number of cycles applied with stress amplitude S_i. The structure is nominally considered to have failed when $D = 1$, although in practice large factors of safety are usually applied. Various methods of computing the fatigue damage D sustained over the design lifetime of an offshore structure are reviewed in the following sections.

6.7.2 Regular Wave Analysis

With this approach the long term environmental conditions are modeled as a succession of regular waves of varying height H and period T. The fatigue damage D is estimated to be

$$D = (n_t/K)\int_0^\infty \int_0^\infty S^r(H,T)p(H,T)dHdT, \tag{6.108}$$

where $S(H,T)$ is the stress amplitude produced by a regular wave with properties (H,T), and the function $p(H,T)$ is the long term jpdf of H and T. Furthermore, n_t is the total number of waves expected over the design life of the structure. The jpdf $p(H,T)$ must be estimated from environmental measurements; in practice, the double integral which appears in equation (6.108) might typically be replaced by a summation over discrete combinations of H and T given in the form of a "frequency of occurrence" scatter diagram.

The advantage of the regular wave approach is its relative simplicity; the main disadvantage is that the environmental model may not give a realistic representation of the wave forces and hence the stresses produced by a truly random sea. The fact that the damage is proportional to S raised to the power r (typically $r \approx 3.5$) implies that the result given by equation (6.107) or equation (6.108) is particularly sensitive to the computed stress levels; for this reason a random environmental model is normally preferred, and the incorporation of such a model within a fatigue analysis is described in the following two sections.

6.7.3 Narrow Band Random Analysis

If the stress produced by the random seastate is "narrow banded" then by definition the stress time history has the appearance of a sine wave of slowly varying frequency and amplitude — for each upwards crossing of zero, the time history displays a single peak (see Figure 6.14).

The amplitudes of the peaks (or maxima), S say, will be random with some probability density function $p_m(S)$; with this notation the expected value of the

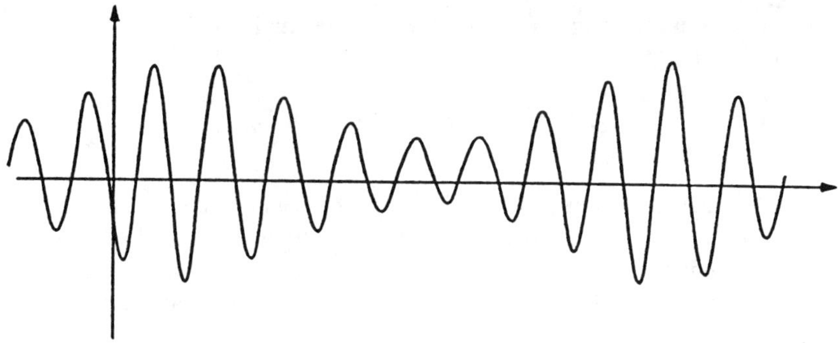

FIGURE 6.14
A sample time history for a narrow band process.

damage D in a storm of duration T can be estimated via equation (6.107) to have the form

$$E[D] = (\nu_0^+ T/K)E[S^r] = (\nu_0^+ T/K)\int_0^\infty S^r p_m(S)dS, \tag{6.109}$$

where $\nu_0{}^+$ is the mean rate at which the stress time history crosses zero with positive slope (i.e., the mean rate of occurrence of the peaks). Equation (6.109) relates to the damage produced by a single stationary seastate of duration T; if the seastate is characterised by the significant wave height H_s and the average period T_1, then the damage can be written in the form $E[D(H_s,T_1)]$. The total damage occurring over the design life of the structure can then be expressed as

$$E[D_L] = (T_L/T)\int_0^\infty \int_0^\infty E[D(H_s, T_1)]p(H_s, T_1)dH_s dT_1, \tag{6.110}$$

where T_L is the design life and $p(H_s,T_1)$ is the long term jpdf of H_s and T_1, as described in Section 6.3.

The form of the peak distribution $p_m(S)$ which appears in equation (6.109) will depend upon the equations of motion of the system and the statistical properties of the wave loading. If the system is linear and the wave loading is taken to be Gaussian (which will *not* be the case in the presence of drag forces), then the stresses induced in the system will be Gaussian. If, as assumed at the beginning of this section, the stress is further taken to be narrow banded, then $p_m(S)$ will have the form of a Rayleigh distribution [24]. In this case equation (6.109) can be evaluated in closed form to yield

$$E[D] = (\nu_0^+ T/K)(\sqrt{2}\sigma_S)^r \Gamma(1 + r/2), \tag{6.111}$$

where Γ is the gamma function and σ_S is the standard deviation of the stress time history $S(t)$. If significant drag forces are present then the stress time history can be highly non-Gaussian, and in this case equation (6.111) tends to yield a severe under-prediction of the fatigue damage; a more accurate prediction may require time domain simulation of the system response in order to estimate $p_m(S)$ numerically. Furthermore, if the stress time history is Gaussian but wide banded, then equation (6.109) no longer applies and the analysis which is outlined in the following section must be employed.

6.7.4 Wide Band Random Analysis

If the stress in a component is "wide banded" then by definition the time history displays multiple peaks for each upwards crossing of zero (see Figure 6.15).

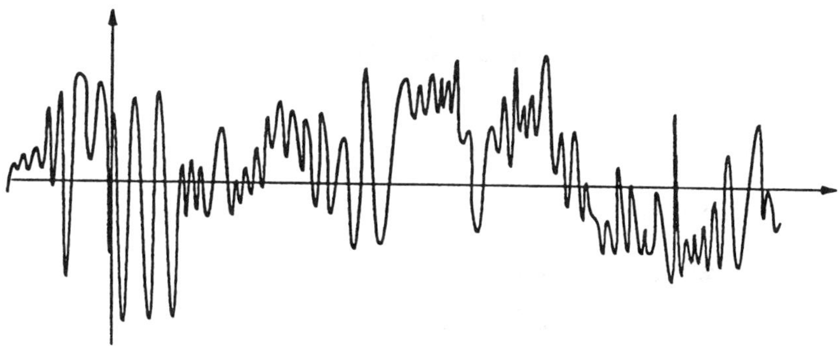

FIGURE 6.15
A sample time history for a broad band process.

In this case it is not so obvious how the time history can be divided into discrete "stress cycles" for use in Miner's Law. Methods of producing an "equivalent" series of stress cycles include: (i) the peak counting method, (ii) the range counting method, and (iii) the rain-flow counting method. In *the peak counting method*, a stress cycle is attributed to each peak which lies above zero, with the amplitude of the cycle being placed equal to the value of the peak. In *the range counting method*, two "half-cycles" are associated with each peak (whether the peak is positive or negative): the amplitude of the first half-cycle is placed equal to half the stress range between the peak and the preceding trough, while the amplitude of the second half-cycle is placed equal to half the stress range between the peak and the following trough. Clearly (i) and (ii) will yield similar results for a narrow band process, but quite different results may be obtained for a wide band process: the range counting method will produce more cycles than the peak counting method, and the amplitudes of the cycles will generally be smaller. For a Gaussian process it can be shown that the fatigue damage predicted by the peak counting method is identical to equation (6.111), whereas the range counting method leads (approximately) to the same result apart from the presence of a multiplying factor α^{r-1} [67]. Here α (<1) is the ratio: (rate of zero crossings with positive slope)/(rate of occurrence of peaks). The range counting method will therefore tend to predict a lower fatigue damage than the peak counting method. In the non-Gaussian case, the two methods are normally used in conjunction with a time domain simulation of the stress time history to yield a numerical estimate of the fatigue damage.

The *rain-flow counting method* [70] is a relatively complex method of converting a wide band random stress history into an "equivalent" sequence of stress cycles. Unlike the peak counting method and the range counting method, the rain-flow approach associates a mean stress level with each cycle, and small stress reversals are recognized as being superimposed upon larger stress cycles. It is not possible to produce a closed form equivalent of equation (6.111) for this approach, and the method must normally be applied numerically to computed (simulated) or measured time histories. It is generally thought that rain-flow counting yields the most realistic estimate of fatigue damage; typically the peak counting method will yield a higher estimate of the damage while the range counting method will predict less damage.

6.8 Reliability Methods

The stochastic response analysis methods which are presented in Section 6.4 and the fatigue analysis which is presented in Section 6.7 are based on the assumption that the statistical uncertainty which arises in the design of an offshore structure is due predominantly to the random environmental conditions. However, as mentioned in Section 6.4.4, significant uncertainties may exist regarding the properties of the structure itself. In this case it is necessary to take a wider view of the structural reliability which encompasses an "ensemble" of possible structures. The question of reliability can be posed in terms of a "safety margin" M of the form

$$M = R_{acc} - R, \tag{6.112}$$

where R is a variable representing the response of the structure and R_{acc} is the maximum acceptable ("safe") value of the response. For example, R might represent the stress in some typical component, while R_{acc} is the maximum allowable stress; alternatively, R might represent the lifetime fatigue damage sustained by the structure, in which case R_{acc} would be unity. In practice M would actually be a composite safety margin which encompasses all of the possible modes of failure. The structure is in a safe (or unfailed) condition for $M>0$, and thus the structural reliability can be measured by the probability that M remains positive over the design lifetime.

To relate equation (6.112) to the analysis of Section 6.4.5, equation (6.26) gives the probability that the system response will exceed a critical level b; this result can be reinterpreted by considering b to be the maximum response R, in which case 1-$P(R,T)$ is effectively the cumulative probability function of R. In the absence of any structural uncertainty, this result can be used immediately to calculate the probability that $M>0$ for any specified R_{acc}. In the presence of structural uncertainty, 1-$P(R,T)$ is the *conditional* probability function of R for a fixed set of structural properties; in principle, this result can be used in conjunction with the statistics of the structural properties to yield a revised probability that $M>0$. The key phrase here is "in principle", because this type of analysis faces a number of practical difficulties, not least of which are: (i) the statistical properties of the structure (and for that matter the environment) may not be known with enough certainty to warrant a full reliability analysis, and (ii) even were the statistical properties of all random quantities are fully defined, the complexity of the reliability calculation might be beyond that which is computationally feasible. To overcome these difficulties a number of approximate methods have been developed: many of the techniques described in the earlier chapters of the present book take the definition of the safety margin M as the starting point for a reliability analysis, and hence these methods are in principle applicable to offshore structures. It can be seen from the present chapter that M is an extremely complex function for a typical offshore structure, and it is clear that the advanced reliability methods described earlier in this text are essential mathematical tools. It can be noted that ongoing developments in the application of reliability theory to offshore structures are reported in the conference series "Integrity of Offshore Structures" [71].

6.9 Concluding Remarks and References

This chapter has presented a review of the methods which are currently available for the stochastic dynamic analysis of offshore structures. In addition to this overview, techniques for the analysis of fixed structures and large floating systems have been presented in some detail to give an indication of the type of mathematical model which might be adopted. There is no doubt that offshore structural dynamics is a challenging discipline, and the continuing needs of the industry provide the motivation for further research work in this area. The analysis of non-linear multidegree-of-freedom systems, and the development of efficient techniques for dealing with structural uncertainties in large complex systems are but two areas in which current analysis methods could ideally be enhanced.

References

[1] Patel, M.H., *Dynamics of Offshore Structures*, Benthem Press, London, 1989.

[2] Langley, R.S., On Quasi-Static Approximations to Non-Stationary Random Vibration, *Journal of Sound and Vibration*, 113(2), 1987, pp. 365–375.

[3] Sarpkaya, T. and Isaacson, M., *Mechanics of Wave Forces on Offshore Structures*, Van Nostrand Reinhold Company, London, 1981.

[4] Longuet-Higgins, M.S., The Effect of Non-Linearities on Statistical Distributions in the Theory of Sea Waves, *Journal of Fluid Mechanics*, 17(3), 1963, pp. 459–480.

[5] Faltinsen, O.M., *Sea Loads on Ships and Offshore Structures*, Cambridge University Press, Cambridge, 1990.

[6] Harris, R.I., The Nature of the Wind, *Proceeding of a seminar held on the 18th June 1970 at the Institute of Civil Engineering: The Modern design of wind sensitive structures*, CIRIA, London, 1971.

[7] Ochi, M.S. and Shin, Y.S., Wind Turbulent Spectra for Design Consideration of Offshore Structures, *Offshore Technology Conference, OTC 5736*, 1988, pp. 461–467.

[8] Kareem, A., Wind-induced Response Analysis of Tension Leg Platforms, *Journal of Structural Engineering*, 11(1), 1985, pp. 37–55.

[9] van Oortmerssen, I.G., *The Motions of a Moored Ship in Waves*, The Netherlands Ship Model Basin Publication No. 510, Wageningen, 1976.

[10] Hearn, G.E. and Donati, E., A New Fluid-structure Interaction Analysis Based on Higher Order Boundary Elements, *International Journal for Numerical Methods in Fluids*, 8(2), 1988, pp. 199–225.

[11] Morison, J.R., O'Brien, M.P., Johnson, J.W., and Schaaf, S.A., The Force Exerted by Surface Waves on Piles, *Petroleum Transactions, TP2846*, 189, 1950, pp. 149–154.

[12] Newman, J.N., *Marine Hydrodynamics*, MIT Press, Cambridge, Mass., 1977.

[13] Wu, S.C. and Tung, C.C., Random Response of Offshore Structures to Wave and Current Forces, Sea Grant Publication UNC-SG-75-22, Department of Civil Engineering, North Carolina State University, 1975.

[14] Gudmestad, O. and Connor, J.J., Linearisation Methods and the Influence of Current on the Nonlinear Hydrodynamic Drag Force, *Applied Ocean Research*, 5(4), 1983.

[15] Hamilton, J., Three-dimensional Fourier Analysis of Drag Force for Compliant Offshore Structures, *Applied Ocean Research*, 2(4), 1980.

[16] Langley, R.S., The Linearisation of Three Dimensional Drag Force in Random Seas with Current, *Applied Ocean Research*, 6(3), 1984, pp. 126–131.

[17] Eatock Taylor, R. and Rajagopalan, A., Superharmonic Drag Resonance Effects in Drag Dominated Structures, *Integrity of Offshore Structures*, Applied Science Publishers, 1981.

[18] Koliopulos, P.K., Quasi-Static and Dynamic Response Statistics of Linear SDOF Systems under Morison-Type Wave Forces, *Engineering Structures*, 10(1), 1990, pp. 24–36.

[19] Eatock Taylor, R., *Non-linear Frequency Domain Analysis*, Computational Mechanics Publications, 1994.

[20] Molin, B. and Bureau, G., A Simulation Model for the Dynamic Behavior of Tankers Moored to Single Point Moorings, *International Symposium on Ocean Engineering Ship Handling*, Paper No. 4, Sweden, 1980.

[21] Battjes, J.A., Probabilistic Aspects on Ocean Waves, *Safety of Structures under Dynamic Loading,* Tapir Publishers, 1978, pp. 387–439.

[22] Bishop, R.E.D. and Price, W.G., *Hydroelasticity of Ships*, Cambridge University Press, Cambridge, 1979.

[23] Kirk, C.L., Dynamic Response of Marine Risers by Single Wave and Spectral Analysis Methods, *Applied Ocean Research*, 7(1), 1985, pp. 213.

[24] Lin, Y.K., *Probabilistic Theory of Structural Dynamics*, McGraw-Hill Book Company, New York, 1967.

[25] Naess, A., On the Long-Term Statistics of Extremes, *Applied Ocean Research*, 6, 1984, pp. 227–228.

[26] Langley, R.S., Techniques for Assessing the Lifetime Reliability of Engineering Structures Subjected to Stochastic Loads, *Engineering Structures*, 9, 1987, pp. 95–103.

[27] Inglis, R.B., Pijfers, J.G.L., and Vugts, J.H., A Unified Probabilistic Approach to Predicting the Response of Offshore Structures, Including the Extreme Response, *Behavior of Offshore Structures*, Elsevier Science Publications, Amsterdam, 1985, pp. 95–109.

[28] Roberts, J.B., First Passage Time for Randomly Excited Non-Linear Oscillators, *Journal of Sound and Vibration*, 109, 1986, pp. 33–50.

[29] Langley, R.S., A First Passage Approximation for Normal Stationary Random Processes, *Journal of Sound and Vibration*, 122(2), 1988, pp. 261–275.

[30] Naess, A., The Effect of the Markov Chain Condition on the Prediction of Extreme Values, *Journal of Sound and Vibration*, 94, 1984, pp. 87–103.

[31] Spanos, P.D., Stochastic Linearisation in Structural Dynamics, *Applied Mechanics Reviews*, 34, 1981, pp. 1–8.

[32] Langley, R.S., Second Order Frequency Domain Analysis of Moored Vessels, *Applied Ocean Research*, 9(1), 1987, pp. 7–18.

[33] Newland, D.E., *Random Vibrations, Spectral and Wavelet Analysis*, Longman Scientific and Technical, Harlow, 1993.

[34] Spanos, P.D., Filter Approaches to Wave Kinematics Approximation, *Applied Ocean Research*, 8(1), 1986, pp. 2–7.

[35] Pradlwarter, H.J., A Selective MC Simulation Technique for Nonlinear Structural Reliability, *Proceedings of the ASCE Speciality Conference*, Denver, Colorado, July 8-10, ASCE New York, 1992, pp. 8–10.

[36] Nigam, N.C., *Introduction to Random Vibrations*, MIT Press, Cambridge, Mass., 1983.

[37] Schetzen, M., *The Volterra and Wiener Theories of Non-Linear Systems*, John Wiley & Sons, New York, 1980.

[38] Donely, M.G. and Spanos, P.D., *Dynamic Analysis of Non-linear Structures by the Method of Statistical Quadratization*, Springer-Verlag Press, New York, 1990.

[39] Roberts, J.B., Nonlinear Analysis of Slow Drift Oscillations of Moored Vessels in Random Seas, *Journal of Ship Research*, 25(2), 1981, pp. 130–140.

[40] Roberts, J.B., A Stochastic Theory for Nonlinear Ship Rolling in Irregular Seas, *Journal of Ship Research*, 26(4), 1982, pp. 229–245.

[41] Pierson, W.J. and Holmes, P., Irregular Wave Forces on a Pile "91.ww 4.1", *Water, Wave and Harbor Division ASCE*, New York, 1965.

[42] Tickell, R.G., Continuous Random Wave Loading on Structural Members, *The Structural Engineer*, 55, 1977, pp. 209–222.

[43] Najafian, G. and Burrows, R., Probabilistic Modeling of Quasi-static Response of Offshore Structures subject to Nonlinear Wave Loading: Pt II Approximate Approaches, *Applied Ocean Research*, 16, 1994, pp. 205–221.

[44] Burrows, R., Quasistatic Response of Offshore Structures using Probabilistic Methods, *Applied Mathematical Modeling*, 1, 1977, pp. 325–332.

[45] Borgman, L.E., Statistical Models for Ocean Waves and Wave Forces, *Advances in Hydroscience*, 8, pp. 139–181.

[46] Browers, J.J.H. and Verbreek, P.H.J, Expected Fatigue Damage and Expected Extreme Response for Morison-Type Wave Loading, *Applied Ocean Research*, 5(3), 1983, pp. 129–133.

[47] Cartwright, D.E. and Longuet-Higgins, M.S., The Statistical Distribution of the Maxima of a Random Function, *Proceedings of the Royal Society of London, Series A*, 1956.

[48] Tung, C.C., Peak Distribution of Random Wave-Current Force, *Journal of Engineering Mechanics Division ASCE*, No. EM5, 100, 1974.

[49] Pinkster, J.A., *Low Frequency Second Order Wave Exciting Forces on Floating Structures*, Publication No. 650, Netherlands Ship Model Basin, Wageningen - Netherlands, 1981.

[50] Naess, A. and Ness, G.M., Sum-Frequency Response Statistics of Tethered Platforms in Random Waves, *Applied Ocean Research*, 14(1), 1992, pp. 23–32.

[51] Vinje, T., On the Statistical Distribution of Second-Order Forces and Motions, *Ship Building Progress*, 30, 1983, pp. 58–68.

[52] Langley, R.S., The Statistics of Second Order Wave Forces, *Applied Ocean Research*, 6(4), 1984, pp. 182–186.

[53] Kac, M. and Siegert, A.J.F., On the Theory of Noise in Radio Receivers with Square Law Detectors, *Journal of Applied Physics*, 283, 18, 1947.

[54] Naess, A., Statistical Analysis of Second Order Response of Marine Structures, *Journal of Ship Research*, 29(4), 1985, pp. 274–284.

[55] Naess, A., The Statistical Distribution of Second-Order Slowly Varying Forces and Motions, *Applied Ocean Research*, 8(2), 1986, pp. 110–118.

[56] Langley, R.S., A Statistical Analysis of Low Frequency Second-Order Forces and Motions, *Applied Ocean Research*, 9(3), 1987, pp. 163–170.

[57] Faltinsen, O.D and Løken, A.E., Slow Drift Oscillations of a Ship in Irregular Waves, *Applied Ocean Research*, 1(1), 1979, pp. 21–31.

[58] Naess, A. and Johnsen, J.M., An Efficient Numerical Method for Calculating the Statistical Distribution of Combined First-Order and WaveDrift Response, *International Conference on Offshore Mechanics and Arctic Engineering Conference*, Stavanger, Norway, 1991, pp. 59–70.

[59] Langley, R.S. and McWilliam, S., A Statistical Analysis of First and Second Order Vessel Motions induced by Waves and Wind Gusts, *Applied Ocean Research*, 15(1), 1993, pp. 13–23.

[60] Kato, S., Kinoshita, T., and Takase, S., Statistical Theory of Total Second Order Responses of Moored Vessels in Random Seas, *Applied Ocean Research*, 12(1), 1990, pp. 2–13.

[61] Langley, R.S., On the Time Domain Simulation of Second Order Wave Forces and Motions, *Applied Ocean Research*, 8(3), 1986, pp. 134–143.

[62] McWilliam, S. and Langley, R.S., Extreme Values of First- and Second-order Wave-induced Vessel Motions, *Applied Ocean Research*, 15, 1993, pp. 169–181.

[63] Marthinsen, T., Dynamics of Slow-Drift Oscillations with Non-Linear Restoring Forces, *Non-Linear Water Waves, IUTAM Symposium*, Tokyo, Japan, 1987, pp. 459–466.

[64] Johnsen, J.M. and Naess, A., Time Variant Wave Drift Damping and its Effect on the Response Statistics of Moored Offshore Structures, *First International Offshore and Polar Engineering Conference*, Edinburgh, 1991, pp. 452–459.

[65] Roberts, J.B. and Spanos, P.D., *Random Vibration and Statistical Linearisation*, John Wiley & Sons, New York, 1990.

[66] McWilliam, S. and Langley, R.S., Response Statistics of Non-Linearly Moored Vessels under the Action of First and Second Order Wave Forces, *Applied Ocean Research*, 16, 1994, pp. 295–311.

[67] Madsen, H.O., Krenk, S., and Lind, N.C., *Methods of Structural Safety*, Prentice-Hall, New Jersey, 1986.

[68] Det Norske Veritas (DnV), *Rules for the Design, Construction and Inspection of Offshore Structures*, Oslo, 1977.

[69] Miner, M.A., Cumulative Damage in Fatigue, *Transactions of the American Society of Mechanical Engineers*, 67, A159, 1945.

[70] Wirsching and Shehata, Fatigue under Wide Band Random Loading using the Rain-Flow Method, *Journal of Engineering Materials and Technology, ASME*, 1977, pp. 205–211.

[71] Faulkner, D., Cowling, M.J., Incecik, A., and Das, P.K. (Editors) *Integrity of Offshore Structures - 5*, Engineering Materials Advisory Services Ltd., Warley, 1993.

Index